375/

International A/AS Level
Physics

Chris Mee • Mike Crundell • Brian Arnold • Wendy Brown

HODDER
EDUCATION
AN HACHETTE UK COMPANY

D1354840

Orders: please contact Bookpoint Ltd, 130 Milton Park, Abingdon, Oxon OX14 4SB. Telephone: (44) 01235 827720. Fax: (44) 01235 400454. Lines are open 9.00–5.00, Monday to Saturday, with a 24-hour message answering service. Visit our website at www.hoddereducation.co.uk.

© Chris Mee, Mike Crundell, Brian Arnold, Wendy Brown 2008
First published in 2008 by
Hodder Education, an Hachette UK Company,
338 Euston Road
London NW1 3BH

Impression number   5 4
Year   2011 2010 2009

Cover photo © Victor Habbick Visions/Science Photo Library
Typeset in 11/13pt Times New Roman by Charon Tec Ltd (A Macmillan Company)
Printed in Dubai

A catalogue record for this title is available from the British Library

ISBN 978 0 340 94564 3

Contents

Preface

This book is a revision of *AS/A2 Physics*, amended and updated so that it is suitable for international students following the University of Cambridge International Examinations (CIE) Physics Syllabus 9702. The book has been fully endorsed by CIE and is listed as a recommended textbook for this syllabus. New chapters on measurement techniques, direct and remote sensing, and communication of information have been added.

All the assessment objectives and learning outcomes of Syllabus 9702 are covered in this book, with some re-ordering of topics in the interests of a logical teaching order. In a few cases, the material of the book goes slightly beyond the requirements of the syllabus in order to arrive at a satisfactory termination of that topic. The learning outcomes addressed in each section, or group of sections, of a chapter are listed using CIE wording, so that students can identify the exact coverage of the syllabus.

Key points, definitions and equations are highlighted in yellow panels. After each section or group of sections, there is a brief summary of the ground covered. Mathematical derivations have been kept to a minimum, but where some guidance may be helpful, this is provided in a *Maths Note* box. To achieve familiarity through practice, worked examples are provided at frequent intervals. These are followed by similar questions for students to attempt, under the heading *Now it's your turn*. In addition, there are groups of problems and questions at the end of each section or group of sections, and a selection of exam-style questions at the end of each chapter. Answers to these questions can be found at the back of the book.

Although the book is specifically directed towards the CIE syllabus, its coverage of the subject will make it suitable for international students working for the physics examinations of other awarding bodies.

Chris Mee
Mike Crundell
Brian Arnold
Wendy Brown
January 2008

1. Physical quantities and units

Physics is the study of how the world behaves and how the laws of nature operate. Accurate measurement is very important in the development of any science, and this is particularly true of physics. Throughout history, as methods of taking measurements have improved, then new ideas have developed.

In general, physicists begin by observing, measuring and collecting data. These data are then analysed to discover whether they fit into a pattern. If there is a pattern and this pattern can be used to explain other events, it becomes a theory. If the theory predicts events successfully, the theory becomes a law. The process is known as the *scientific method* (see Figure 1.1).

This chapter looks at the enormous range of numbers which are met in physics, the units in which quantities are measured and how these are expressed.

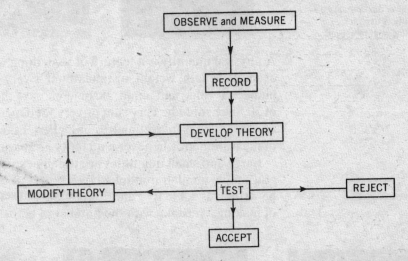

Figure 1.1 Block diagram to illustrate the scientific method

1.1 Quantities and units

At the end of Section 1.1 you should be able to:
- show an understanding that all physical quantities consist of a numerical magnitude and a unit
- use prefixes and symbols to indicate decimal sub-multiples and multiples of units
- recall the following base quantities and their units: mass (kg), length (m), time (s), current (A), temperature (K), amount of substance (mol)

- express derived units as products or quotients of the base units and use these units as appropriate
- use base units to check the homogeneity of physical equations
- understand and use the conventions for labelling graph axes and table columns
- make reasonable estimates of physical quantities.

Figure 1.2 Brahe (1546–1601) measured the elevations of stars; these days a modern theodolite is used for measuring angular elevation.

A **physical quantity** is a feature of something which can be measured, for example, length, weight, or time of fall. Every physical quantity has a numerical value and a unit. If someone says they have a waist measurement of 50, they could be very slim or very fat depending on whether the measurement is in centimetres or inches! Take care – it is vital to give the unit of measurement whenever a quantity is measured or written down.

Large and small quantities are usually expressed in scientific notation, i.e. as a simple number multiplied by a power of ten. For example, 0.00034 would be written as 3.4×10^{-4} and 154 000 000 as 1.54×10^8. There is far less chance of making a mistake with the number of zeros!

Figure 1.3 The elephant is large in comparison with the boy but small compared with the jumbo jet.

SI units

In very much the same way that languages have developed in various parts of the world, many different systems of measurement have evolved. Just as languages can be translated from one to another, units of measurement can also be converted between systems. Although some conversion factors are easy to remember, some are very difficult. It is much better to have just one system of units. For this reason, scientists around the world use the **Système International** (SI) which is based on the metric system of measurement. Each quantity has just one unit and this unit can have **multiples** and **sub-multiples** to cater for larger or smaller values. The unit is given a **prefix** to denote the multiple or submultiple (see Table 1.1). For example, one thousandth of a metre is known as a millimetre (mm) and 1.0 millimetre equals 1.0×10^{-3} metres (m).

Figure 1.4 The mass of this jewel could be measured in kilograms, pounds, carats, grains, etc.

prefix	symbol	multiplying factor
peta	P	10^{15}
tera	T	10^{12}
giga	G	10^{9}
mega	M	10^{6}
kilo	k	10^{3}
deci	d	10^{-1}
centi	c	10^{-2}
milli	m	10^{-3}
micro	μ	10^{-6}
nano	n	10^{-9}
pico	p	10^{-12}

Table 1.1 The more commonly used prefixes

Beware when converting units for areas and volumes!

$$1\,mm = 10^{-3}\,m$$
Squaring both sides $\quad 1\,mm^2 = (10^{-3})^2\,m^2 = 10^{-6}\,m^2$
and $\quad 1\,mm^3 = (10^{-3})^3\,m^3 = 10^{-9}\,m^3$
Note also that $\quad 1\,cm^2 = (10^{-2})^2\,m^2 = 10^{-4}\,m^2$
and $\quad 1\,cm^3 = (10^{-2})^3\,m^3 = 10^{-6}\,m^3$

A distance of thirty metres should be written as 30 m and not 30 ms or 30 m s. The letter s is *never* included in a unit for the plural. If a space is left between two letters, the letters denote different units. So, 30 m s would mean thirty metre seconds and 30 ms means 30 milliseconds.

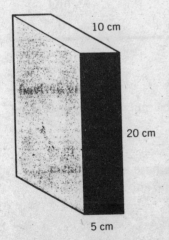

10 cm

20 cm

5 cm

This box has a volume of $1.0 \times 10^3\,cm^3$ or $1.0 \times 10^6\,mm^3$ or $1.0 \times 10^{-3}\,m^3$.

Example

Calculate the number of micrograms in 1.0 milligram.

$$1.0\,g = 1.0 \times 10^3\,mg$$
$$\text{and } 1.0\,g = 1.0 \times 10^6 \text{ micrograms (μg)}$$
$$\text{so, } 1.0 \times 10^3\,mg = 1.0 \times 10^6\,μg$$
$$\text{and } 1.0\,mg = (1.0 \times 10^6)/(1.0 \times 10^3) = \mathbf{1.0 \times 10^3\,μg}$$

▶▶

> ### Now it's your turn
>
> **1** Calculate the area, in cm², of the top of a table with sides of 1.2 m and 0.9 m.
>
> **2** Determine the number of cubic metres in one cubic kilometre.

Base units

If a quantity is to be measured accurately, the unit in which it is measured must be defined as precisely as possible.

SI is founded upon seven fundamental or **base units**.

The base quantities and the units with which they are measured are listed in Table 1.2. For completeness, the candela has been included, but this unit will not be used in the A/AS course.

Table 1.2 The base quantities

quantity	unit	symbol
mass	kilogram	kg
length	metre	m
time	second	s
electric current	ampere (amp)	A
thermodynamic temperature	kelvin	K
amount of substance	mole	mol
luminous intensity	candela	cd

Derived units

All quantities, apart from the base quantities, can be measured using **derived units**.

Derived units consist of some combination of the base units. The base units may be multiplied together or divided by one another, but never added or subtracted.

See Table 1.3 for examples of derived units. Some quantities have a named unit. For example, the unit of force is the newton, symbol N, but the newton can be expressed in terms of base units. Quantities which do not have a named unit are expressed in terms of other units. For example, specific latent heat (Chapter 4) is measured in joules per kilogram ($J\,kg^{-1}$).

Table 1.3 Some examples of derived units which may be used in the A/AS course

quantity	unit	derived unit
frequency	hertz (Hz)	s^{-1}
speed	$m\,s^{-1}$	$m\,s^{-1}$
acceleration	$m\,s^{-2}$	$m\,s^{-2}$
force	newton (N)	$kg\,m\,s^{-2}$
energy	joule (J)	$kg\,m^2\,s^{-2}$
power	watt (W)	$kg\,m^2\,s^{-3}$
electric charge	coulomb (C)	$A\,s$
potential difference	volt (V)	$kg\,m^2\,s^{-3}\,A^{-1}$
electrical resistance	ohm (Ω)	$kg\,m^2\,s^{-3}\,A^{-2}$
specific heat capacity	$J\,kg^{-1}\,K^{-1}$	$m^2\,s^{-2}\,K^{-1}$

Example

What are the base units of speed?

Speed is defined as $\dfrac{distance}{time}$ and so the unit is $\dfrac{m}{s}$.

Division by a unit is shown using a negative index, that is s^{-1}.
The base units of speed are **$m\,s^{-1}$**.

Now it's your turn

Use the information in Tables 1.2 and 1.3 to determine the base units of the following quantities.

1 Density $\left(= \dfrac{mass}{volume} \right)$ $kg\ m^{-3}$

2 Force ($= mass \times acceleration$) $kg\ m\ s^{-2}$

Equations

It is possible to work out the total number of oranges in two bags if one bag contains four and the other five (the answer is nine!). This exercise would, of course, be nonsense if one bag contained three oranges and the other four mangoes. In the same way, for any equation to make sense, each term involved in the equation must have the same base units. A term in an equation is a group of numbers and symbols, and each of these terms (or groups) is added to, or subtracted from, other terms. For example, in the equation

$$v = u + at$$

the terms are v, u, and at.

In any equation where each term has the same base units, the equation is said to be **homogeneous** or 'balanced'.

In the example on page 5, each term has the base units m s^{-1}. If the equation is not homogeneous, then it is incorrect and is not valid. When an equation is known to be homogeneous, then the balancing of base units provides a means of finding the units of an unknown quantity.

Example

Use base units to show that the following equation is homogeneous.

work done = gain in kinetic energy + gain in gravitational potential energy

The terms in the equation are work, (gain in) kinetic energy, and (gain in) gravitational potential energy.

work done = force × distance moved

and so the base units are kg m s^{-2} × m = **kg m^2 s^{-2}**.

kinetic energy = $\frac{1}{2}$ × mass × (speed)^2

Since any pure number such as $\frac{1}{2}$ has no unit, the base units are kg × (m s^{-1})^2 = **kg m^2 s^{-2}**.

potential energy = mass × gravitational field strength g × distance

The base units are kg × m s^{-2} × m = **kg m^2 s^{-2}**.

Conclusion: **All terms have the same base units and the equation is homogeneous.**

Now it's your turn

1 Use base units to check whether the following equations are balanced:
   (a) *pressure = depth × density × gravitational field strength* yes
   (b) *energy = mass × (speed of light)^2*

   yes

2 The thermal energy Q needed to melt a solid of mass m without any change of temperature is given by the equation

$$Q = mL$$

where L is a constant. Find the base units of L.

Conventions for symbols and units

You may have noticed that when symbols and units are printed, they appear in different styles of type. The symbol for a physical quantity is printed in *italic* (sloping) type, whereas its unit is in roman (upright) type. For example, velocity v is italic, but its unit m s^{-1} is roman. Of course, you will not be able to make this distinction in handwriting.

At A/AS level and beyond, there is a special convention for labelling columns of data in tables and graph axes. The symbol is printed first (in italic), separated by a forward slash (the printing term is a solidus) from the unit (in roman). Then the data is presented in a column, or along an axis, as pure numbers. This is illustrated in Figure 1.5, which shows a table of data and the resulting graph for the velocity v of a particle at various times t.

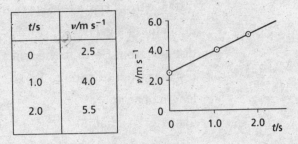

t/s	$v/m\ s^{-1}$
0	2.5
1.0	4.0
2.0	5.5

Figure 1.5 The convention for labelling tables and graphs

If you remember that a physical quantity contains a pure number and a unit, the reason for this style of presentation becomes clear. By dividing a physical quantity such as time (a number and a unit) by the appropriate unit, you are left with a pure number. It is then algebraically correct for the data in tables, and along graph axes, to appear as pure numbers.

You may also see examples in which the symbol for the physical quantity is followed by the slash, and then by a power of 10, and then the unit, for example $t/10^2s$. This means that the column of data has been divided by 100, to save repeating lots of zeros in the table. If you see a table or graph labelled $t/10^2s$ and the figures 1, 2, 3 in the table column or along the graph axis, this means that the experimental data was obtained at values of t of 100 s, 200 s, 300 s.

Try to get out of the habit of heading table columns and graphs in ways such as 't in s', '$t(s)$' or even of recording each reading in the table as 1.0 s, 2.0 s, 3.0 s.

Order of magnitude of quantities

It is often useful to be able to estimate the size, or **order of magnitude**, of a quantity. Strictly speaking, the order of magnitude is the power of ten to which the number is raised. The ability to estimate is particularly important in a subject like physics where quantities have such widely different values. A *short* distance for an astrophysicist is a light-year (about 9.5×10^{15} m) whereas a *long* distance for a nuclear physicist is 6×10^{-15} m (the approximate diameter of a nucleus)! Table 1.4 gives some values of distance which may be met in the A/AS Physics course.

The ability to estimate orders of magnitude is valuable when planning and carrying out experiments or when suggesting theories. Having an idea of the expected result provides a useful check that a silly error has not been made.

Figure 1.6 The ratio of the mass of the humpback whale to the mass of the mouse is about 10^4. That is minute compared to the ratio of the mass of a galaxy to the mass of a nucleus (10^{68})!

This is also true when using a calculator. For example, the acceleration of free fall at the Earth's surface is about $10\,\text{m s}^{-2}$. If a value of $9800\,\text{m s}^{-2}$ is calculated, then this is obviously wrong and a simple error in the power of ten is likely to be the cause. Similarly, a calculation in which the cost of boiling a kettle of water is found to be several dollars, rather than a few cents, may indicate that the energy has been measured in watt-hours rather than kilowatt-hours.

Table 1.4 Some values of distance

	distance/m
distance from Earth to edge of observable Universe	1.4×10^{26}
diameter of a galaxy	1.2×10^{21}
distance from Earth to the Sun	1.5×10^{11}
distance from London to Paris	3.5×10^{5}
length of a car	4
diameter of a hair	5×10^{-4}
diameter of an atom	3×10^{-10}
diameter of a nucleus	6×10^{-15}

Example

It is worthwhile to remember the sizes of some common objects so that comparisons can be made. For example, a jar of peanut butter has a mass of about $500\,\text{g}$ and a carton of orange juice has a volume of $1000\,\text{cm}^3$ (1 litre).

Now it's your turn

Estimate the following quantities:
(a) the mass of an orange,
(b) the mass of an adult human,
(c) the height of a room in a house,
(d) the diameter of a pencil,
(e) the volume of a small bean,
(f) the volume of a human head,
(g) the speed of a jumbo jet,
(h) the temperature of the human body.

Section 1.1 Summary

- All physical quantities have a magnitude (size) and a unit.
- The SI base units of mass, length, time, electric current, thermodynamic temperature and amount of substance are the kilogram, metre, second, ampere, kelvin and mole respectively.
- Units of all mechanical, electrical, magnetic and thermal quantities may be derived in terms of these base units.

- Physical equations must be homogeneous (balanced). Each term in an equation must have the same base units.
- The convention for printing headings in tables of data, and for labelling graph axes, is the symbol for the physical quantity (in *italic*), followed by a forward slash, followed by the abbreviation for the unit (in roman). In handwriting, one cannot distinguish between italic and roman type.
- The order of magnitude of a number is the power of ten to which the number is raised. The order of magnitude can be used to make a check on whether a calculation gives a sensible answer.

Section 1.1 Questions

1 Write down, using scientific notation, the values of the following quantities:
(a) 6.8 pF
(b) 32 μC
(c) 60 GW

2 How many electric fires, each rated at 2.5 kW, can be powered from a generator providing 2.0 MW of electric power?

3 An atom of gold, Figure 1.7, has a diameter of 0.26 nm and the diameter of its nucleus is 5.6×10^{-3} pm. Calculate the ratio of the diameter of the atom to that of the nucleus.

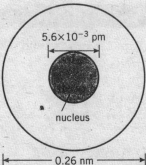

Figure 1.7 |← 0.26 nm →|

4 Determine the base units of the following quantities:
(a) energy (= *force* × *distance*)
(b) specific heat capacity
(*thermal energy change = mass × specific heat capacity × temperature change*)

5 Show that the left-hand side of the equation

pressure + $\frac{1}{2}$ *density* × (*speed*)^2 = *constant*

is homogeneous and find the base units of the constant on the right-hand side.

6 The period T of a pendulum of mass M is given by the expression

$$T = 2\pi\sqrt{\frac{I}{Mgh}}$$

where g is the gravitational field strength and h is a length.
Determine the base units of the constant I.

1.2 Scalars and vectors

At the end of Section 1.2 you should be able to:
- distinguish scalar and vector quantities and give examples of each
- add and subtract coplanar vectors
- represent a vector as two perpendicular components
- use a vector triangle to represent forces in equilibrium.

Figure 1.8 Although the athlete runs 10 km in the race, his final distance from the starting point may well be zero!

All physical quantities have a magnitude and a unit. For some quantities, magnitude and units do not give us enough information to fully describe the quantity. For example, if we are given the time for which a car travels at a certain speed, then we can calculate the distance travelled. However, we cannot find out how far the car is from its starting point unless we are told the direction of travel. In this case, the speed and direction must be specified.

A quantity which can be described fully by giving its magnitude is known as a **scalar quantity**. A **vector quantity** has magnitude and direction.

Some examples of scalar and vector quantities are given in Table 1.5.

quantity	scalar	vector
mass	✓	
weight		✓
speed	✓	
velocity		✓
force		✓
pressure	✓	
electric current		✓
temperature	✓	

Table 1.5 Some scalars and vectors

Example

A 'big wheel' at a theme park has a diameter of 14 m and people on the ride complete one revolution in 24 s. Calculate:
(a) the distance a rider moves in 3.0 minutes,
(b) the distance of the rider from the starting position.

(a) In 3.0 minutes, the rider completes $\frac{3.0 \times 60}{24} = 7.5$ revolutions.

distance travelled = 7.5 × circumference of wheel
$$= 7.5 \times 2\pi \times 7.0$$
$$= \textbf{330 m}$$

(b) 7.5 revolutions completed. Rider is $\frac{1}{2}$ revolution from starting point.
The rider is at the opposite end of a diameter of the big wheel.
So, the distance from starting position = **14 m**.

Now it's your turn

1 State whether the following quantities are scalars or vectors:
   (a) time of departure of a train,
   (b) gravitational field strength,
   (c) density of a liquid.

> **2** A student has a part-time job, earning 5 dollars per hour. He works
> for 4 hours and then spends 12 dollars. Calculate and decide
> whether each quantity can be described as a scalar or a vector:
> **(a)** the total of the amounts that have changed hands,
> **(b)** the balance.

Vector representation

When you hit a tennis ball, you have to judge the direction you want it to
move in, as well as how hard to hit it. The force you exert is therefore a vector
quantity and cannot be represented by a number alone. One way to represent
a vector is by means of an arrow. The direction of the arrow is the direction of
the vector quantity. The length of the arrow, drawn to scale, represents its
magnitude. This is illustrated in Figure 1.9.

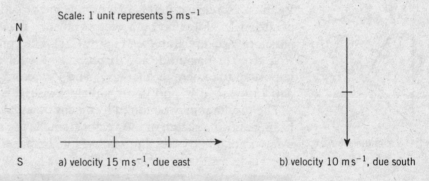

Scale: 1 unit represents 5 m s^{-1}

a) velocity 15 m s^{-1}, due east

b) velocity 10 m s^{-1}, due south

Figure 1.9 Representation of a vector quantity

Addition of vectors

The addition of two scalar quantities which have the same unit is no problem.
The quantities are added using the normal rules of addition. For example, a
beaker of volume 250 cm^3 and a bucket of volume 9.0 litres have a total
volume of 9250 cm^3.

Adding together two vectors is more difficult because they have direction
as well as magnitude. If the two vectors are in the same direction, then they
can simply be added together. Two objects of weight 50 N and 40 N have a
combined weight of 90 N because both weights act in the same direction
(vertically downwards). Figure 1.10 shows the effect of adding two forces of
magnitudes 30 N and 20 N which act in the same direction or in opposite
directions. The angle between the forces is 0° when they act in the same
direction and 180° when they are in opposite directions. For all other angles
between the directions of the forces, the combined effect, or **resultant**, is
some value between 10 N and 50 N.

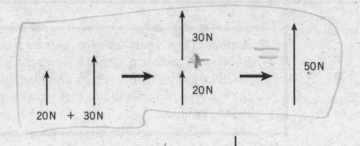

this same direction =

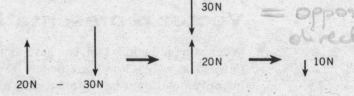

= *opposite direction*

20N + 30N

20N − 30N

Figure 1.10 Vector addition

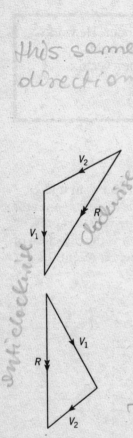

anticlockwise *clockwise*

Figure 1.11 Vector triangles

In cases where the vectors do not act in the same or opposite directions, the resultant is found by means of a **vector triangle**. Each one of the two vectors V_1 and V_2 is represented in magnitude and direction by the side of a triangle. Note that both vectors must be in either a clockwise or an anticlockwise direction (see Figure 1.11). The combined effect, or resultant R, is given in magnitude and direction by the third side of the triangle. It is important to remember that, if V_1 and V_2 are drawn clockwise, then R is anticlockwise; if V_1 and V_2 are anticlockwise, R is clockwise.

The resultant may be found by means of a scale diagram. Alternatively, having drawn a sketch of the vector triangle, the problem may be solved using trigonometry (see the Maths Note on page 17).

Example

A ship is travelling due north with a speed of $12\,\text{km h}^{-1}$ relative to the water. There is a current in the water flowing at $4.0\,\text{km h}^{-1}$ in an easterly direction relative to the shore. Determine the velocity of the ship relative to the shore by:
(a) scale drawing,
(b) calculation.

(a) By scale drawing (Figure 1.12):

Scale: 1 cm represents $2\,\text{km h}^{-1}$

resultant R

The velocity relative to the shore is:

$6.3 \times 2 =$ **12.6 km h^{-1}** in a direction **18° east of north**.

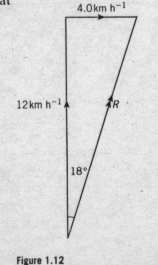

Figure 1.12

(b) By calculation:

Referring to the diagram (Figure 1.13) and using Pythagoras' theorem,

$$R^2 = 12^2 + 4^2 = 160$$
$$R = \sqrt{160} = 12.6$$
$$\tan \alpha = \frac{4}{12} = 0.33$$
$$\alpha = 18.4°$$

The velocity of the ship relative to the shore is **12.6 km h^{-1}** in a direction **18.4° east of north**.

Figure 1.13

Now it's your turn

1 A swimmer who can swim in still water at a speed of 4 km h^{-1} is swimming in a river. The river flows at a speed of 3 km h^{-1}. Calculate the speed of the swimmer relative to the river bank when she swims:
  (a) downstream,
  (b) upstream.

2 Draw to scale a vector triangle to determine the resultant of the two forces shown in Figure 1.14. Check your answer by calculating the resultant.

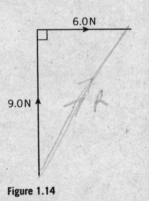

Figure 1.14

The use of a vector triangle for finding the resultant of two vectors can be demonstrated by means of a simple laboratory experiment. A weight is attached to each end of a flexible thread and the thread is then suspended over two pulleys, as shown in Figure 1.15. A third weight is attached to a point P near the centre of the thread. The string moves over the pulleys and then comes to rest. The positions of the threads are marked on a piece of paper held on a board behind the threads. This is easy to do if light from a small lamp is shone at the board. Having noted the sizes W_1 and W_2 of the weights on the ends of the thread, a vector triangle can then be drawn on the paper, as shown in Figure 1.16. The resultant of W_1 and W_2 is found to be equal in magnitude but opposite in direction to the weight W_3. If this were not so, there would be a resultant force at P and the thread and weights would move. The use of a vector triangle is justified. The three forces W_1, W_2 and W_3 are in equilibrium. The condition for the vector diagram of these forces to represent the equilibrium situation is that the three vectors should form a **closed triangle**.

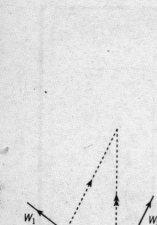

Figure 1.16 The vector triangle

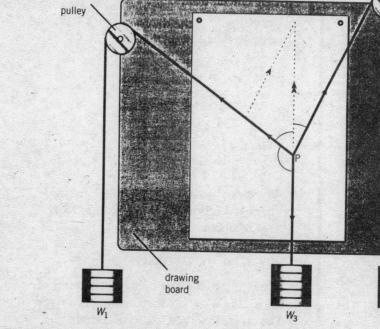

Figure 1.15 Apparatus to check the use of a vector triangle

We have considered only the addition of two vectors. When three or more vectors need to be added, the same principles apply, provided the vectors are **coplanar** (all in the same plane). The vector triangle then becomes a vector polygon: the resultant forms the missing side to close the polygon.

To subtract two vectors, reverse the direction (that is, change the sign) of the vector to be subtracted, and add.

Resolution of vectors

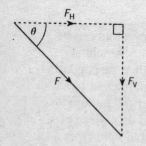

Figure 1.17 Resolving a vector into components

On pages 11–14 we saw that two vectors may be added together to produce a single resultant. This resultant behaves in the same way as the two individual vectors. It follows that a single vector may be split up, or **resolved**, into two vectors, or **components**. The combined effect of the components is the same as the original vector. In later chapters, we will see that resolution of a vector into two perpendicular components is a very useful means of solving certain types of problem.

Consider a force of magnitude F acting at an angle of θ below the horizontal (see Figure 1.17). A vector triangle can be drawn with a component F_H in the horizontal direction and a component F_V acting vertically. Remembering that F, F_H and F_V form a right-angled triangle, then

$$F_H = F \cos \theta$$

and $\quad F_V = F \sin \theta$

The force F has been resolved into two perpendicular components, F_H and F_V. The example chosen is concerned with forces, but the method applies to all types of vector quantity.

Example

A glider is launched by an aircraft with a cable, as shown in Figure 1.18. At one particular moment, the tension in the cable is 620 N and the cable makes an angle of 25° with the horizontal. Calculate:

(a) the force pulling the glider horizontally,

(b) the vertical force exerted by the cable on the nose of the glider.

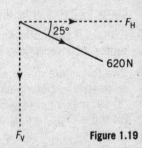

Figure 1.18

(a) horizontal component $F_H = 620 \cos 25$
$= \mathbf{560\,N}$

(b) vertical component $F_V = 620 \sin 25 = \mathbf{260\,N}$

Figure 1.19

Now it's your turn

1 An aircraft is travelling 35° east of north at a speed of 310 km h^{-1}. Calculate the speed of the aircraft in:
   (a) the northerly direction,
   (b) the easterly direction.

2 A cyclist is travelling down a hill at a speed of 9.2 m s^{-1}. The hillside makes an angle of 6.3° with the horizontal. Calculate, for the cyclist:
   (a) the vertical speed,
   (b) the horizontal speed.

Section 1.2 Summary

- A scalar quantity has magnitude only.
- A vector quantity has magnitude and direction.
- A vector quantity may be represented by an arrow, with the length of the arrow drawn to scale to give the magnitude.
- The combined effect of two (or more) vectors is called the resultant.
- Coplanar vectors may be added (or subtracted) using a vector diagram.
- The resultant may be found using a scale drawing of the vector diagram, or by calculation.
- A single vector may be divided into two separate components.
- The dividing of a vector into components is known as the resolution of the vector.
- In general, a vector is resolved into two components at right-angles to each other.

Section 1.2 Questions

1 State whether the following quantities are scalars or vectors:
(a) movement of the hands of a clock, vector
(b) frequency of vibration, S
(c) flow of water in a pipe. V

2 Speed and velocity have the same units. Explain why speed is a scalar quantity whereas velocity is a vector quantity.

3 A student states that a bag of sugar has a weight of 10 N and that this weight is a vector quantity. Discuss whether the student is correct when stating that weight is a vector.

4 Explain how an arrow may be used to represent a vector quantity.

5 Two forces are of magnitude 450 N and 240 N respectively. Determine:
(a) the maximum magnitude of the resultant force, 690 N
(b) the minimum magnitude of the resultant force, 210 N
(c) the resultant force when the forces act at right-angles to each other.
Use a vector diagram and then check your result by calculation.

6 A boat can be rowed at a speed of 7.0 km h^{-1} in still water. A river flows at a constant speed of 1.5 km h^{-1}. Use a scale diagram to determine the angle to the bank at which the boat must be rowed in order that the boat travels directly across the river.

7 Two forces act at a point P as shown in Figure 1.20. Draw a vector diagram, to scale, to determine the resultant force. Check your work by calculation.

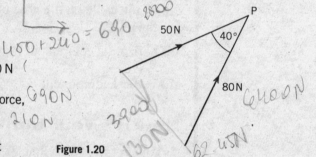

450 + 240 = 690   2800

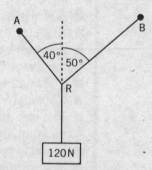

Figure 1.20

Exam-style Questions

1 (a) (i) Explain what is meant by a *base unit*.
     (ii) Give four examples of base units.
(b) State what is meant by a *derived unit*.
(c) (i) For any equation to be valid, it must be homogeneous. Explain what is meant by a *homogeneous* equation.
     (ii) The pressure p of an ideal gas of density ρ is given by the equation

$$p = \tfrac{1}{3}\rho\langle c^2 \rangle$$

where $\langle c^2 \rangle$ is the mean-square-speed (i.e. it is a quantity measured as [speed]^2). Use base units to show that the equation is homogeneous.

2 (a) Determine the base units of:
     (i) work done,
     (ii) the moment of a force.
(b) Explain why your answers to (a) mean that caution is required when the homogeneity of an equation is being tested.

3 (a) Distinguish between a *scalar* and a *vector* quantity.
(b) A mass of weight 120 N is hung from two strings as shown in Figure 1.21.

Figure 1.21

Determine, by scale drawing or by calculation, the tension in:
(i) RA,   (ii) RB.

(c) Use your answers in (b) to determine the horizontal component of the tension in:
  (i) RA,   **(ii)** RB.
  Comment on your answer

4 A fielder in a cricket match throws the ball to the wicket-keeper. At one moment of time, the ball has a horizontal velocity of $16\,\mathrm{m\,s^{-1}}$ and a velocity in the vertically upward direction of $8.9\,\mathrm{m\,s^{-1}}$.
  (a) Determine, for the ball:
    (i) its resultant speed,
    (ii) the direction in which it is travelling relative to the horizontal.

(b) During the flight of the ball to the wicket-keeper, the horizontal velocity remains unchanged. The speed of the ball at the moment when the wicket-keeper catches it is $19\,\mathrm{m\,s^{-1}}$. Calculate, for the ball just as it is caught:
  (i) its vertical speed,
  (ii) the angle that the path of the ball makes with the horizontal.
(c) Suggest with a reason whether the ball, at the moment it is caught, is rising or falling.

Maths Note

Sine rule
For any triangle (Figure 1.22, top),

$$\frac{a}{\sin A} = \frac{b}{\sin B} = \frac{c}{\sin C}$$

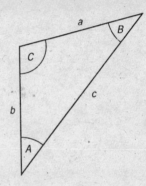

Cosine rule
For any triangle,

$$a^2 = b^2 + c^2 - 2bc \cos A$$

$$b^2 = a^2 + c^2 - 2ac \cos B$$

$$c^2 = a^2 + b^2 - 2ab \cos C$$

Pythagoras' theorem
For a right-angled triangle (Figure 1.22, bottom),

$$h^2 = o^2 + a^2$$

Also for a right-angled triangle:

$$\sin \theta = \frac{o}{h}$$

$$\cos \theta = \frac{a}{h}$$

$$\tan \theta = \frac{o}{a}$$

Figure 1.22

2. Measurement techniques

This book is about experimental, rather than purely theoretical, physics. The ideas you will meet in it can all be tested by experiment. Because of this, it is important that you should have an informed and confident approach to work in the laboratory. You should be able to make a sensible choice of what instrument to use to measure a particular physical quantity. You should also be aware of the sources of error and uncertainty in your experimental work. This chapter looks at some of the measurement techniques available in your school or college physics laboratory. You will also learn how errors and uncertainties in your readings affect the final value of the quantity to be measured.

2.1 Measurements

At the end of Section 2.1 you should be able to:
- use techniques for the measurement of length, volume, angle, mass, time, temperature and electrical quantities, all to appropriate ranges of magnitude. In particular, you should be able to:
  - measure lengths using a ruler, vernier scale and micrometer
  - measure weight and hence mass using spring and lever balances
  - measure an angle using a protractor
  - measure time intervals using clocks, stopwatches and the calibrated time-base of a cathode-ray oscilloscope (c.r.o.)
  - measure temperature using a thermometer as a sensor
  - use ammeters and voltmeters with appropriate scales
  - use a galvanometer in null methods
  - use a cathode-ray oscilloscope (c.r.o.)
  - use a calibrated Hall probe
- use both analogue scales and digital displays
- use calibration curves.

All experiments designed to obtain a quantitative result for a physical quantity involve measurements. These measurements must be of some combination of the base quantities length, mass, time, temperature and current. (To complete the list, we should include quantity of substance and luminous intensity, but these are not encountered in experimental work in A/AS Physics.) In the following sections we shall look at the methods available for measuring the base quantities in a school or college laboratory. By understanding the principles of the available methods, we shall be able to make an informed decision about the choice of a particular technique, with

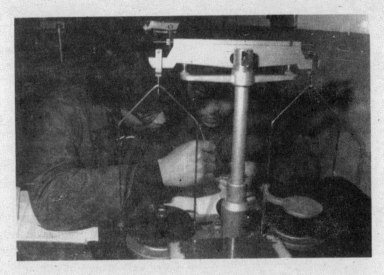

Figure 2.1 Students doing an experiment in a physics laboratory

respect to making the experiment as precise and reproducible as possible, and avoiding sources of systematic error. For all of the quantities, the effective choice will be limited by what instruments are available in your laboratory. However, in one type of exam question, on planning and design, you may be asked to devise an experiment and draw on your theoretical, rather than practical, knowledge of various types of apparatus.

At A/AS level, students generally assume that the calibration of the instruments they use is correct. However, it is worth thinking about how to compare the calibration of one instrument against another, even if this is a check you will very seldom make. In a planning/design question, you might be asked to suggest a method of calibration. Generally it is easy to *compare* the calibration of different instruments, but not so easy to determine which of two instruments giving different readings is correct.

On the pages that follow, we shall spend most time on methods for measuring length, because length-measuring instruments of several different types will be available in your laboratory.

Methods of measuring length

The metre rule

The simplest length-measuring instrument to be found in your laboratory is a metre (or half-metre) rule. It has the great advantages of being cheap, convenient and simple to use. A relatively unskilled student should have no difficulty in taking a reading with an uncertainty of 0.5 mm. However, you should be aware of three possible sources of error in using a metre rule.

The first may arise if the end of the rule is worn, giving rise to a zero error (Figure 2.2). For this reason, it is bad practice to place the zero end of the rule against one end of the object to be measured and to take the reading at the other end. You should place the object against the rule so that a reading is made at each end of the object. The length of the object is then obtained by subtraction of the two readings. A zero error like this is a systematic error,

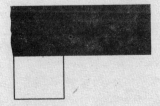

Figure 2.2 Zero error with a metre rule

because it is involved every time a reading is taken from the zero end of the rule. (A more detailed explanation of systematic errors will be found in Section 2.2.) In general, the zero reading of any instrument may be subject to an error. We shall meet this type of error again in the micrometer screw gauge and in the ammeter.

The calibration of the metre rule may give rise to another systematic error because the markings are incorrect. Try comparing the 30 cm graduated length of one plastic rule with the same nominal length on another. You are quite likely to find a discrepancy of one or two millimetres. One of the reasons why wooden or plastic metre rules are cheap is that the manufacturer does not claim any great accuracy for the scale markings. The calibration may be checked by laying the rule alongside a more accurate rule, such as a steel 30 cm or metre rule, and noting any discrepancy. If you compare an engineer's steel rule with a plastic rule, you will see at once that the engraved marks on the steel rule are much finer than the impressed marks on the plastic. Of course, the extra care which has been taken in engraving the steel rule has to be paid for. A one-metre steel rule is many times more expensive than a plastic metre rule.

Another source of error with the metre rule is parallax error. If the object to be measured is not on the same level as the graduated surface of the rule, the angle at which the scale is viewed will affect the result (Figure 2.3). This is a random error (see Section 2.2), because the angle of view may be different for different readings. It may be reduced by arranging the rule so that there is no gap between the scale and the object. Parallax error is also important in reading any instrument in which a needle moves over a scale. A rather sophisticated way of eliminating parallax error is to place a mirror alongside the scale. When the needle and scale are viewed directly, the needle and its image in the mirror coincide. This ensures that the scale reading is always taken at the same viewing angle.

The smallest division on the metre rule is 1 mm. If you take precautions to avoid parallax error, you should be able to estimate a reading to about 0.5 mm. If you are measuring the length of an object by taking a reading at each end, the uncertainties add to give a total uncertainty of 1 mm. The range of the metre rule is from 1 mm to 1000 mm. To measure a length of more than 1 m with a metre rule will introduce a further uncertainty, of perhaps 1 mm or 2 mm, because of the difficulty of making a reference mark at the 1 m end of the rule and moving the rule so that the zero exactly corresponds with this reference. It is usually better to use a steel tape to measure lengths of more than 1 m.

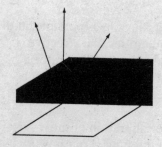

Figure 2.3 Parallax error with a metre rule

The micrometer screw gauge

The type of micrometer screw gauge available in a school or college laboratory may be used to measure the dimensions of objects up to a maximum of about 50 mm. Measurements can easily be made with an uncertainty of 10 μm or less. The principle of the instrument is the magnification of linear motion using the circular motion of a screw. The instrument consists of a U-shaped piece of steel with a fixed, plane, end-piece A (see Figure 2.4). Opposite this is a screw with a corresponding end-piece B. The position of the screw can be adjusted using the ratchet C which is connected to the thimble D. There are graduations along the barrel of the instrument (the bearing in which the screw turns), and round the

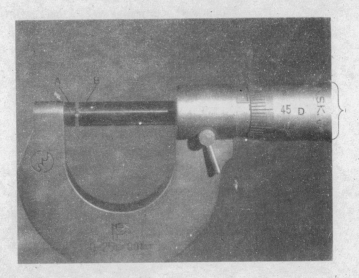

Figure 2.4 Micrometer screw gauge

circumference of the thimble. The purpose of the ratchet is to ensure that the same torque (that is, the amount of twist) is applied to the thimble for each reading. If this torque is exceeded, the ratchet slips. The object to be measured is placed in the jaws of the gauge between end-pieces A and B, and B is screwed down on to the object, using the ratchet C, until the ratchet slips.

The screw advances exactly 1 mm for two revolutions. That is, the pitch of the screw is 0.5 mm or 500 µm. If you look at the graduations on the barrel of the screw bearing you will see that there are divisions every 0.5 mm. The reading on the barrel corresponds to the position of the edge of the thimble (see Figure 2.5). When taking a reading it is important to check which half of the millimetre the edge of the barrel is in. The graduations round the circumference of the thimble run from 0 to 50. Each division corresponds to one-hundredth of a mm, or 10 µm. The reading on the thimble is added to the reading on the barrel. Thus, Figure 2.5 shows a reading of 9.5 mm on the barrel plus 0.36 mm on the thimble, 9.86 mm in total. You can easily read to the nearest division on the thimble; that is, to the nearest 0.01 mm (10 µm).

The micrometer screw gauge is very likely to have a systematic zero error. Every time you use the gauge, you should check the zero error by moving face B so that it makes contact with face A. The screw must be tightened with the ratchet C, so that a reproducible zero is obtained. Then take the reading on the barrel and the thimble. This gives the zero error, which must be allowed for in all subsequent readings. Figure 2.6a shows a screw gauge with a zero error of +0.12 mm. If this were the error which applied when the reading of 9.86 mm was obtained in Figure 2.5, the true length of the object would be 9.86 mm – 0.12 mm = 9.74 mm. Figure 2.6b shows a zero error of –0.08 mm. In this case, the true length of the object in Figure 2.5 would be 9.86 mm + 0.08 mm = 9.94 mm.

In the case of a wooden or plastic metre rule, it is good practice to check the calibration against an engineer's steel rule, if one is available in your laboratory. It would be unusual to do the same with a micrometer screw gauge, but if there is doubt about the calibration of a particular gauge, it can be checked by measuring the dimensions of a series of gauge blocks. A gauge

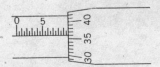

Figure 2.5 Screw gauge scales with a reading of 9.86 mm

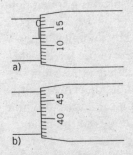

Figure 2.6 a) Zero error of +0.12 mm b) zero error of –0.08 mm

block is a rectangular steel block with faces which are accurately plane and parallel. The length of the gauge block is known to an uncertainty of less than $1\,\mu\text{m}$. However, not many school laboratories possess gauge blocks.

The vernier caliper

A vernier caliper is a versatile instrument for measuring the dimensions of an object, the diameter of a hole, or the depth of a hole. Its range is up to about 100 mm, and it can be read to 0.1 mm or 0.05 mm depending on the type of vernier with which it is fitted. It consists of a steel mm scale A with two reference posts at the zero mark (see Figure 2.7):

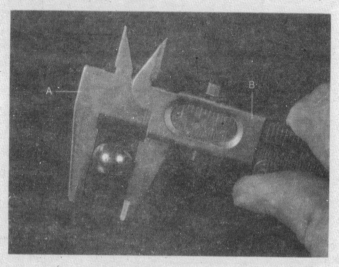

Figure 2.7 Vernier caliper

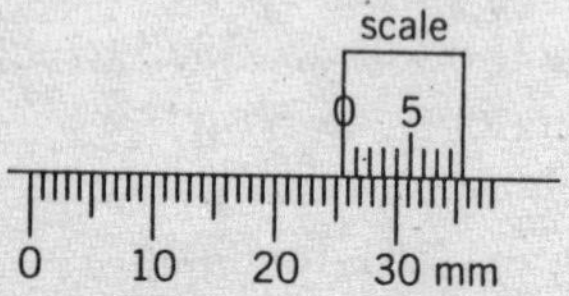

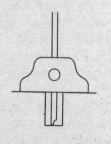

Figure 2.8 Vernier scale with a reading of 25.4 mm

A sliding part B moves along the scale. The slider has the vernier scale engraved on it. The zero of the vernier corresponds with reference posts on the sliding part. One set of reference posts, those with the straight parts on the inside, is used like the jaws of a screw gauge: the object to be measured is placed between the jaws, and the sliding part B is moved along until the object is gripped tightly. A reading to the nearest mm is taken on the fixed scale, at the zero end of the vernier scale. The reading to a tenth of a mm is obtained by finding where a graduation of the vernier scale coincides with a graduation of the fixed scale. Figure 2.8 shows the scale of a vernier caliper giving a reading of 25.4 mm.

The second set of jaws has the straight parts on the outside. These can be used to measure the diameter of a hole. The jaws are placed inside the hole and are moved apart until they are in contact with the edges of the hole. The scale and vernier can then be read.

A pin at the end of the sliding part of the caliper can be used to measure the depth of a hole; for example, a hole which has been drilled in, but not right through, a wooden board. The end of the fixed scale is placed on the board, across the hole, and the pin moved into the hole until it reaches the bottom (see Figure 2.9). The reading of the scale and vernier gives the depth of the hole.

As with the micrometer screw gauge, the vernier caliper should be checked for a systematic zero error before taking a reading.

Figure 2.9 Measurement of the depth of a blind hole

Examples

1 Figure 2.10a shows the scale of a micrometer screw gauge when the zero error is being checked, and Figure 2.10b the scale when the gauge is tightened on an object. What is the length of the object?

From Figure 2.10a, the zero error is +0.12 mm. The reading in Figure 2.10b is 15.62 mm. The length of the object is thus (15.62 − 0.12) mm = **15.50 mm**.

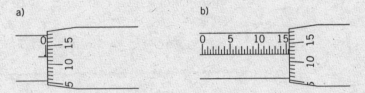

Figure 2.10

2 Figure 2.11 shows the scale of a vernier caliper. What is the reading?

The zero of the vernier scale is between the 5.5 cm and 5.6 cm divisions of the fixed scale. There is coincidence between the third graduation of the vernier scale and one of the graduations of the fixed scale. The reading is thus **5.53 cm** or **55.3 mm**.

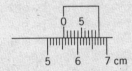

Figure 2.11

Now it's your turn

Figures 2.12a and 2.12b show the scales of a micrometer screw gauge when the zero is being checked, and again when measuring the diameter of an object. What is the diameter?

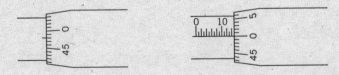

Figure 2.12

Choice of method

A summary of the range and reading uncertainty of length-measuring instruments is given in Table 2.1.

Table 2.1 Length-measuring instruments

instrument	range	reading uncertainty	notes
metre rule	1 m	0.5 mm	check zero, calibration errors
micrometer screw gauge	50 mm	0.01 mm	check zero error
vernier caliper	100 mm	0.1 mm	versatile: inside and outside diameters, depth

In deciding which instrument to use in a particular experiment, you should consider first of all the nature of the length measurement you have to make. For example, if you need to find the diameter of a steel sphere, the screw gauge and caliper techniques are obvious candidates. You should then consider whether you need the greater accuracy of the micrometer. In a particular experiment, the uncertainty in the diameter of the sphere may be the dominant uncertainty (see Section 2.2), and in such a case the fact that the accuracy available with the screw gauge is ten times that for the vernier caliper will sway the argument. In an experiment which may last some time, you should also think about the resources of your laboratory. Is it sensible to use what may be one of only a small number of available screw gauges, when they may also be in demand by other students for other experiments? Sometimes, in design questions, you are asked to think about the cost of setting up an experiment. We have mentioned the difference in cost of a steel metre rule compared with wooden or plastic rules. It would be foolish economics to supply a steel rule for each of a number of students, when it would be perfectly adequate to provide each of them with a wooden rule and have one steel rule available in the laboratory for calibration purposes.

Application: measurement of pressure difference

A difference in gas pressure may be measured by comparing the heights of liquid in the two arms of a U-tube. Figure 2.13 shows a U-tube connected to a

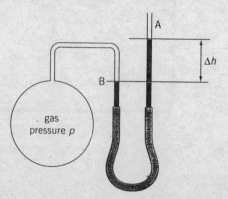

Figure 2.13

container of gas. The pressure above the liquid in tube A is atmospheric pressure p_{atm}. The pressure above the liquid in tube B, and hence the pressure of gas in the container, is p. The relation between the pressures is

$$p = p_{atm} + \Delta h \rho g$$

where Δh is the difference in vertical height between the levels of the liquid in the two arms of the tube, ρ is the density of the liquid, and g is the acceleration of free fall. To find the pressure of the gas, all we need to do is to measure the difference Δh between the liquid levels, assuming that p_{atm}, ρ (and g) are known.

In some laboratories, a U-tube mounted on a board to which a vertical millimetre scale is attached may be available. This device is called a manometer. It is then a simple matter to measure Δh. If the manometer contains oil or water, the liquid in the tube will have a curved surface which is concave downwards (Figure 2.14a). This surface is called the meniscus. Use a set-square to find the reading on the vertical scale corresponding to the *bottom* of the meniscus in each side of the U-tube. The surface of the liquid in a manometer filled with mercury will be convex (Figure 2.14b), and in this case you should read to the *top* of the meniscus. Again, use a set-square to avoid parallax error. If a manometer is not available, you will have to arrange your own system of U-tube and scale; for example, a half-metre rule. Make sure that the scale is clamped vertically.

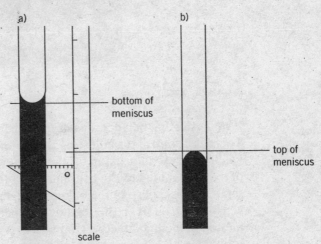

Figure 2.14

Methods of measuring mass

The method of measuring mass is with a balance. In fact, balances compare the *weight* of the unknown mass with the weight of a standard mass. But because weight is proportional to mass, equality between the unknown weight and the weight of the standard mass means that the unknown mass is equal to the standard mass.

In your laboratory, you may have access to a number of different types of balance, including the top-pan balance, the lever balance and the spring balance. It is important that you should familiarise yourself with the use of all types that are available to you, so that you do not restrict your choice to one particular type. Note also that some types of spring balance may be calibrated in force units (that is, in newton) rather than in mass units (kilogram).

The top-pan balance

The top-pan balance (Figure 2.15) is a direct-reading instrument, based on a pressure sensor, or sometimes a spring. The unknown mass is placed on the pan, and its weight applies a force to the sensor. The mass corresponding to this force is displayed on a digital read-out.

Figure 2.15 Top-pan balance

When using the balance, ensure that the initial (unloaded) reading is zero. There is a control for adjusting the zero reading. The balance may have a tare facility, for use in backing off the mass of an empty container so that the mass of material added to the container is obtained directly. This works in the same way as adjusting the balance for zero error.

The uncertainty in the reading of a particular top-pan balance will be quoted in the manufacturer's manual. As with other digital instruments, it is likely to be expressed as a percentage uncertainty of the reading shown on the scale, together with the uncertainty in the final figure of the display.

The spring balance and the lever balance

Spring balances (Figure 2.16) are based on Hooke's law (see page 86): the extension of a loaded spring is proportional to the load. The extension is measured directly, by a marker moving along a straight scale, or by a pointer moving over a circular scale. As with any instrument using a scale and pointer, you should take care not to introduce a parallax error when you take readings. Position yourself so that your line of sight is perpendicular to the scale. Before placing the object of unknown mass on the pan, check for zero error. There is likely to be a zero-error adjustment screw on the balance.

Figure 2.16 Spring balance

Lever balances (Figure 2.17) are based on the principle of moments. In one common type, the unknown mass is placed on a pan, and balance is achieved by sliding a mass along a bar, calibrated in mass units, until the bar is horizontal. This represents the condition in which the torque of the load is equal and opposite to the torque of the sliding mass and the bar. A reading is taken from the edge of the sliding mass on the divisions marked on the bar. In this case, parallax error is less likely to be serious. Again, check for zero error before taking a reading.

Figure 2.17 Lever balance

Figure 2.18 Lever balance with circular scale

Another type of lever balance has a pointer moving along a circular scale (Figure 2.18). A weight on the pointer arm is placed in one of two positions in order to change the range (for example, from 0–100 g to 0–1 kg).

   Both of these types of balance are used more for the convenience of obtaining a rapid, approximate reading, rather than for an accurate determination. An indication of the uncertainty involved in readings with a particular balance can be obtained from the smallest division on the scale.

Example

The mass of a quantity of chemical is determined using a lever balance. Over the range of masses involved, the separation between mass graduations on the bar is 2 g. The mass of the empty container is 56 g, and the mass of the container plus the chemical is 104 g. Find the mass of the chemical, and the uncertainty in this value.

By subtraction, the mass of the chemical is 104 − 56 = **48 g**. The uncertainty in each reading is likely to be the difference between mass graduations on the beam, that is ±2 g. Each of the two readings has an uncertainty of ±2 g: the uncertainty in the mass of the chemical is thus **±4 g**.

Now it's your turn

The mass of a chemical used to make up a solution is determined as follows. A dish containing the chemical is placed on the pan of a spring balance. The pointer reading on the scale is shown in Figure 2.19a. The chemical is then tipped into a known volume of water, and the empty dish replaced on the pan, giving the pointer reading shown in Figure 2.19b. What is the mass of chemical, and what is the uncertainty in this value?

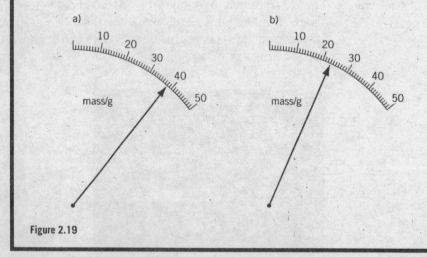

Figure 2.19

Choice of method

As stated above, the top-pan and spring balance are direct-reading instruments. This means that readings can be obtained quickly and conveniently. The lever balance requires adjustment of the sliding mass, but this takes only a very short time. There may be some rules in your laboratory about which type of balance should be used for which task. In general, the pan of the balance should be kept clean and dry. Do not weigh out loose chemicals on the pan; always use a container, the mass of which you have determined beforehand, or for which you have made allowance using the tare control.

In general, choose a balance of sensitivity appropriate to the experiment you are carrying out.

Application: current balance

A U-shaped magnet is placed on a top-pan balance (Figure 2.20). A wire is clamped so that it runs along the channel of the magnet. The wire is connected in a circuit with a d.c. supply, a rheostat, an ammeter and a switch. When the supply is switched on, the balance reading is seen to change, because a force is exerted on the wire in the magnetic field. By Newton's third law, a force is also exerted on the magnet. This is detected by the change Δm in the mass reading. This change must be converted into force F by multiplying Δm by g. The variation with current I of the magnetic force F may be determined. The equation

$$F = BIl \qquad \text{(see page 322)}$$

where l is the length of the wire in the magnetic field, may be verified. The direction of the force, as predicted theoretically by Fleming's left-hand rule, may also be verified by checking whether the mass reading increases or decreases for a given current direction.

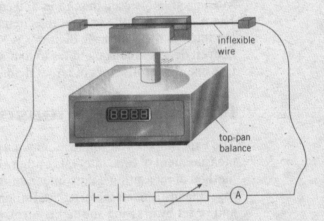

Figure 2.20 Current balance experiment

Measuring an angle

Angles are measured using an instrument called a protractor. It looks like a semi-circular, or sometimes circular, ruler, with its scale marked out in angular measure, invariably degrees rather than radians. The centre of the circle is clearly marked.

To measure the angle between two lines, the centre of the circle of the protractor is placed exactly over the point of intersection of the lines and one line is aligned with the 0° direction of the protractor (Figure 2.21). The angle between the lines is then given by the reading on the scale at which the second line passes through the circumference of the circle.

If the direction of a single line needs to be defined, this is always referred to the direction of the Ox axis, the horizontal axis pointing towards the right, as zero.

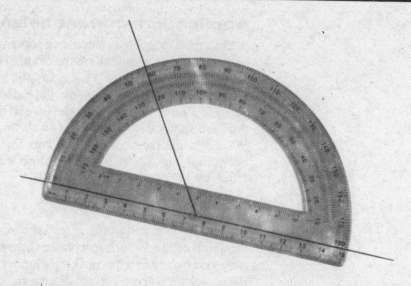

Figure 2.21 Using a protractor

Most protractors used in schools and colleges have a diameter of about 10 cm. The size of the scale at the circumference of the circle is then such that the interval between scale divisions is 1°. It is easy to make a reading to the nearest degree, and sometimes to half a degree, if the line being measured is fine enough. Protractors of larger diameter may be marked in half degrees.

Methods of measuring time

The experiments you will meet in your practical physics course deal with the measurement of time intervals, rather than with absolute time. The basic method of measuring a time interval is with a stopclock or stopwatch. In each case, the instrument is started and stopped by pressing a lever or a button, and re-set by pressing another control. You should familiarise yourself with the way of operating the instrument before you start a timing experiment in earnest. Remember that the reaction time of the experimenter (a few tenths of a second) is likely to be much greater than the uncertainty of the instrument itself. If you do not reduce the effects of reaction time, an unacceptable systematic error may be built in to the experiment. As explained in Section 2.2, one way of reducing the effect of reaction time is to time enough events (for example, the swings of a pendulum) to make the interval being measured very much longer than the experimenter's reaction time. A good technique is to count the events (the swings), commencing by counting down to zero, and starting the timer at the zero count. Wherever possible, work with at least 20 seconds' worth of events (oscillations), and repeat each set of timings three times. (Sometimes, when carrying out experiments on damped oscillations, you will have to be satisfied with fewer swings, but try not to go below intervals of 10 seconds.)

The stopclock

A mechanical, spring-powered stopclock will have an analogue display; that is, a hand (or hands) which move round a dial (Figure 2.22). Such an instrument is likely to read to the nearest one-fifth of a second.

Figure 2.22 Analogue stopclock

The stopwatch or digital timer

This instrument has a digital display (Figure 2.23). It is based on the oscillations of a quartz crystal. The read-out will probably be to the nearest one-hundredth of a second. In addition to the start, stop and re-set controls, digital stopwatches often have a 'lap' facility, which allows one reading to be held in the display while the watch is still running. Because of this complexity, it is vital that you know the functions of all the controls.

Your own wristwatch may well have a built-in stopwatch, which may be just as precise as the watches available in the laboratory. However, the start and stop controls on wrist stopwatches are sometimes rather small, and it is important that you should not fumble a start or stop signal.

Figure 2.23 Digital timer

Choice of method

Often, students are attracted to a digital stopwatch because it reads to one-hundredth of a second. However, in all experiments in which the start and stop signals are applied manually, such accuracy is unnecessary and inappropriate. The reaction time of the experimenter, which is likely to be a few tenths of a second, will cancel out the accuracy of the watch. It would be misleading, and bad practice, to enter times such as '21.112 s' in a table of results if the systematic error due to reaction time had not been fully accounted for. Thus, if you are doing an experiment on the timing of oscillations and there are no digital timers available, you will be at no disadvantage if you have to use an analogue stopclock.

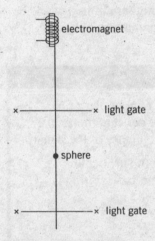

electromagnet

× ——————— × light gate

• sphere

× ——————— × light gate

Figure 2.24 Determination of the acceleration of free fall

Application: determination of the acceleration of free fall

A steel sphere is released from an electromagnet and falls under gravity. As it falls, it passes through light gates which switch an electronic timer on and off (Figure 2.24). The acceleration of free fall can be determined from the values of the time intervals and distances.

This is an experiment in which electronic switching is essential in order to reduce the potentially very large error caused by the reaction time of the experimenter. Here, timing to one-hundredth of a second is essential.

Application: measurement of frequency using a cathode-ray oscilloscope

A cathode-ray oscilloscope (c.r.o.) has a calibrated time-base, so that measurements from the screen of the c.r.o. can be used to give values of time intervals. One application is to measure the frequency of a periodic signal, for example the sine-wave output of a signal generator. The signal is connected to the Y-input of the c.r.o., and the Y-amplifier and time-base controls are adjusted until a trace of at least one, but fewer than about five, complete cycles of the signal is obtained on the screen. The distance L on the graticule (the scale on the screen) corresponding to one complete cycle is measured (Figure 2.25). It is good practice to measure the length of, say, four cycles, and then divide by four so as to obtain an average value of L. The graticule will probably be divided into centimetre and perhaps millimetre or two-millimetre divisions. If the time-base setting is x (which will be in units of seconds, milliseconds or microseconds per centimetre), the time T for one cycle is given by $T = Lx$. The frequency f of the

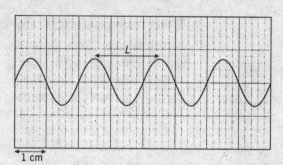

Figure 2.25 Measurement of frequency using a c.r.o.

1 cm

signal is then obtained from $f = 1/T$. The uncertainty of the determination will depend on how well you can estimate the measurement of the length of the cycle from the graticule. Remembering that the trace has a finite width, you can probably measure this length to an uncertainty of about ±2 mm.

As with most instruments you will use, your laboratory time will be so limited that you will probably have to take the time-base settings on trust. However, it is worth thinking about possible methods of checking the calibration. You could try checking against a calibrated signal generator: but who is to say which of the signal generator or c.r.o. has the correct calibration? Another method would be to connect a microphone to the Y-input, and sound a tuning fork of known frequency near the microphone.

Example

The output of a signal generator is connected to the Y-input of a c.r.o. When the time-base control is set at 0.50 milliseconds per centimetre, the trace shown in Figure 2.26 is obtained. What is the frequency of the signal?

Two complete cycles of the trace occupy 6.0 cm on the graticule. The length of one cycle is therefore 3.0 cm. The time-base setting is 0.50 ms cm^{-1}, so 3.0 cm is equivalent to $3.0 \times 0.50 = 1.5$ ms. The frequency is thus $1/1.5 \times 10^{-3} = $ **670 Hz**.

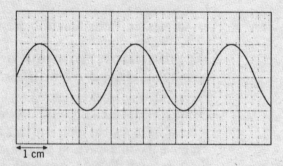

1 cm

Figure 2.26

Now it's your turn

The same signal is applied to the Y-input of the c.r.o. as in the example above, but the time-base control is changed to 2.0 milliseconds per centimetre. How many complete cycles of the trace will appear on the screen, which is 8.0 cm wide?

Methods of measuring temperature

The SI unit of temperature, the kelvin, is based on the ideal gas (or thermo-dynamic) scale of temperature. The scale may be arrived at using an instrument called a constant-volume gas thermometer (see page 136). The equation relating the Celsius temperature scale to the thermodynamic scale is

$$\theta = T - 273.15$$

where θ is in degrees Celsius and T is in kelvin.

Fortunately, in your practical course you will come across nothing more complicated than a liquid-in-glass (probably a mercury-in-glass) thermometer. You may, however, do experiments on thermocouple thermometers and resistance thermometers, and assess their suitability for use as a thermometer.

The mercury-in-glass thermometer

Liquid-in-glass thermometers are based on the thermal expansion of a liquid. A quantity of liquid is contained in a bulb at the end of a thin capillary tube. The space above the liquid contains an inert gas at low pressure. If the bulb is placed in a beaker of water which is gradually heated, the liquid expands and the thread of liquid occupies more and more of the capillary. The capillary is graduated: the position of the end of the thread gives the temperature.

Most thermal physics experiments which you will carry out will involve the measurement of temperatures between 0 °C (the temperature of melting ice) and 100 °C (the temperature of steam above boiling water at a pressure of 1 atmosphere). The most useful thermometer covering this range is a mercury-in-glass thermometer (Figure 2.27) with graduations from −10 °C to 110 °C, in 1 °C intervals. You will find it easy to take readings to the nearest half degree, and perhaps to 0.2 °C. There are a number of precautions you should take when using the thermometer. Always allow time for the thermometer to reach thermal equilibrium with its surroundings. If you are measuring the temperature of a beaker of liquid which is being heated, the liquid must be thoroughly stirred before taking the reading. (Because of convection currents, there is a temperature difference of several degrees between the top and the bottom of the liquid.) The thermometer is calibrated for use at a standard depth of immersion; this may be stated on the stem. If it is not, try to ensure that the thermometer is always immersed in the liquid to the same depth. The length of the bulb plus about 20 mm is a good rule of thumb.

There are some points to be made about safety. Thermometers are relatively fragile instruments. Because of their shape, they have a tendency to roll along the bench-top. Make sure that your thermometer does not roll off and fall to the ground. If a thermometer does break and the mercury in it comes out, do not be tempted to play with it. Mercury is a poison. To reduce the risk of breakage, do not use the thermometer as a stirrer, unless it is of a robust type designated as a 'stirring thermometer'. If you have to fit a thermometer through a rubber bung, make sure that the hole in the bung is large enough, and lubricate the rubber thoroughly with soap. Wear gloves and grip the thermometer so that, if it breaks, your wrist will not be cut.

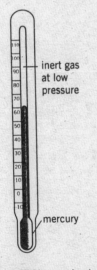

Figure 2.27 Mercury-in-glass thermometer

inert gas at low pressure

mercury

Example

The temperature of a mixture of ice, salt and water is measured using a mercury-in-glass thermometer. When thermal equilibrium has been reached, the mercury thread in the thermometer is as shown in Figure 2.28. What is the temperature of the mixture? What is the uncertainty in this value?

By interpolation between the scale divisions, the temperature reading is **−2.5°C**. The uncertainty is probably about **±0.2 deg C**.

Figure 2.28

Now it's your turn

The temperature of a solidifying liquid is measured using a liquid-in-glass thermometer. When thermal equilibrium has been reached, the liquid thread in the thermometer is as shown in Figure 2.29. What is the solidification temperature? What is the uncertainty in this value?

Figure 2.29

The thermocouple thermometer

A thermocouple thermometer consists of two wires made of different metals or alloys, joined at one end. The other ends of the wires are connected to the terminals of a millivoltmeter. This may be a digital instrument, which is calibrated in °C (see Figure 2.30). The thermocouple may also be connected to a datalogger. The junction is placed in thermal contact with the object, the temperature of which is required.

Figure 2.30 Thermocouple thermometer

The thermocouple thermometer actually measures the difference in temperature between the junction of the two metals (the hot junction) and a cold junction. In some applications, the cold junction is placed in an ice–water mixture, so as to achieve a known reference (see Figure 6.9, page 140).

The thermocouple may be connected to a millivoltmeter which has not been calibrated in temperature units. In this case, you will have to make use of the known variation of thermoelectric e.m.f. for that particular pair of metals with temperature. You will need to draw a **calibration curve**, a graph of temperature against e.m.f., so that you can read off the temperature corresponding to an e.m.f. reading. Note that this graph is often a curve rather than a straight line, as shown in Figure 2.31. In some cases the curvature is so much that a particular e.m.f. can be obtained for two different temperatures. Clearly, this restricts the temperature range over which the thermocouple is useful.

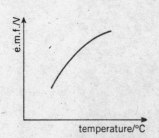

Figure 2.31 Calibration curve for a thermocouple

Choice of method

The heat capacity of the bulb of a liquid-in-glass thermometer is much greater than that of the hot junction of a thermocouple. For this reason, the thermocouple is particularly useful if a rapidly varying temperature is to be measured, or if the object, the temperature of which is required, has a small heat capacity.

Mercury-in-glass thermometers are available to cover the temperature range from about −40 °C to 350 °C. Thermocouples using different pairs of metal or alloy wires can cover a much larger range.

The choice of a particular thermometer in a given application will depend on the range of temperatures to be covered, the heat capacity of the object, and whether the temperature is varying rapidly.

Methods of measuring current and potential difference

Your physics laboratory will probably have a selection of instruments for measuring current and potential difference (voltage). The two main types are analogue meters, in which a pointer moves over a scale (Figure 2.32), and digital, in which the value is displayed on a read-out consisting of a series of integers (Figure 2.33).

Analogue meters

The normal analogue meter is restricted to the measurement of the relevant quantity over a single range. For example, a 0–1 A d.c. ammeter will measure direct currents in the range from zero to 1 A. A 0–30 V d.c. voltmeter will measure steady potential differences in the range from zero to 30 V. Some analogue meters have a dual-range facility, with a common negative terminal and two positive terminals, each of which is associated with a separate scale on the instrument. Thus, one scale might be 0–3 A, and the other 0–10 A. Each of the positive terminals is marked with the scale to which it refers. Be careful to take the reading on the scale corresponding to the pair of terminals you have selected.

Figure 2.32 Analogue meter

Figure 2.33 Digital meter

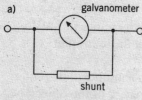

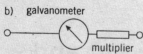

Figure 2.34 a) Galvanometer with shunt, for current measurements
b) galvanometer with multiplier, for voltage measurements

Analogue meters are subject to zero error. Before switching on the circuit, check whether the needle is exactly at the zero mark. If it is not, return the needle to zero by adjusting the screw at the needle pivot. There is also the possibility of parallax error. The needle should be read from a position directly above it and the scale, and not from one side. Sometimes a strip of mirror is provided close to the scale so that the experimenter can align the needle with its image in the mirror, ensuring that viewing is vertical. The uncertainty associated with a current or voltage reading from an analogue meter is usually taken to be (±) half the smallest scale reading.

A **galvanometer** is a sensitive current-measuring analogue meter. It may be converted into an ammeter by the connection of a suitable resistor in parallel with the meter (Figure 2.34a). Such a resistor is called a *shunt*. The meter may be converted into a voltmeter by the connection of a suitable resistor in series with the meter (Figure 2.34b). Such a resistor is called a *multiplier*.

The manufacturers provide shunts and multipliers which are clearly labelled with the conversion function and full-scale deflection, for attachment to the basic galvanometer (Figure 2.35). All you need do is to select the shunt or multiplier required for your experiment and make sure that you apply the correct factor when reading the scale.

A galvanometer with a centre-zero scale shows negative currents when the needle is to the left-hand side of the zero mark and positive currents when it is to the right. This type of meter is often used as a null indicator; that is, to detect when the current in a part of a circuit is zero.

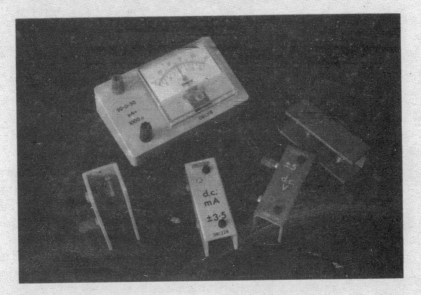

Figure 2.35 Shunts and multipliers for use with a galvanometer

Digital meters

The use of a digital meter may save you the trouble of selecting an instrument with a suitable range for your application. Most have an auto-ranging function; that is, the instrument selects the most sensitive range for the particular value of current or voltage being measured. All the experimenter has to do is check whether there is a zero error and adjust if necessary, note whether the display indicates 'A' or 'mA', and observe the position of the decimal point!

The uncertainty in the reading of a digital meter is expressed in terms of the overall uncertainty and the uncertainty in the last digit. When in use, you will note that the last digit of the display fluctuates from one figure to another. You can try to estimate the mean of the fluctuations, but if this fluctuation occurs, there is clearly uncertainty in the last digit of the value.

Multimeters

Multimeters, or multifunction instruments, are available in both analogue and digital forms (Figure 2.36). Such meters may include switched options for the measurement of direct and alternating currents and voltages, and of resistance, with several ranges for each quantity being measured. If you use a multimeter, make sure that you are familiar with the controls, so that you can set the instrument to measure the quantity you require.

Figure 2.36 Multimeter

Choice of method

Much will depend on the selection of meters available in your laboratory. Before you set up your circuit, make a rough calculation to determine the ranges of currents and voltages that you will have to measure. This is a vital part of the planning process, and will help to make sure that you select the appropriate instrument from those that are available.

In some laboratories, multimeters are provided for use primarily as test instruments, to be available to anyone who wishes to make a rapid check on currents, potential differences or resistances in a circuit. If this is the rule in your laboratory, it is bad practice to use a multimeter in a long experiment, when a single-function and single-range instrument would do the job equally successfully.

Remember that, to measure a current in a component in a circuit, an ammeter should be connected in series with the component. To measure the potential difference across the component, a voltmeter should be connected in parallel with the component. The arrangement is shown in Figure 2.37.

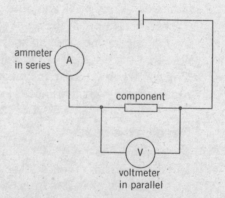

Figure 2.37 An ammeter is connected in series with the component, a voltmeter in parallel

Application: measurement of voltage using a cathode-ray oscilloscope

The cathode-ray oscilloscope, with its calibrated Y-amplifier, may be used to measure the amplitude of an alternating voltage signal. (We have already seen how the time-base of the c.r.o. may be used to measure time.) The signal is connected to the Y-input, and the Y-amplifier and time-base settings are adjusted until a suitable trace is obtained (Figure 2.38). The amplitude A of the trace is measured. If the Y-amplifier setting is Q (in units of volts per centimetre), the peak value V_0 of the signal is given by $V_0 = AQ$. The peak-to-peak value is $2V_0$, and the r.m.s. (root-mean-square) voltage is $V_0/\sqrt{2}$. (Remember that the reading obtained on an analogue or digital voltmeter is the r.m.s. value.)

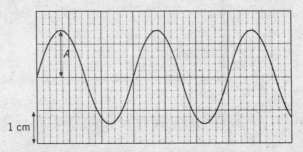

Figure 2.38 Measurement of alternating voltage

Example

The output from a signal generator is connected to the Y-input of a c.r.o. When the Y-amplifier control is set to 5.0 millivolts per centimetre, the trace shown in Figure 2.39 is obtained. Find the peak voltage of the signal, and the r.m.s. voltage.

Measure the amplitude of the trace on the graticule: this is 1.4 cm. The Y-amplfier setting is $5.0\,\text{mV cm}^{-1}$. 1.4 cm is thus equivalent to $1.4 \times 5.0 = 7.0\,\text{mV}$. The peak voltage of the signal is **7.0 mV**. The r.m.s. voltage is given by $7.0/\sqrt{2}$ = **4.9 mV r.m.s.**

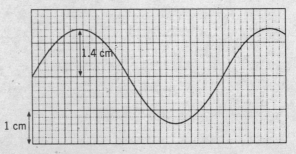

Figure 2.39

Now it's your turn

The output from a signal generator is connected to the Y-input of a c.r.o. When the Y-amplifier control is set to 20 millivolts per centimetre, the trace shown in Figure 2.40 is obtained. Find:
(a) the peak-to-peak voltage of the signal,
(b) the r.m.s. voltage.

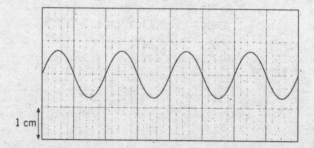

1 cm

Figure 2.40

Measuring magnetic flux density

The flux density of a magnetic field may be measured using a **Hall probe**, a device which makes use of the Hall effect (see page 332). The Hall probe units in use in school or college laboratories consist of a thin slice of a semiconductor material which is placed with its plane at right angles to the direction of the magnetic field. The control unit is arranged to pass a certain current through the semiconductor slice; the Hall voltage, which is proportional to the magnetic flux density, is read off on an analogue or digital meter, which will have already been calibrated in units of magnetic flux density (tesla). The arrangement is illustrated in Figure 2.41.

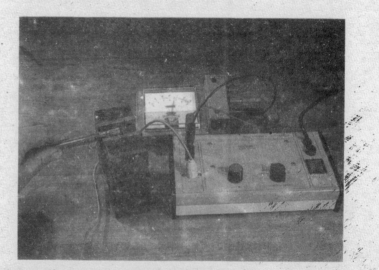

Figure 2.41 Hall probe apparatus

Section 2.1 Summary

- Methods available for the measurement of length include:
  – metre rule (range 1 m, reading uncertainty 0.5 mm)
  – micrometer screw gauge (range 50 mm, reading uncertainty 0.01 mm)
  – vernier caliper (range 100 mm, reading uncertainty 0.1 mm).
- Methods available for the measurement of mass include:
  – top-pan balance
  – spring balance
  – lever balance.
- Methods available for the measurement of time include:
  – stopclock (reading uncertainty 0.2 s)
  – stopwatch (reading uncertainty 0.01 s)
  – cathode-ray oscilloscope.
- Methods available for the measurement of temperature include:
  – liquid-in-glass thermometer
  – thermocouple thermometer.
- Methods available for the measurement of current and potential difference include:
  – analogue meter
  – digital meter
  – multimeter
  – cathode-ray oscilloscope.
- Methods available for the measurement of magnetic flux density include the Hall probe.

2.2 Errors and uncertainties

At the end of Section 2.2 you should be able to:
- show an understanding of the distinction between precision and accuracy
- show an understanding of the distinction between systematic errors (including zero errors) and random errors
- assess the uncertainty in a derived quantity by simple addition of actual, fractional or percentage uncertainties.

Accuracy and precision

Accuracy is the degree to which a measurement approaches the 'true value'. Accuracy depends on the equipment used, the skill of the experimenter and the techniques used. Reducing error or uncertainty in a measurement improves its accuracy. **Precision** is that part of accuracy which is within the control of the experimenter. The experimenter may choose different measuring instruments and may use them with different levels of skill, thus affecting the precision of measurement.

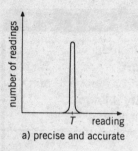

a) precise and accurate

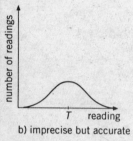

b) imprecise but accurate

Figure 2.42

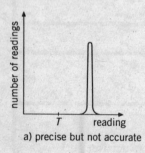

a) precise but not accurate

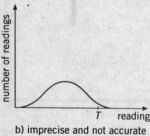

b) imprecise and not accurate

Figure 2.43

If we want to measure the diameter of a steel sphere or a marble, we could use a metre rule, or a vernier caliper, or a micrometer screw gauge. The choice of measuring instrument would depend on the precision to which we want the measurement to be made. For example, the metre rule could be used to measure to the nearest millimetre, the vernier caliper to the nearest tenth of a millimetre, and the micrometer screw gauge to the nearest one-hundredth of a millimetre. We could show the readings as follows.

metre rule: 1.2   ± 0.1 cm

vernier caliper: 1.21 ± 0.01 cm

micrometer screw gauge: 1.212 ± 0.001 cm

The degree of precision to which the measurement has been made increases as we move from the metre rule to the vernier caliper and finally to the micrometer screw gauge. Note that the number of significant figures quoted for the measurement increases as the precision increases. In fact, the number of significant figures in a measurement gives an indication of the precision of the measurement.

When a measurement is repeated many times with a precise instrument, the readings are all close together, as shown in Figure 2.42a. Using a measuring instrument with less precision means that there would be a greater spread of readings, as shown in Figure 2.42b, resulting in greater uncertainty.

Precision is partly to do with the accuracy of an observation or measurement. A reading may be very precise but it need not be accurate. Accuracy is concerned with how close a reading is to its true value. For example, a micrometer screw gauge may be precise to ±0.001 cm but, if there is a large zero error, then the readings from the scale for the diameter of a sphere or marble would not be accurate. The distinction between precision and accuracy is illustrated in Figure 2.43. On each of the graphs the value T is the true value of the quantity.

Uncertainty

In the list above, each of the measurements is shown with its precision. For example, using the metre rule, the measurement of the diameter is 1.2 cm with a precision of 0.1 cm. In reality, precision is not the only factor affecting the accuracy of the measurement.

The total range of values within which the measurement is likely to lie is known as its **uncertainty**.

For example, a measurement of 46.0 ± 0.5 cm implies that the most likely value is 46.0 cm, but it could be as low as 45.5 cm or as high as 46.5 cm. The uncertainty in the measurement is ±0.5 cm or ±(0.5/46) × 100% = ±1%.

It is important to understand that, when writing down measurements, the number of significant figures of the measurement indicates its uncertainty. Some examples of uncertainty are given in Table 2.2.

Table 2.2 Examples of uncertainty

instrument	uncertainty	typical reading
stopwatch with 0.2 s divisions	± 0.1 s	16.2 s
thermometer with 1 deg C intervals	± 0.5 deg C	22.5 °C
ammeter with 0.1 A divisions	± 0.05 A	2.15 A

Note that while a particular temperature is shown as a number with the unit °C, a temperature *interval* is correctly shown as a number with the unit deg C. However, most people use the unit °C for both a particular temperature and a temperature interval.

It should be remembered that the uncertainty in a reading is not wholly confined to the reading of its scale or to the skill of the experimenter. Any measuring instrument has a built-in uncertainty. For example, a metal metre rule expands as its temperature rises. At only one temperature will readings of the scale be precise. At all other temperatures, there will be an uncertainty due to the expansion of the scale. Knowing by how much the rule expands would enable this uncertainty to be removed and hence improve precision.

Manufacturers of digital meters quote the uncertainty for each meter. For example, a digital voltmeter may be quoted as $\pm 1\%$ ± 2 digits. The $\pm 1\%$ applies to the total reading shown on the scale and the ± 2 digits is the uncertainty in the final display figure. This means that the uncertainty in a reading of 4.00 V would be $(\pm 4.00 \times 1/100) \pm 0.02 = \pm 0.06$ V. This uncertainty would be added to any further uncertainty due to a fluctuating reading.

The uncertainty in a measurement is sometimes referred to as being its *error*. This not strictly true. Error would imply that a mistake has been made. There is no mistake in taking the measurement, but there is always some doubt or some uncertainty as to its value.

Example

A student takes a large number of imprecise readings for the current in a wire. He uses an ammeter with a zero error of $-\Delta I$, meaning that all scale readings are too small by ΔI. The true value of the current is I. Sketch a distribution curve of the number of readings plotted aginst the measured value of the current. Label any relevant values.

This is the case illustrated in **Figure 2.43b**. The peak of the curve is centred on a value of $I - \Delta I$.

Now it's your turn

1 A large number of precise readings for the diameter D of a wire is made using a micrometer screw gauge. The gauge has a zero error $+E$, which means that all readings are too large. Sketch a distribution curve of the number of readings plotted against the measured value of the diameter.

> **2** The manufacturer of a digital ammeter quotes its uncertainty as ±1.5% ±2 digits.
>
> **(a)** Determine the uncertainty in a constant reading of 2.64 A.
>
> **(b)** The meter is used to measure the current from a d.c. power supply. The current is found to fluctuate randomly between 1.58 A and 2.04 A. Determine the most likely value of the current, with its uncertainty.

Choice of instruments

The precision of an instrument required for a particular measurement is related to the measurement being made. Obviously, if the diameter of a hair is being measured, a high-precision micrometer screw gauge is required, rather than a metre rule. Similarly, a galvanometer should be used to measure currents of the order of a few milliamperes, rather than an ammeter. Choice is often fairly obvious where single measurements are being made, but care has to be taken where two readings are subtracted. Consider the following example.

The distance of a lens from a fixed point is measured using a metre rule. The distance is 95.2 cm (see Figure 2.44). The lens is now moved closer to the fixed point and the new distance is 93.7 cm. How far has the lens moved? The answer is obvious: (95.2 − 93.7) = 1.5 cm. But how accurate is the measurement?

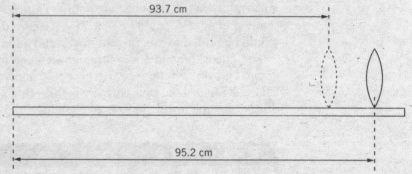

Figure 2.44

We have seen that the uncertainty in each measurement using a metre rule is, optimistically, ±1 mm ($\pm\frac{1}{2}$ mm at the zero end of the rule plus $\pm\frac{1}{2}$ mm when finding the position of the centre of the lens). This means that each separate measurement of length has an uncertainty of about (1/940 × 100)%, i.e. about 0.1%. That appears to be good! However, the uncertainty in the distance moved is ±2 mm, because both distances have an uncertainty, and these uncertainties add up, so the uncertainty is ±(2/15 × 100)% = ±13%. This uncertainty is, quite clearly, unacceptable. Another means by which the distance moved could be measured must be devised to reduce the uncertainty.

During your A/AS course, you will meet with many different measuring instruments. You must learn to recognise which instrument is most appropriate for particular measurements. A stopwatch may be suitable for measuring the period of oscillation of a pendulum but you would have difficulty using it to find the time taken for a stone to fall vertically from rest through a distance of 1 m. Choice of appropriate instruments is likely to be examined when you are planning experiments.

Example

Suggest appropriate instruments for the measurement of the dimensions of a single page of this book.

The obvious instrument to measure the height and width of a page is a 30 cm ruler, which can be read to ±1 mm. The width, the smaller dimension, is about 210 mm, so the actual uncertainty is 210 mm ±1 mm and the percentage uncertainty is about ±0.5%. It is not sensible to try to measure the thickness of a single page, even with a micrometer screw gauge, as the percentage error will be very high. Instead, use the screw gauge to measure the thickness of a large number of pages (but don't include the covers!). 400 pages are about 18 mm thick. The uncertainty in this measurement, using a screw gauge, is ±0.01 mm, giving a percentage uncertainty of about ±0.05% in the thickness of all 400 pages. This is also the percentage uncertainty in the thickness of a single page. If an uncertainty of ±0.5% is acceptable, a vernier caliper should be used instead of the screw gauge.

Now it's your turn

1 Suggest appropriate instruments for the measurement of:
   (a) the discharge current of a capacitor (of the order of 10^{-6} A),
   (b) the time for a feather to fall in air through a distance of about 40 cm,
   (c) the time for a ball to fall vertically through a distance of about 40 cm,
   (d) the length of a pendulum having a period of about 1 s,
   (e) the temperature of some water as it cools to room temperature,
   (f) the temperature of a roaring Bunsen flame,
   (g) the weight of 20 small glass beads,
   (h) the weight of a house brick.

2 The diameter of a ball is measured using a metre rule and a set square, as illustrated in Figure 2.45. The readings on the rule are 16.8 cm and 20.4 cm. Each reading has an uncertainty of ±1 mm. Calculate, for the diameter of the ball:
   (a) its actual uncertainty, **(b)** its percentage uncertainty.

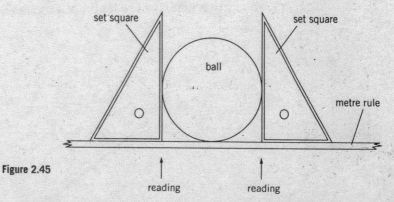

Figure 2.45

Suggest an alternative, but more precise, method by which the diameter could be measured.

Systematic and random uncertainty (error)

Not only is the choice of instrument important so that any measurement is made with acceptable precision but, also, the techniques of measurement must optimise accuracy. That is, your experimental technique must reduce as far as possible any uncertainties in readings. These uncertainties may be classed as either **systematic** or **random**.

Systematic uncertainty (error)

A systematic uncertainty will result in all readings being either above or below the accepted value. This uncertainty cannot be eliminated by repeating readings and then averaging. Instead systematic uncertainty can be reduced only by improving experimental techniques. Examples of systematic uncertainty are:

- **Zero error** on an instrument. The scale reading is not zero before measurements are taken – see Figure 2.46. Check before starting the experiment.
- **Wrongly calibrated scale.** In school laboratories we assume that measuring devices are correctly calibrated, and would not be expected to check the calibration in an experiment. However, if you have doubts, you can check the calibration of an ammeter by connecting several in series in the circuit, or of a voltmeter by connecting several in parallel. Rules can be checked by laying several of them alongside each other. Thermometers can be checked by placing several in well-stirred water. These checks will not enable you to say which of the instruments are calibrated correctly, but they will show you if there is a discrepancy.
- **Reaction time** of experimenter. When timings are carried out manually, it must be accepted that there will be a delay between the experimenter observing the event and starting the timing device. This delay, called the reaction time, may be as much as a few tenths of a second. To reduce the effect, you should arrange that the intervals you are timing are much greater than the reaction time. For example, you should time sufficient swings of a pendulum for the total time to be of the order of at least ten seconds, so that a reaction time of a few tenths of a second is less important.

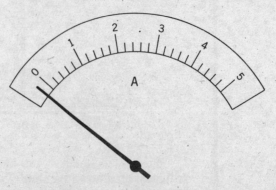

Figure 2.46 This ammeter has a zero error of about −0.1 A.

Random uncertainty (error)

Random uncertainty results in readings being scattered around the accepted value. Random uncertainty may be reduced by repeating a reading and averaging, and by plotting a graph and drawing a best-fit line. Examples of random errors are:

- reading a scale, particularly if this involves the experimenter's judgement about interpolation between scale readings
- timing oscillations without the use of a reference marker, so that timings may not always be made to the same point of the swing
- taking readings of a quantity that varies with time, involving the difficulty of reading both a timer scale and another meter simultaneously
- reading a scale from different angles, so introducing a variable parallax error. (In contrast, if a scale reading is always made from the same non-normal angle, this will introduce a systematic error.)

Example

The current in a resistor is to be measured using an analogue ammeter. State one source of (a) systematic uncertainty, (b) random uncertainty. In both cases, suggest how the uncertainty may be reduced.

(a) Systematic uncertainty could be a zero error on the meter, or a wrongly calibrated scale. This can be reduced by checking for a zero reading before starting the experiment, or using two ammeters in series to check that the readings agree.
(b) Random uncertainty could be a parallax error caused by taking readings from different angles. This can be reduced by the use of a mirror behind the scale and viewing normally.

Now it's your turn

1 The length of a pencil is measured with a 30 cm rule. Suggest one possible source of (a) a systematic uncertainty, (b) a random uncertainty. In each case, suggest how the uncertainty may be reduced.

2 The diameter of a wire is to be measured to a precision of ±0.01 mm.
  (a) Name a suitable instrument.
  (b) Suggest a source of systematic uncertainty.
  (c) Explain why it is good practice to average a set of diameter readings, taken spirally along the length of the wire.

Combining uncertainties

There are two simple rules for obtaining an estimate of the overall uncertainty in a final result. The rules are:

1 For quantities which are added or subtracted to give a final result, add the actual uncertainties.
2 For quantities which are multiplied together or divided to give a final result, add the fractional uncertainties.

Suppose that we wish to obtain the value of a physical quantity x by measuring two other quantities, y and z. The relation between x, y and z is known, and is ·

$$x = y + z$$

If the uncertainties in y and z are Δy and Δz respectively, the uncertainty Δx in x is given by

$$\Delta x = \Delta y + \Delta z$$

If the quantity x is given by

$$x = y - z$$

the uncertainty in x is again given by

$$\Delta x = \Delta y + \Delta z$$

Examples

1 I_1 and I_2 are two currents coming into a junction in a circuit. The current I going out of the junction is given by

$$I = I_1 + I_2$$

In an experiment, the values of I_1 and I_2 are determined as 2.0 ± 0.1 A and 1.5 ± 0.2 A respectively. What is the value of I? What is the uncertainty in this value?

Using the given equation, the value of I is given by $I = 2.0 + 1.5$ = **3.5 A**. The rule for combining the uncertainties gives $\Delta I = 0.1 + 0.2 = 0.3$ A. The result for I is thus **(3.5 ± 0.3) A**.

2 In an experiment, a liquid is heated electrically, causing the temperature to change from 20.0 ± 0.2 °C to 21.5 ± 0.5 °C. Find the change of temperature, with its associated uncertainty.

The change of temperature is $21.5 - 20.0 =$ **1.5 °C**. The rule for combining the uncertainties gives the uncertainty in the temperature change as $0.2 + 0.5 = 0.7$ °C. The result for the temperature change is thus **(1.5 ± 0.7) °C**.

Note that this second example shows that a small difference between two quantities may have a large uncertainty, even if the uncertainty in measuring each of the quantities is small. This is an important factor in considering the design of experiments, where the difference between two quantities may introduce an unacceptably large error.

Now it's your turn

Two set-squares and a ruler are used to measure the diameter of a cylinder. The cylinder is placed·between the set-squares, and the set-squares are aligned with the ruler, in the manner of the jaws of calipers. The readings on the ruler at opposite ends of a diameter are 4.15 cm and 2.95 cm. Each reading has an uncertainty of ±0.05 cm. What is the diameter of the cylinder? What is the uncertainty in the diameter?

Now suppose that we wish to find the uncertainty in a quantity x, whose relation to two measured quantities, y and z, is

$$x = Ayz$$

where A is a constant. The uncertainty in the measurement of y is $\pm\Delta y$, and that in z is $\pm\Delta z$. The **fractional uncertainty** in x is given by

$$\frac{\Delta x}{x} = \frac{\Delta y}{y} + \frac{\Delta z}{z}$$

To combine the uncertainties when the quantities are raised to a power, for example

$$x = Ay^a z^b$$

where A is a constant, the rule is

$$\frac{\Delta x}{x} = a\left(\frac{\Delta y}{y}\right) + b\left(\frac{\Delta z}{z}\right)$$

Example

A value of the acceleration of free fall g was determined by measuring the period of oscillation T of a simple pendulum of length l. The relation between g, T and l is

$$g = 4\pi^2\left(\frac{l}{T^2}\right)$$

In the experiment, l was measured as 0.55 ± 0.02 m and T was measured as 1.50 ± 0.02 s. Find the value of g, and the uncertainty in this value.

Substituting in the equation, $g = 4\pi^2(0.55/1.50^2) = $ **9.7 m s^{-2}**. The fractional uncertainties are $\Delta l/l = 0.02/0.55 = 0.036$ and $\Delta T/T = 0.02/1.50 = 0.013$.

Applying the rule to find the fractional uncertainty in g

$$\frac{\Delta g}{g} = \frac{\Delta l}{l} + \frac{2\Delta T}{T} = 0.036 + 2 \times 0.013 = 0.062$$

The actual uncertainty in g is given by (value of g) $\times$ (fractional uncertainty in g) $= 9.7 \times 0.062 = 0.60$ m s^{-2}. The experimental value of g, with its uncertainty, is thus **(9.7 ± 0.6) m s^{-2}**.

Note that it is not good practice to determine g from the measurement of the period of a pendulum of fixed length. It would be much better to take values of T for a number of different lengths l, and to draw a graph of T^2 against l. The gradient of this graph is $4\pi^2/g$.)

▶▶

Now it's your turn

A value of the volume V of a cylinder is determined by measuring the radius r and the length L. The relation between V, r and L is

$$V = \pi r^2 L$$

In an experiment, r was measured as 3.30 ± 0.05 cm, and L was measured as 25.4 ± 0.4 cm. Find the value of V, and the uncertainty in this value.

If you find it difficult to deal with the fractional uncertainty rule, you can easily estimate the uncertainty by substituting extreme values into the equation. For $x = Ay^a z^b$, taking account of the uncertainties in y and z, the lowest value of x is given by

$$x_{low} = A(y - \Delta y)^a (z - \Delta z)^b$$

and the highest by

$$x_{high} = A(y + \Delta y)^a (z + \Delta z)^b$$

If x_{low} and x_{high} are worked out, the uncertainty in the value of x is given by $(x_{high} - x_{low})/2$.

Example

Apply the extreme value method to the data for the simple pendulum experiment in the Example on page 49.

Because of the form of the equation for g, the lowest value for g will be obtained if the lowest value of l and the highest value for T are substituted. This gives

$$g_{low} = 4\pi^2 (0.53/1.52^2) = 9.1 \, \text{m s}^{-2}$$

The highest value for g is obtained by substituting the highest value for l and the lowest value for T. This gives

$$g_{high} = 4\pi^2 (0.57/1.48^2) = 10.3 \, \text{m s}^{-2}$$

The uncertainty in the value of g is thus $(g_{high} - g_{low})/2$ $= (10.3 - 9.1)/2 = \textbf{0.6 m s}^{-2}$, as before.

Now it's your turn

Apply the extreme value method to the data for the volume of the cylinder, at the top of this page.

If the expression for the quantity under consideration involves both products (or quotients) and sums (or differences), then the best approach is the extreme value method.

Section 2.2 Summary

- Accuracy is concerned with how close a reading is to its true value.
- Precision is the part of accuracy which can be controlled by the experimenter.
- Uncertainty indicates the range of values within which a measurement is likely to lie.
- A systematic uncertainty (or systematic error) is often due to instrumental causes, and results in all readings being above or below the true value. It cannot be eliminated by averaging.
- A random uncertainty (or random error) is due to the scatter of readings around the true value. It may be reduced by repeating a reading and averaging, or by plotting a graph and taking a best-fit line.
- Combining uncertainties:
  – for expressions of the form $x = y + z$ or $x = y - z$, the overall uncertainty is $\Delta x = \Delta y + \Delta z$
  – for expressions of the form $x = Ay^a z^b$, the overall fractional uncertainty is $\Delta x/x = a(\Delta y/y) + b(\Delta z/z)$

Exam-style Questions

1 You are asked to measure the internal diameter of a glass capillary tube (diameter about 2 mm). You are also to investigate the uniformity of the tube along its length. Suggest suitable methods.

2 The value of the acceleration of free fall varies slightly at different places on the Earth's surface. Discuss whether this means that **(a)** a top-pan balance, **(b)** a spring balance, **(c)** a lever balance, should be re-calibrated when they are moved to different locations. If you needed to, how would you calibrate a balance?

3 The shutter on a particular camera has settings which allow it to be open for (nominally) 1 s, 0.5 s, 0.25 s, 0.125 s, 0.067 s, 0.033 s, 0.017 s, 0.008 s, 0.004 s, 0.002 s and 0.001 s. Suggest a method (or methods) of calibrating the exposure times over this range.

4 Explain the factors you would consider when deciding whether to use a liquid-in-glass or a thermoelectric thermometer in particular experimental situations.

5 Summarise the advantages and disadvantages of analogue and digital ammeters.

6 Explain how to use a cathode-ray oscilloscope to measure the characteristics of the sinusoidal output from a signal generator.

3. Kinematics

'Kinematics' is the description of how objects move. The aim of this chapter is to describe motion in terms of quantities such as position, speed, velocity and acceleration. To make matters easy, we shall talk about the motion of a particle, rather than of a real body. The simplification of dealing with a particle – a body with no size – is one that is adopted in many branches of physics.

3.1 Average speed

At the end of Sections 3.1–3.4 you should be able to:
- define displacement, speed, velocity and acceleration
- use graphical methods to represent displacement, speed, velocity and acceleration
- find displacement from the area under a velocity–time graph
- use the slope of a displacement–time graph to find the velocity
- use the slope of a velocity–time graph to find the acceleration.

When talking about motion, we shall discuss the way in which the position of a particle varies with time. Think about a particle moving along a straight line. In a certain time, the particle will cover a certain distance. The **average speed** of the particle is defined as the distance covered divided by the time taken. Written as a word equation, this is

$$average\ speed = \frac{distance\ covered}{time\ taken}$$

The unit of speed is the metre per second (m s^{-1}).

One of the most fundamental of physical constants is the speed of light in a vacuum. It is important because it is used in the definition of the metre, and because, according to the theory of relativity, it defines an upper limit to attainable speeds. The range of speeds that you are likely to come across is enormous; some are summarised in Table 3.1.

It is important to recognise that speed has a meaning only if it is quoted relative to a fixed reference. In most cases, speeds are quoted relative to the surface of the Earth, which – although it is moving relative to the solar system – is often taken to be fixed. Thus, when we say that a bird can fly at a certain average speed, we are relating its speed to the Earth. However, a passenger on a ferry may see that a seagull, flying parallel to the boat, appears to be practically stationary. If this is the case, the seagull's speed relative to the

Table 3.1 Examples of speeds

	speed/m s^{-1}
light	3.0×10^8
electron around nucleus	2.2×10^6
Earth around Sun	3.0×10^4
jet airliner	2.5×10^2
typical car speed (80 km per hour)	22
sprinter	10
walking speed	1.5
snail	1×10^{-3}

boat is zero. However, if the speed of the boat through the water is $8\,\text{m s}^{-1}$, then the speed of the seagull relative to Earth is also $8\,\text{m s}^{-1}$. When talking about relative speeds we must also be careful about directions. It is easy if the motions are in the same direction, as in the example of the ferry and the seagull. The addition of velocity vectors is considered in Section 1.2.

Examples

1 The radius of the Earth is $6.4 \times 10^6\,\text{m}$; one revolution about its axis takes 24 hours ($8.6 \times 10^4\,\text{s}$). Calculate the average speed of a point on the equator relative to the centre of the Earth.

In 24 hours, the point on the equator completes one revolution and travels a distance of $2\pi \times$ the Earth's radius, that is $2\pi \times 6.4 \times 10^6 = 4.0 \times 10^7\,\text{m}$. The average speed is (distance covered)/(time taken), or $4.0 \times 10^7/8.6 \times 10^4 =$ **$4.7 \times 10^2\,\text{m s}^{-1}$**.

2 How far does a cyclist travel in 11 minutes if his average speed is $22\,\text{km h}^{-1}$?

First convert the average speed in km h^{-1} to a value in m s^{-1}. $22\,\text{km}$ ($2.2 \times 10^4\,\text{m}$) in 1 hour ($3.6 \times 10^3\,\text{s}$) is an average speed of $6.1\,\text{m s}^{-1}$. 11 minutes is $660\,\text{s}$. Since average speed is (distance covered)/(time taken), the distance covered is given by (average speed) $\times$ (time taken), or $6.1 \times 660 =$ **$4000\,\text{m}$**.

Note the importance of working in consistent units: this is why the average speed and the time were converted to m s^{-1} and s respectively.

3 A train is travelling at a speed of $25\,\text{m s}^{-1}$ along a straight track. A boy walks along the corridor in a carriage towards the rear of the train, at a speed of $1\,\text{m s}^{-1}$ relative to the train. What is his speed relative to Earth?

In one second, the train travels $25\,\text{m}$ forwards along the track. In the same time the boy moves $1\,\text{m}$ towards the rear of the train, so he has moved $24\,\text{m}$ along the track. His speed relative to Earth is thus $25 - 1 =$ **$24\,\text{m s}^{-1}$**.

Now it's your turn

1 The speed of an electron in orbit about the nucleus of a hydrogen atom is $2.2 \times 10^6 \, \text{m s}^{-1}$. It takes $1.5 \times 10^{-16} \, \text{s}$ for the electron to complete one orbit. Find the radius of the orbit.

2 The average speed of an airliner on a domestic flight is $220 \, \text{m s}^{-1}$. How long will it take to fly between two airports on a flightpath $700 \, \text{km}$ long?

3 Two cars are travelling in the same direction on a long, straight road. The one in front has an average speed of $25 \, \text{m s}^{-1}$ relative to Earth; the other's is $31 \, \text{m s}^{-1}$, also relative to Earth. What is the speed of the second car relative to the first when it is overtaking?

3.2 Speed and velocity

In ordinary language, there is no difference between the terms *speed* and *velocity*. However, in physics there is an important distinction between the two. **Velocity** is used to represent a vector quantity: the magnitude of how fast a particle is moving, and the direction in which it is moving. **Speed** does not have an associated direction. It is a scalar quantity (see Section 1.2).

So far, we have talked about the total distance travelled by a body. Like speed, distance is a scalar quantity, because we do not have to specify the direction in which the distance is travelled. However, in defining velocity we introduce a quantity called **displacement**. Displacement of a particle is its change of position. Consider a cyclist travelling $500 \, \text{m}$ along a straight road, and then turning round and coming back $300 \, \text{m}$. The total distance travelled is $800 \, \text{m}$, but the displacement is only $200 \, \text{m}$, since the cyclist has ended up $200 \, \text{m}$ from the starting point.

The **average velocity** is defined as the displacement divided by the time taken.

$$average \; velocity = \frac{displacement}{time \; taken}$$

Because distance and displacement are different quantities, the average speed of motion will sometimes be different from the magnitude of the average velocity. If the time taken for the cyclist's trip in the example above is $120 \, \text{s}$, the average speed is $800/120 = 6.7 \, \text{m s}^{-1}$, whereas the magnitude of the average velocity is $200/120 = 1.7 \, \text{m s}^{-1}$. This may seem confusing, but the difficulty arises only when the motion involves a change of direction and we take an average value. If we are interested in describing the motion of a particle at a particular moment in time, the speed at that moment is the same as the magnitude of the velocity at that moment.

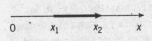

Figure 3.1

We now need to define average velocity more precisely, in terms of a mathematical equation, instead of our previous word equation. Suppose that at time t_1 a particle is at a point x_1 on the x-axis (Figure 3.1). At a later time t_2, the particle has moved to x_2. The displacement (the change in position) is $(x_2 - x_1)$, and the time taken is $(t_2 - t_1)$. The average velocity $\bar{v}$ is then

$$\bar{v} = \frac{x_2 - x_1}{t_2 - t_1}$$

The bar over v is the symbol meaning 'average'. As a shorthand, we can write $(x_2 - x_1)$ as Δx, where Δ (the Greek capital letter delta) means 'the change in'. Similarly, $t_2 - t_1$ is written as Δt. This gives us

$$\bar{v} = \frac{\Delta x}{\Delta t}$$

If x_2 were less than x_1, $(x_2 - x_1)$ and Δx would be negative. This would mean that the particle had moved to the left, instead of to the right as in Figure 3.1. The sign of the displacement gives the direction of particle motion. If Δx is negative, then the average velocity v is also negative. The sign of the velocity, as well as the sign of the displacement, indicates the direction of the particle's motion. This is because both displacement and velocity are vector quantities.

3.3 Describing motion by graphs

Position–time graphs

Figure 3.2 is a graph of position x against time t for a particle moving in a straight line. This curve gives a complete description of the motion of the particle. We can see from the graph that the particle starts at the origin O (at which $x = 0$) at time $t = 0$. From O to A the graph is a straight line: the particle is covering equal distances in equal periods of time. This represents a period of *uniform velocity*. The average velocity during this time is $(x_1 - 0)/(t_1 - 0)$. Clearly, this is the gradient of the straight-line part of the graph between O and A. Between A and B the particle is slowing down, because the distances travelled in equal periods of time are getting smaller. The average velocity during this period is $(x_2 - x_1)/(t_2 - t_1)$. On the graph, this is represented by the gradient of the straight line joining A and B. At B, for a moment, the particle is at rest, and after B it has reversed its direction and is heading back towards the origin. Between B and C the average velocity is $(x_3 - x_2)/(t_3 - t_2)$. Because x_3 is less than x_2, this is a negative quantity, indicating the reversal of direction.

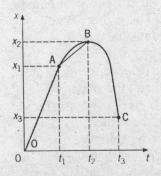

Figure 3.2

Calculating the average velocity of the particle over the relatively long intervals t_1, $(t_2 - t_1)$ and $(t_3 - t_2)$ will not, however, give us the complete description of the motion. To describe the motion exactly, we need to know the particle's velocity at every instant. We introduce the idea of **instantaneous velocity**. To define instantaneous velocity we make the intervals of time over which we measure the average velocity shorter and shorter. This has the effect of approximating the curved displacement–time graph by a series of short straight-line segments. The approximation becomes better the shorter the time interval, as illustrated in Figure 3.3. Eventually, in the case of extremely small time intervals (mathematically we would say 'infinitesimally small'), the straight-line segment has the same direction as the tangent to the curve. This limiting case gives the instantaneous velocity as the slope of the tangent to the displacement–time curve.

Figure 3.3

Displacement–time and velocity–time graphs

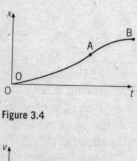

Figure 3.4

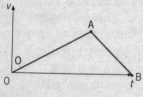

Figure 3.5

Figure 3.4 is a sketch graph showing how the displacement of a car, travelling along a straight test track, varies with time. We interpret this graph in a descriptive way by noting that between O and A the distances travelled in equal intervals of time are progressively increasing: that is, the velocity is increasing as the car is accelerating. Between A and B the distances for equal time intervals are decreasing; the car is slowing down. Finally, there is no change in position, even though time passes, so the car must be at rest. We can use Figure 3.4 to deduce the details of the way in which the car's instantaneous velocity v varies with time. To do this, we draw tangents to the curve in Figure 3.4 at regular intervals of time, and measure the slope of each tangent to obtain values of v. The plot of v against t gives the graph in Figure 3.5. This confirms our descriptive interpretation: the velocity increases from zero to a maximum value, and then decreases to zero again. We will look at this example in more detail on pages 65–66, where we shall see that the area under the velocity–time graph in Figure 3.5 gives the displacement x.

3.4 Acceleration

We have used the word *accelerating* in describing the increase in velocity of the car in the previous section. **Acceleration** is a measure of the rate at which the velocity of the particle is changing. **Average acceleration** is defined by the word equation

$$average\ acceleration = \frac{change\ in\ velocity}{time\ taken}$$

The unit of acceleration is the unit of velocity (the metre per second) divided by the unit of time (the second), giving the metre per second per second (m s^{-2}). In symbols, this equation is

$$\bar{a} = \frac{v_2 - v_1}{t_2 - t_1} = \frac{\Delta v}{\Delta t}$$

where v_1 and v_2 are the velocities at time t_1 and t_2 respectively. To obtain the **instantaneous acceleration**, we take extremely small time intervals, just as we did when defining instantaneous velocity. Because it involves a change in velocity (a vector quantity), acceleration is also a vector quantity: we need to specify both its magnitude and its direction.

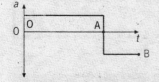

Figure 3.6

We can deduce the acceleration of a particle from its velocity–time graph by drawing a tangent to the curve and finding the slope of the tangent. Figure 3.6 shows the result of doing this for the car's motion described by Figure 3.4 (the displacement–time graph) and Figure 3.5 (the velocity–time graph). The car accelerates at a constant rate between O and A, and then decelerates (that is, slows down) uniformly between A and B.

An acceleration with a very familiar value is the acceleration of free fall near the Earth's surface (see page 61): this is $9.81\ \text{m s}^{-2}$, often approximated to $10\ \text{m s}^{-2}$. To illustrate the range of values you may come across, some accelerations are summarised in Table 3.2.

Table 3.2 Examples of accelerations

	acceleration/m s^{-2}
due to circular motion of electron around nucleus	9×10^{26}
car crash	1×10^3
free fall on Earth	10
family car	2
free fall on Moon	2
at equator, due to rotation of Earth	3×10^{-2}
due to circular motion of Earth around Sun	6×10^{-5}

Examples

1 A sports car accelerates along a straight test track from rest to $70\,km\,h^{-1}$ in 6.3 s. What is its average acceleration?

First convert the data into consistent units. 70 km ($7.0 \times 10^4\,m$) in 1 hour ($3.6 \times 10^3\,s$) is $19\,m\,s^{-1}$. Since average acceleration is (change of velocity)/(time taken), the acceleration is $19/6.3 = \mathbf{3.0\ m\ s^{-2}}$.

2 A railway train, travelling along a straight track, takes 1.5 minutes to come to rest from a speed of $115\,km\,h^{-1}$. What is its average acceleration while braking?

$115\,km\,h^{-1}$ is $31.9\,m\,s^{-1}$, and 1.5 minutes is 90 s. The average acceleration is (change of velocity)/(time taken) $= -31.9/90 = \mathbf{-0.35\ m\ s^{-2}}$.

Note that the acceleration is a negative quantity because the change of velocity is negative: the final velocity is less than the initial. A negative acceleration is often called a deceleration.

Now it's your turn

1 A sprinter, starting from the blocks, reaches his full speed of $9.0\,m\,s^{-1}$ in 1.5 s. What is his average acceleration?

2 A car is travelling at a speed of $25\,m\,s^{-1}$. At this speed, it is capable of accelerating at $1.8\,m\,s^{-2}$. How long would it take to accelerate from $25\,m\,s^{-1}$ to the speed limit of $31\,m\,s^{-1}$?

Sections 3.1–3.4 Summary

- Speed is a scalar quantity, and is described by magnitude only. Velocity is a vector, and requires magnitude and direction. Speed is the magnitude of velocity. Displacement is the change in position of a particle, and is a vector quantity.
- Average speed is defined by: (*distance covered*)/(*time taken*)
- Average velocity is defined by: (*displacement*)/(*time taken*) or $\Delta x/\Delta t$
- The instantaneous velocity is the average velocity measured over an infinitesimally short time interval.
- Average acceleration is defined by: (*change in velocity*)/(*time taken*) or $\Delta v/\Delta t$
- Acceleration is a vector. Instantaneous acceleration is the average acceleration measured over an infinitesimally short time interval.
- Velocity may be deduced from the gradient of a displacement–time graph.
- Acceleration may be deduced from the gradient of a velocity–time graph.

Sections 3.1–3.4 Questions

1 At an average speed of 24 km h⁻¹, how many kilometres will a cyclist travel in 75 minutes?

2 An aircraft travels 1600 km in 2.5 hours. What is its average speed, in m s⁻¹?

3 Does a car speedometer register speed or velocity? Explain.

4 An aircraft travels 1400 km at a speed of 700 km h⁻¹, and then runs into a headwind that reduces its speed over the ground to 500 km h⁻¹ for the next 800 km. What is the total time for the flight? What is the average speed of the aircraft?

5 A sports car can stop in 6.1 s from a speed of 110 km h⁻¹. What is its acceleration?

6 Can the velocity of a particle change if its speed is constant? Can the speed of a particle change if its velocity is constant? If the answer to either question is 'yes', give examples.

3.5 Uniformly accelerated motion

At the end of Sections 3.5–3.7 you should be able to:

- derive, from the definitions of velocity and acceleration, equations which represent uniformly accelerated motion in a straight line
- solve problems using equations which represent uniformly accelerated motion in a straight line, including the motion of bodies falling in a uniform gravitational field without air resistance
- recall that the weight of a body is equal to the product of its mass and the acceleration of free fall
- describe an experiment to determine the acceleration of free fall using a falling body
- describe qualitatively the motion of bodies falling in a uniform gravitational field with air resistance
- describe and explain motion due to a uniform velocity in one direction and a uniform acceleration in a perpendicular direction.

Having defined displacement, velocity and acceleration, we shall use the definitions to derive a series of equations, called the *kinematic equations*, which can be used to give a complete description of the motion of a particle in a straight line. The mathematics will be simplified if we deal with situations in which the acceleration does not vary with time; that is, the acceleration is uniform (or constant). This approximation applies for many practical cases. However, there are two important types of motion for which the kinematic equations do not apply: circular motion, and the oscillatory motion called simple harmonic motion. We shall deal with these separately in Chapter 10.

Think about a particle moving along a straight line with constant acceleration a. Suppose that its initial velocity, at time $t = 0$, is u. After a further time t its velocity has increased to v. From the definition of acceleration as (change in velocity)/(time taken), we have $a = (v - u)/t$ or, re-arranging,

$$v = u + at$$

From the definition of average velocity $\bar{v}$ as (distance travelled)/(time taken), over the time t the distance travelled s will be given by the average velocity multiplied by the time taken, or

$$s = \bar{v}t$$

The average velocity $\bar{v}$ is written in terms of the initial velocity u and final velocity v as

$$\bar{v} = \frac{u + v}{2}$$

and, using the previous equation for v,

$$\bar{v} = (u + u + at)/2 = u + at/2$$

Substituting this we have

$$s = ut + \tfrac{1}{2}at^2$$

The right-hand side of this equation is the sum of two terms. The ut term is the distance the particle would have travelled in time t if it had been travelling with a constant speed u, and the $\tfrac{1}{2}at^2$ term is the additional distance travelled as a result of the acceleration.

The equation relating the final velocity v, the initial velocity u, the acceleration a and the distance travelled s is

$$v^2 = u^2 + 2as$$

If you wish to see how this is obtained from previous equations, see the Maths Note below.

Maths Note

From $v = u + at$,

$$t = (v - u)/a$$

Substitute this in $s = ut + \tfrac{1}{2}at^2$,

$$s = u(v - u)/a + \tfrac{1}{2}a(v - u)^2/a^2$$

Multiplying both sides by $2a$ and expanding the terms,

$$2as = 2uv - 2u^2 + v^2 - 2uv + u^2$$

or $v^2 = u^2 + 2as$

The four equations relating the various quantities which define the motion of the particle in a straight line in uniformly accelerated motion are

$$v = u + at$$
$$s = ut + \frac{1}{2}at^2$$
$$v^2 = u^2 + 2as$$
$$\bar{v} = (u + v)/2$$

In these equations u is the initial velocity, v is the final velocity, $\bar{v}$ is the average velocity, a is the acceleration, s is the distance travelled, and t is the time taken.

In solving problems involving kinematics, it is important to understand the situation before you try to substitute numerical values into an equation. Identify the quantity you want to know, and then make a list of the quantities you know already. This should make it obvious which equation is to be used.

Free fall acceleration

A very common example of uniformly accelerated motion is when a body falls freely near the Earth's surface. Because of the gravitational attraction of the Earth, all objects fall with the same uniform acceleration. This acceleration is called the **acceleration of free fall**, and is represented by the symbol g. It has a value of $9.81\ \mathrm{m\ s^{-2}}$, and is directed downwards. For completeness, we ought to qualify this statement by saying that the fall must be in the absence of air resistance, but in most situations this can be assumed to be true.

One reason why the acceleration of free fall is important is that it provides the link between the mass m and the weight W of a body. The relationship is

$$W = mg$$

We will meet this equation again in Chapter 5. At present you need only remember the relationship.

The acceleration of free fall may be determined by an experiment in which the time of fall t of a body between two points a distance s apart is measured. If the body falls from rest, we can use the second of the equations for uniformly accelerated motion in the form

$$g = 2s/t^2$$

to calculate the value of g. Note that, because the time of fall is likely to be only a few tenths of a second, precise timing to one-hundredth of a second is required. An experiment involving the switching of light gates by the falling object has been described in Chapter 2 (Figure 2.24). The light gates are connected to an electronic timer.

Until the sixteenth century, the idea of the acceleration of a falling body was not fully appreciated. It was commonly thought that heavier bodies fell faster than light ones. This idea was a consequence of the effect of air resistance on light objects with a large surface area, such as feathers. However, Galileo Galilei (1564–1642) suggested that, in the absence of resistance, all bodies would fall

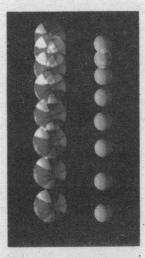

Figure 3.7 Strobo-flash photograph of objects in free fall

Figure 3.8 Galileo in his study

Figure 3.9 Leaning Tower of Pisa

with the same constant acceleration. He showed mathematically that, for a body falling from rest, the distance travelled is proportional to the square of the time. Galileo tested the relation experimentally by timing the fall of objects from various levels of the Leaning Tower of Pisa (Figure 3.9). This is the relation we have derived as $s = ut + \frac{1}{2}at^2$. For a body starting from rest, $u = 0$ and $s = \frac{1}{2}at^2$. That is, the distance is proportional to time squared.

We have mentioned that, in most situations, air resistance can be neglected. In fact, there are some applications in which this resistance is most important. One is the case of the fall of a parachutist, where air resistance plays a vital part. The velocity of a body falling through a resistive fluid (a liquid or a gas) does not increase indefinitely, but eventually reaches a maximum velocity, called the **terminal velocity**. The force due to air resistance increases with speed. When this resistive force has reached a value equal and opposite to the weight of the falling body, the body no longer accelerates and continues at uniform velocity. This is a case of motion with non-uniform acceleration. The acceleration starts off with a value of g, but decreases to zero at the time when the terminal velocity is achieved. Thus, raindrops and parachutists are normally travelling at a constant speed by the time they approach the ground (Figure 3.10).

Figure 3.10 A parachutist about to land

Examples

1 A car increases its speed from 25 m s^{-1} to 31 m s^{-1} with a uniform acceleration of 1.8 m s^{-2}. How far does it travel while accelerating?

In this problem we want to know the distance s. We know the initial speed $u = 25$ m s^{-1}, the final speed $v = 31$ m s^{-1}, and the acceleration $a = 1.8$ m s^{-2}. We need an equation linking s with u, v and a; this is

$$v^2 = u^2 + 2as$$

Substituting the values, we have $31^2 = 25^2 + 2 \times 1.8\,s$. Re-arranging, $s = (31^2 - 25^2)/(2 \times 1.8) =$ **93 m**.

2 The average acceleration of a sprinter from the time of leaving the blocks to reaching his maximum speed of $9.0\,\mathrm{m\,s^{-1}}$ is $6.0\,\mathrm{m\,s^{-2}}$. For how long does he accelerate? What distance does he cover in this time?

In the first part of this problem, we want to know the time t. We know the initial speed $u = 0$, the final speed $v = 9.0\,\mathrm{m\,s^{-1}}$, and the acceleration $a = 6.0\,\mathrm{m\,s^{-2}}$. We need an equation linking t with u, v and a; this is

$$v = u + at$$

Substituting the values, we have $9.0 = 0 + 6.0t$. Re-arranging, $t = 9.0/6.0 = \mathbf{1.5\,s}$.

For the second part of the problem, we want to know the distance s. We know the initial speed $u = 0$, the final speed $v = 9.0\,\mathrm{m\,s^{-1}}$, and the acceleration $a = 6.0\,\mathrm{m\,s^{-2}}$; we have also just found the time $t = 1.5\,\mathrm{s}$. There is a choice of equations linking s with u, v, a and t. We can use

$$s = ut + \tfrac{1}{2}at^2$$

Substituting the values, $s = 0 + \tfrac{1}{2} \times 6.0 \times (1.5)^2 = \mathbf{6.8\,m}$.

Another relevant equation is $\bar{v} = \Delta x/\Delta t$. Here the average velocity $\bar{v}$ is given by $\bar{v} = (u + v)/2 = 9.0/2 = 4.5\,\mathrm{m\,s^{-1}}$. $\Delta x/\Delta t$ is the same as s/t, so $4.5 = s/1.5$, and $s = 4.5 \times 1.5 = \mathbf{6.8\,m}$ as before.

3 A cricketer throws a ball vertically upward into the air with an initial velocity of $18.0\,\mathrm{m\,s^{-1}}$. How high does the ball go? How long is it before it returns to the cricketer's hands?

In the first part of the problem, we want to know the distance s. We know the initial velocity $u = 18.0\,\mathrm{m\,s^{-1}}$ upwards and the acceleration $a = g = 9.81\,\mathrm{m\,s^{-2}}$ downwards. At the highest point the ball is momentarily at rest, so the final velocity $v = 0$. The equation linking s with u, v and a is

$$v^2 = u^2 + 2as$$

Substituting the values, $0 = (18.0)^2 + 2(-9.81)s$. Thus $s = -(18.0)^2/2(-9.81) = \mathbf{16.5\,m}$. Note that here the ball has an upward velocity but a downward acceleration, and that at the highest point the velocity is zero but the acceleration is not zero.

In the second part we want to know the time t for the ball's up-and-down flight. We know u and a, and also the overall displacement $s = 0$, as the ball returns to the same point at which it was thrown. The equation to use is

$$s = ut + \tfrac{1}{2}at^2$$

Substituting the values, $0 = 18.0t + \tfrac{1}{2}(-9.81)t^2$. Doing some algebra, $t(36.0 - 9.81t) = 0$. There are two solutions, $t = 0$ and $t = 36.0/9.81 = 3.7\,\mathrm{s}$. The $t = 0$ value corresponds to the time when the displacement was zero when the ball was on the point of leaving the cricketer's hands. The answer required here is $\mathbf{3.7\,s}$.

▶▶

3.6 Graphs of the kinematic equations

It is often useful to represent the motion of a particle graphically, instead of by means of a series of equations. In this section we bring together the graphs which correspond to the equations we have already derived. We shall see that there are some important links between the graphs.

First, think about a particle moving in a straight line with constant velocity. Constant velocity means that the particle covers equal distances in equal intervals of time. A graph of displacement x against time t is thus a straight line, as in Figure 3.11. Here the particle has started at $x = 0$ and at time $t = 0$. The slope of the graph is equal to the magnitude of the velocity, since, from the definition of average velocity, $\bar{v} = (x_2 - x_1)/(t_2 - t_1) = \Delta x/\Delta t$. Because this graph is a straight line, the average velocity and the instantaneous velocity are the same. The equation describing the graph is $x = vt$.

Now think about a particle moving in a straight line with constant acceleration. The particle's velocity will change by equal amounts in equal intervals of time. A graph of the magnitude v of the velocity against time t will be a straight line, as in Figure 3.12. Here the particle has started with velocity u at time $t = 0$. The slope of the graph is equal to the magnitude of the acceleration. The graph is a straight line showing that the acceleration is a constant. The equation describing the graph is $v = u + at$.

An important feature of the velocity–time graph is that we can deduce the displacement of the particle by calculating the area between the graph and the t-axis, between appropriate limits of time. Suppose we want to obtain the displacement of the particle between times t_1 and t_2 in Figure 3.12. Between these times the average velocity $\bar{v}$ is represented by the horizontal line AB. The area between the graph and the t-axis is equal to the area of the rectangle whose top edge is AB, or average velocity $\bar{v}$. This area is $\bar{v}\Delta t$. But, by the definition of average velocity ($\bar{v} = \Delta x/\Delta t$), $\bar{v}\Delta t$ is equal to the displacement Δx during the time interval Δt.

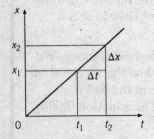

Figure 3.11

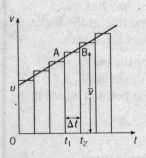

Figure 3.12

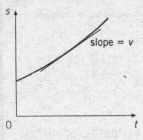

Figure 3.13

We can deduce the graph of displacement s against time t from the velocity–time graph by calculating the area between the graph and the t-axis for a succession of values of t. As shown in Figure 3.12, we can split the area up into a number of rectangles. The displacement at a certain time is then just the sum of the areas of the rectangles up to that time. Figure 3.13 shows the result of plotting the displacement s determined in this way against time t. It is a curve with a slope which increases the higher the value of t, indicating that the particle is accelerating. The slope at a particular time gives the magnitude of the instantaneous velocity. The equation describing Figure 3.13 is $s = ut + \frac{1}{2}at^2$.

Example

The displacement–time graph for a car on a straight test track is shown in Figure 3.14. Use this graph to draw velocity–time and acceleration–time graphs for the test run.

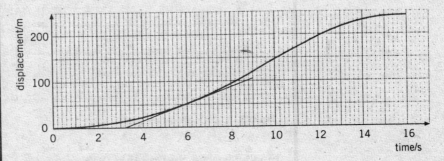

Figure 3.14 Displacement–time graph

We have already met this graph when we discussed the concepts of velocity and acceleration (Figure 3.4, page 56). In Figure 3.14 it has been re-drawn to scale, and figures have been put on the displacement and time axes. We find the magnitude of the velocity by measuring the gradient of the displacement–time graph. As an example, a tangent to the graph has been drawn at $t = 6.0\,$s. The slope of this tangent is $18\,$m s^{-1}. If the process is repeated at different times, the following velocities are determined.

t/s	2	4	6	8	10	12	14	16
v/m s^{-1}	6	12	18	24	30	20	10	0

These values are plotted on the velocity–time graph of Figure 3.15. Check some of the values by drawing tangents yourself. *Hint*: When drawing tangents, use a mirror or a transparent ruler.

Figure 3.15 shows two straight-line portions. Initially, from $t = 0$ to $t = 10\,$s, the car is accelerating uniformly, and from $t = 10\,$s to $t = 16\,$s it is decelerating. The acceleration is given by $a = \Delta v/\Delta t = 30/10 = 3\,$m s^{-2} up to $t = 10\,$s. Beyond $t = 10\,$s the acceleration is $-30/6 = -5\,$m s^{-2}. (The minus sign shows that the car is decelerating.)

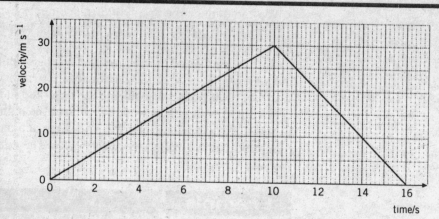

Figure 3.15 Velocity–time graph

The acceleration–time graph is plotted in Figure 3.16.

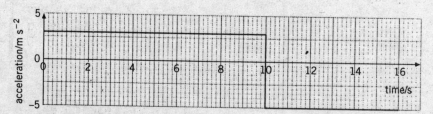

Figure 3.16 Acceleration–time graph

Finally, we can confirm that the area under a velocity–time graph gives the displacement. The area under the line in Figure 3.15 is

$$\left(\tfrac{1}{2} \times 10 \times 30\right) + \left(\tfrac{1}{2} \times 6 \times 30\right) = 240 \text{ m}$$

the value of s at $t = 16\,$s on Figure 3.14.

Now it's your turn

In a test of a sports car on a straight track, the following readings of velocity v were obtained at the times t stated.

t/s	0	5	10	15	20	25	30	35
v/m s^{-1}	0	15	23	28	32	35	37	38

(a) On graph paper, draw a velocity–time graph and use it to determine the acceleration of the car at time $t = 5\,$s.
(b) Find also the total distance travelled between $t = 0$ and $t = 30\,$s.

Note: These figures refer to a case of non-uniform acceleration, which is more realistic than the previous example. However, the same rules apply: the acceleration is given by the slope of the velocity–time graph at the relevant time, and the distance travelled can be found from the area under the graph.

3.7 Two-dimensional motion under a constant force

Figure 3.17 Cricketer bowling the ball

So far we have been dealing with motion along a straight line; that is, one-dimensional motion. We now think about the motion of particles moving in paths in two dimensions. We shall need to make use of ideas we have already learnt regarding vectors in Chapter 1. The particular example we shall take is where a particle moves in a plane under the action of a constant force.

An example is the motion of a ball thrown at an angle to the vertical (Figure 3.17), or an electron moving at an angle to an electric field. In the case of the ball, the constant force acting on it is its weight. For the electron, the constant force is the force provided by the electric field.

This topic is often called **projectile motion**. Galileo first gave an accurate analysis of this motion. He did so by splitting the motion up into its vertical and horizontal components, and considering these separately. The key is that the two components can be considered independently.

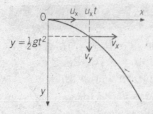

Figure 3.18

As an example, think about a particle sent off in a horizontal direction and subject to a vertical gravitational force (its weight). As before, air resistance will be neglected. We will analyse the motion in terms of the horizontal and vertical components of velocity. The particle is projected at time $t = 0$ at the origin of a system of x, y co-ordinates (Figure 3.18) with velocity u_x in the x-direction. Think first about the particle's vertical motion (in the y-direction). Throughout the motion, it has an acceleration of g (the acceleration of free fall) in the y-direction. The initial value of the vertical component of velocity is $u_y = 0$. The vertical component increases continuously under the uniform acceleration g. Using $v = u + at$, its value v_y at time t is given by $v_y = gt$. Also at time t, the vertical displacement y downwards is given by $y = \frac{1}{2}gt^2$. Now for the horizontal motion (in the x-direction): here the acceleration is zero, so the horizontal component of velocity remains constant at u_x. At time t the horizontal displacement x is given by $x = u_x t$. To find the velocity of the particle at any time t, the two components v_x and v_y must be added vectorially. The direction of the resultant vector is the direction of motion of the particle. The curve traced out by a particle subject to a constant force in one direction is a **parabola**.

Figure 3.19 Water jets from a garden sprinkler showing a parabola-shaped spray

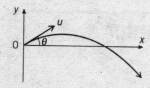

Figure 3.20

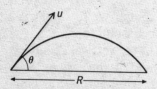

Figure 3.21

If the particle had been sent off with velocity u at an angle θ to the horizontal, as in Figure 3.20, the only difference to the analysis of the motion is that the initial y-component of velocity is $u \sin \theta$. In the example illustrated in Figure 3.20 this is upwards. Because of the downwards acceleration g, the y-component of velocity decreases to zero, at which time the particle is at the crest of its path, and then increases in magnitude again but this time in the opposite direction. The path is again a parabola.

For the particular case of a particle projected with velocity u at an angle θ to the horizontal from a point on level ground (Figure 3.21), the range R is defined as the distance from the point of projection to the point at which the particle reaches the ground again. We can show that R is given by

$$R = \frac{(u^2 \sin 2\theta)}{g}$$

For details, see the Maths Note below.

Maths Note

Suppose that the particle is projected from the origin ($x = 0$, $y = 0$). We can interpret the range R as being the horizontal distance x travelled at the time t when the value of y is again zero. The equation which links displacement, initial speed, acceleration and time is $s = ut + \frac{1}{2}at^2$. Adapting this for the vertical component of the motion, we have

$$0 = (u \sin \theta)t - \tfrac{1}{2}gt^2$$

The two solutions of this equation are $t = 0$ and $t = (2u \sin \theta)/g$. The $t = 0$ case is when the particle was projected; the second is when it returns to the ground at $y = 0$. We use this second value of t with the horizontal component of velocity $u \cos \theta$ to find the distance x travelled (the range R). This is

$$x = R = (u \cos \theta)t = (2u^2 \sin \theta \cos \theta)/g$$

There is a trigonometric relation $\sin 2\theta = 2 \sin \theta \cos \theta$, use of which puts the range expression in the required form

$$R = (u^2 \sin 2\theta)/g$$

We can see that R will have its maximum value for a given speed of projection u when $\sin 2\theta = 1$, or $2\theta = 90°$, or $\theta = 45°$. The value of this maximum range is $R_{max} = u^2/g$.

Examples

1 A stone is thrown from the top of a vertical cliff, 45 m high above level ground, with an initial velocity of 15 m s^{-1} in a horizontal direction (Figure 3.22). How long does it take to reach the ground? How far from the base of the cliff is it when it reaches the ground?

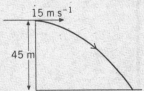

15 m s^{-1}

45 m

Figure 3.22

To find the time t for which the stone is in the air, work with the vertical component of the motion, for which we know that the initial component of velocity is zero, the displacement y = 45 m, and the acceleration a is 9.81 m s^{-2}. The equation linking these is $y = \frac{1}{2}gt^2$. Substituting the values, we have $45 = \frac{1}{2} \times 9.81t^2$. This gives $t = \sqrt{(2 \times 45/9.81)} = \textbf{3.0 s}$.

For the second part of the question, we need to find the horizontal distance x travelled in the time t. Because the horizontal component of the motion is not accelerating, x is given simply by $x = u_x t$. Substituting the values, we have $x = 15 \times 3.0 = \textbf{45 m}$.

2 An electron, travelling with a velocity of 2.0×10^7 m s^{-1} in a horizontal direction, enters a uniform electric field. This field gives the electron a constant acceleration of 5.0×10^{15} m s^{-2} in a direction perpendicular to its original velocity (Figure 3.23). The field extends for a horizontal distance of 60 mm. What is the magnitude and direction of the velocity of the electron when it leaves the field?

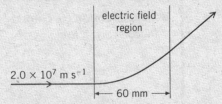

electric field region

2.0×10^7 m s^{-1}

Figure 3.23   ← 60 mm →

The horizontal motion of the electron is not accelerated. The time t spent by the electron in the field is given by $t = x/u_x = 60 \times 10^{-3}/2.0 \times 10^7 = 3.0 \times 10^{-9}$ s. When the electron enters the field, its vertical component of velocity is zero; in time t, it has been accelerated to $v_y = at = 5.0 \times 10^{15} \times 3.0 \times 10^{-9} = 1.5 \times 10^7$ m s^{-1}. When the electron leaves the field, it has a horizontal component of velocity $v_x = 2.0 \times 10^7$ m s^{-1}, unchanged from the initial value u_x. The vertical component is $v_y = 1.5 \times 10^7$ m s^{-1}. The resultant velocity v is given by $v = \sqrt{(v_x^2 + v_y^2)} = \sqrt{[(2.0 \times 10^7)^2 + (1.5 \times 10^7)^2]} = \textbf{2.5} \times \textbf{10}^7 \textbf{m s}^{-1}$. The direction of this resultant velocity makes an angle θ to the horizontal, where θ is given by $\tan \theta = v_y/v_x = 1.5 \times 10^7/2.0 \times 10^7$. The angle θ is **36.9°**.

Now it's your turn

1 A ball is thrown horizontally from the top of a tower 30 m high and lands 15 m from its base (Figure 3.24). What is the ball's initial speed?

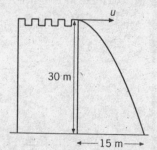

Figure 3.24

2 A football is kicked on level ground at a velocity of 15 m s^{-1} at an angle of 30° to the horizontal (Figure 3.25). How far away is the first bounce?

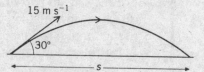

Figure 3.25

Sections 3.5–3.7 Summary

- The equations for a body moving in a straight line with uniform acceleration are:

$$v = u + at \qquad\qquad v^2 = u^2 + 2as$$
$$s = ut + \tfrac{1}{2}at^2 \qquad\qquad \bar{v} = (u + v)/2$$

- Objects falling freely near the surface of the Earth in the absence of air resistance, experience the same acceleration, the acceleration of free fall g, which has the value $g = 9.81$ m s^{-2}.
- The acceleration of free fall provides the link between the mass m and the weight W of a body: $W = mg$
- The gradient of a displacement–time graph gives the velocity of a particle, and the gradient of a velocity–time graph gives its acceleration. The area between a velocity–time graph and the time axis gives the displacement.
- The motion of projectiles is analysed in terms of two independent motions at right angles. The horizontal component of the motion is at a constant velocity, while the vertical motion is subject to a constant acceleration g.

Sections 3.5–3.7 Questions

1 A car accelerates from $5.0 \, m \, s^{-1}$ to $20 \, m \, s^{-1}$ in $6.0 \, s$. Assuming uniform acceleration, how far does it travel in this time?

2 If a raindrop were to fall from a height of $1 \, km$, with what velocity would it hit the ground if there were no air resistance?

3 Traffic police can estimate the speed of vehicles involved in accidents by the length of the marks made by skidding tyres on the road surface. It is known that the maximum deceleration that a car can attain when braking on a normal road surface is about $9 \, m \, s^{-2}$. In one accident, the tyre-marks were found to be $125 \, m$ long. Estimate the speed of the vehicle before braking.

4 On a theme park ride, a cage is travelling upwards at constant speed. As it passes a platform alongside, a passenger drops coin A through the cage floor. At exactly the same time, a person standing on the platform drops coin B from the platform.

Which coin, A or B (if either), reaches the ground first? Which (if either) has the greater speed on impact?

5 William Tell was faced with the agonising task of shooting an apple from his son Jemmy's head. Assume that William is placed $25 \, m$ from Jemmy; his crossbow fires a bolt with an initial speed of $45 \, m \, s^{-1}$. The crossbow and apple are on the same horizontal line. At what angle to the horizontal should William aim so that the bolt hits the apple?

6 The position of a sports car on a straight test track is monitored by taking a series of photographs at fixed time intervals. The following record of position x was obtained at the stated times t.

t/s	0	0.5	1.0	1.5	2.0	2.5	3.0	3.5	4.0	4.5	5.0
x/m	0	0.4	1.8	4.2	7.7	12.4	18.3	25.5	33.9	43.5	54.3

On graph paper, draw a graph of x against t. Use your graph to obtain values for the velocity v of the car at a number of values of t. Draw a second graph of v against t. From this graph, what can you deduce about the acceleration of the car?

Exam-style Questions

1 In a driving manual, it is suggested that, when driving at $13 \, m \, s^{-1}$ (about $45 \, km$ per hour), a driver should always keep a minimum of two car-lengths between the driver's car and the one in front.
   (a) Suggest a scientific justification for this safety tip, making reasonable assumptions about the magnitudes of any quantities you need.
   (b) How would you expect the length of this 'exclusion zone' to depend on speed for speeds higher than $13 \, m \, s^{-1}$?

2 A student, standing on the platform at a railway station, notices that the first two carriages of an arriving train pass her in $2.0 \, s$, and the next two in $2.4 \, s$. The train is decelerating uniformly. Each carriage is $20 \, m$ long. When the train stops, the student is opposite the last carriage. How many carriages are there in the train?

3 A ball is to be kicked so that, at the highest point of its path, it just clears a horizontal cross-bar on a pair of goal-posts. The ground is level and the cross-bar is $2.5 \, m$ high. The ball is kicked from ground level with an initial velocity of $8.0 \, m \, s^{-1}$.

   (a) Calculate the angle of projection of the ball and the distance of the point where the ball was kicked from the goal-line.
   (b) Also calculate the horizontal velocity of the ball as it passes over the cross-bar.
   (c) For how long is the ball in the air before it reaches the ground on the far side of the cross-bar?

4 An athlete competing in the long jump leaves the ground at an angle of $28°$ and makes a jump of $7.40 \, m$.
   (a) Calculate the speed at which the athlete took off.
   (b) If the athlete had been able to increase this speed by 5%, what percentage difference would this have made to the length of the jump?

5 A hunter, armed with a bow and arrow, takes direct aim at a monkey hanging from the branch of a tree. At the instant that the hunter releases the arrow, the monkey takes avoiding action by releasing its hold on the branch. By setting up the relevant equations for the motion of the monkey and the motion of the arrow, show that the monkey was mistaken in its strategy.

4. Work, energy and power

Machines such as wind turbines do work for us. They change energy from one form into some other more useful form. This chapter deals with the subject of energy in its various forms. Not only is the availability of useful forms of energy important, but also the rate at which it can be converted from one form to another. The rate of converting energy or using energy is known as power.

4.1 Work

At the end of Section 4.1 you should be able to:
- show an understanding of the concept of work in terms of the product of a force and displacement in the direction of the force
- calculate the work done in a number of situations, including the work done by a gas which is expanding against a constant external pressure: $W = p\Delta V$

'I'm going to work today.'

'Where do you work?'

'I've done some work in the garden.'

'Lots of work was done lifting the box.'

'I've done my homework.'

The word 'work' is in use in everyday English language but it has a variety of meanings. In physics, the word **work** has a definite meaning. The vagueness of the term 'work' in everyday speech causes problems for some students when they come to give a precise scientific definition of work.

Work is done when a force moves the point at which it acts (the point of application) in the direction of the force.

work done = force × distance moved by force in the direction of the force

It is very important to include direction in the definition of work done. A car can be pushed horizontally quite easily but, if the car is to be lifted off its wheels, much more work has to be done and a machine, such as a car-jack, is used.

When a force moves its point of application in the direction of the force, the force does work and the work done *by* the force is said to be *positive*. Conversely, if the direction of the force is *opposite* to the direction of movement, work is done *on* the force. This work done is then said to be *negative*. This is illustrated in Figure 4.2.

Figure 4.1 The weight-lifter uses a lot of energy to lift the weights but they can be rolled along the ground with little effort.

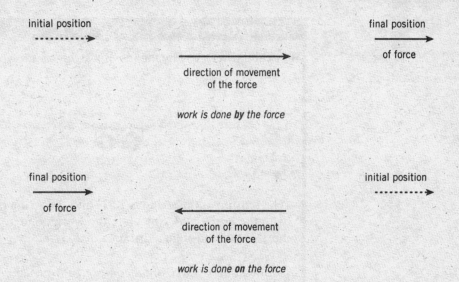

work is done **by** the force

work is done **on** the force

Figure 4.2

An alternative name for distance moved in a particular direction is **displacement**. Displacement is a vector quantity, as is force. However, work done has no direction, only size, and is a scalar quantity. It is measured in joules (J).

> When a force of one newton moves its point of application by one metre in the direction of the force, one joule of work is done.

work done in joules = force in newtons × distance moved in metres in the direction of the force

It follows that a joule (J) may be said to be a newton-metre (N m). If the force and the displacement are not both in the same direction, then the component of the force in the direction of the displacement must be found. Consider a force F acting along a line at an angle θ to the displacement, as shown in Figure 4.4. The component of the force along the direction of the displacement is $F \cos \theta$.

$$\text{work done for displacement } x = F \cos \theta \times x$$
$$= Fx \cos \theta$$

Note that the component $F \sin \theta$ of the force is at right-angles to the displacement. Since there is no displacement in the direction of this component, no work is done.

Figure 4.3 The useful work done by the small tug-boat is found using the component of the tension in the rope along the direction of motion of the ship.

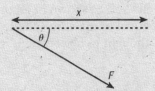

Figure 4.4

Example

A child tows a toy by means of a string as shown in Figure 4.5.

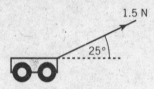

Figure 4.5

The tension in the string is 1.5 N and the string makes an angle of 25° with the horizontal. Calculate the work done in moving the toy horizontally through a distance of 265 cm.

work done = *horizontal component of tension* × *distance moved*

$$= 1.5 \cos 25 \times \frac{265}{100}$$

$$= \mathbf{3.6\ J}$$

Now it's your turn

1 A box weighs 45 N. Calculate the work done in lifting the box through a vertical height of:
   (a) 4.0 m,
   (b) 67 cm.

2 A force of 36 N acts at an angle of 55° to the vertical. The force moves its point of application by 64 cm in the direction of the force. Calculate the work done by:
   (a) the horizontal component of the force,
   (b) the vertical component of the force.

Work done by an expanding gas

A building can be demolished with explosives (Figure 4.6). When the explosives are detonated, large quantities of gas at high pressure are produced. As the gas expands, it does work by breaking down the masonry. In this section, we will derive an equation for the work done when a gas changes its volume.

Consider a gas contained in a cylinder by means of a frictionless piston of area A, as shown in Figure 4.7. The pressure p of the gas in the cylinder is equal to the atmospheric pressure outside the cylinder. This pressure may be thought to be constant.

Figure 4.6 Explosives produce large quantities of high-pressure gas. When the gas expands, it does work in demolishing the building.

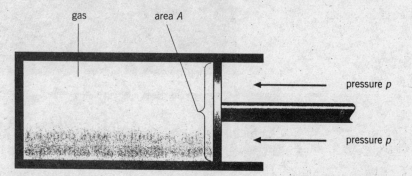

Figure 4.7

Since $pressure = \dfrac{force}{area}$, the gas produces a force F on the piston given by

$$F = pA$$

When the gas expands at constant pressure, the piston moves outwards through a distance x. So,

work done by the gas = force × distance moved
$$W = pAx$$

However, Ax is the increase in volume of the gas ΔV. Hence,

$$W = p\Delta V$$

When the volume of a gas changes at constant pressure,

work done = pressure × change in volume

When the gas *expands*, work is done *by* the gas. If the gas *contracts*, then work is done *on* the gas.

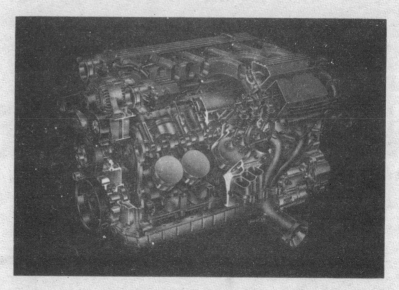

Figure 4.8 It is expanding gases pushing on the pistons which cause work to be done by the engine in a car.

Remember that the unit of work done is the joule (J). The pressure must be in pascals (Pa) or newtons per metre squared (N m^{-2}) and the change in volume in metres cubed (m^3).

Example

A sample of gas has a volume of 750 cm^3. The gas expands at a constant pressure of 1.4×10^5 Pa so that its volume becomes 900 cm^3. Calculate the work done by the gas during the expansion.

change in volume ΔV = (900 − 750)
$$= 150 \, \text{cm}^3$$
$$= 150 \times 10^{-6} \, \text{m}^3$$

work done by gas = $p\Delta V$
$$= (1.4 \times 10^5) \times (150 \times 10^{-6})$$
$$= \mathbf{21 \, J}$$

Now it's your turn

1 The volume of air in a tyre is 9.0×10^{-3} m^3. Atmospheric pressure is 1.0×10^5 Pa. Calculate the work done against the atmosphere by the air when the tyre bursts and the air expands to a volume of 2.7×10^{-2} m^3.

2 High-pressure gas in a spray-can has a volume of 250 cm^3. The gas escapes into the atmosphere through a nozzle, so that its final volume is four times the volume of the can. Calculate the work done by the gas, given that atmospheric pressure is 1.0×10^5 Pa.

Section 4.1 Summary

- When a force moves its point of application in the direction of the force, work is done.
- *Work done* = $Fx \cos \theta$, where θ is the angle between the direction of the force F and the displacement x.
- When a gas expands at constant pressure:
  work done = pressure × change in volume, or $W = p\Delta V$

Section 4.1 Questions

1 A force F moves its point of application by a distance x in a direction making an angle θ with the direction of the force, as shown in Figure 4.9.

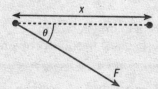

Figure 4.9

The force does an amount W of work. Copy and complete the following table.

F/N	x/m	θ/°	W/J
15	6.0	0	
15	6.0	90	
15	6.0	30	
46		23	6.4
2.4×10^3	1.6×10^2		3.1×10^5
	2.8	13	7.1×10^3

2 An elastic band is stretched so that its length increases by 2.4 cm. The force required to stretch the band increases linearly from 6.3 N to 9.5 N. Calculate:
  (a) the average force required to stretch the elastic band,
  (b) the work done in stretching the band.

3 When water boils at an atmospheric pressure of 101 kPa, 1.00 cm³ of liquid becomes 1560 cm³ of steam. Calculate the work done against the atmosphere when a saucepan containing 550 cm³ of water is allowed to boil dry.

4.2 Energy

At the end of Section 4.2 you should be able to:

- distinguish between gravitational potential energy, electric potential energy and elastic potential energy
- derive, from the defining equation $W = Fs$, the formula $\Delta E_p = mg\Delta h$ for potential energy changes near the Earth's surface
- recall and use the formula $\Delta E_p = mg\Delta h$ for potential energy changes near the Earth's surface
- show an understanding and use the relationship between force and potential energy in a uniform field to solve problems
- derive, from the equations of motion, the formula $E_k = \frac{1}{2} mv^2$
- recall and apply the formula $E_k = \frac{1}{2} mv^2$
- give examples of energy in different forms, examples of its conversion and conservation, and apply the principle of energy conservation to simple examples
- show an appreciation for the implications of energy losses in practical devices and use the concept of efficiency to solve problems.

In order to wind up a spring, work has to be done because a force must be moved through a distance. When the spring is released, it can do work; for example, making a child's toy move. When the spring is wound, it stores the ability to do work.

The ability to do work is called **energy**.

Since work done is measured in joules (J), energy is also measured in joules. Table 4.1 lists some typical values of energy.

Figure 4.10 The spring stores energy as it is stretched, releasing the energy as it returns to its original shape.

Table 4.1 Typical energy values

	order of magnitude of energy/J
radioactive decay of a nucleus	10^{-13}
sound of speech on ear for 1 second	10^{-8}
moonlight on face for 1 second	10^{-3}
beat of the heart	1
burning a match	10^3
large cream cake	10^6
energy released from 100 kg of coal	10^{10}
earthquake	10^{19}
energy received on Earth from the Sun in one year	10^{25}
rotational energy of the Milky Way Galaxy	10^{50}
estimated energy of formation of the Universe	10^{70}

Potential energy

Potential energy is the ability of an object to do work as a result of its position or shape.

We have already seen that a wound-up spring stores energy. This energy is potential energy because the spring is strained. More specifically, the energy may be called **elastic** (or **strain**) **potential energy**. Elastic potential energy is stored in objects which have had their shape changed elastically. Examples include stretched wires, twisted elastic bands and compressed gases.

Another form of potential energy is **electric potential energy**. The law of charges – like charges repel, unlike charges attract – means that work has to be done when charges are moved relative to one another. If, for example, two positive charges are moved closer together, work is done and the electric potential energy of the charges increases. The electric potential energy stored is released when the charges move apart. Conversely, if a positive charge moves closer to a negative charge, energy is released because there is a force of attraction.

Newton's law of gravitation (see Chapter 11) tells us that all masses attract one another. We rely on the force of gravity to keep us on Earth! When two masses are pulled apart, work is done on them and so they gain **gravitational potential energy**. If the masses move closer together, they lose gravitational potential energy. Changes in gravitational potential energy are of particular importance for an object near to the Earth's surface because we frequently do work raising masses and, conversely, the energy stored is released when the mass is lowered again. An object of mass m near the Earth's surface has weight mg, where g is the acceleration of free fall. This weight is the force with which the Earth attracts the mass (and the mass attracts the Earth). If the mass moves a *vertical* distance Δh,

$$work\ done = force \times distance\ moved$$
$$= mg\Delta h$$

Figure 4.11 The cars on the rollercoaster have stored gravitational potential energy. This energy is released as the cars fall.

When the mass is raised, the work done is stored as *gravitational potential energy* and this energy can be recovered when the mass falls.

Change in gravitational potential energy $\Delta E_p = mg\Delta h$

It is important to remember that, for the energy to be measured in joules, the mass m must be in kilograms, the acceleration g in metres second^{-2} and the change in height Δh in metres.

Notice that a zero point of gravitational potential energy has not been stated. We are concerned with *changes* in potential energy when a mass rises or falls.

Example

A shop assistant stacks a shelf with 25 tins of beans, each of mass 460 g. Each tin has to be raised through a distance of 1.8 m. Calculate the gravitational potential energy gained by the tins of beans, given that the acceleration of free fall is 9.8 m s^{-2}.

total mass raised = 25 × 460 = 11 500 g
              = 11.5 kg

increase in potential energy = $m \times g \times \Delta h$
              = 11.5 × 9.8 × 1.8
              = **200 J**

Now it's your turn

The acceleration of free fall is 9.8 m s^{-2}. Calculate the change in gravitational potential energy when:

(a) a person of mass 70 kg climbs a cliff of height 19 m,

(b) a book of mass 940 g is raised vertically through a distance of 130 cm,

(c) an aircraft of total mass 2.5×10^5 kg descends by 980 m.

1.8 m

Figure 4.12

Kinetic energy

Figure 4.13 When the mass falls, it gains kinetic energy and drives the pile into the ground.

As an object falls, it loses gravitational potential energy and in so doing it speeds up. Energy is associated with a moving object. In fact, we know that a moving object can be made to do work as it slows down. For example, a moving hammer hits a nail and, as it stops, does work to drive the nail into a piece of wood.

Kinetic energy is energy due to motion.

Consider an object of mass m moving with a constant acceleration a. In a distance s, the object accelerates from velocity u to velocity v. Then, by referring to the equations of motion (see Chapter 3),

$$v^2 = u^2 + 2as$$

By Newton's law (see Chapter 5), the force F giving rise to the acceleration a is given by

$$F = ma$$

Combining these two equations,

$$v^2 = u^2 + 2\frac{F}{m}s$$

Re-arranging,

$$mv^2 = mu^2 + 2Fs$$
$$2Fs = mv^2 - mu^2$$
$$Fs = \tfrac{1}{2}mv^2 - \tfrac{1}{2}mu^2$$

By definition, the term Fs is the work done by the force moving a distance s. Therefore, since Fs represents work done, then the other terms in the equation, $\tfrac{1}{2}mv^2$ and $\tfrac{1}{2}mu^2$, must also have the units of work done, or energy (see Chapter 1). The magnitude of each of these terms depends on velocity squared and so $\tfrac{1}{2}mv^2$ and $\tfrac{1}{2}mu^2$ are terms representing energy which depends on velocity (or speed). The kinetic energy E_k of an object of mass m moving with speed v is given by

$$E_k = \tfrac{1}{2}mv^2$$

For the kinetic energy to be in joules, mass must be in kilograms and speed in metres per second.

The full name for the term $E_k = \tfrac{1}{2}mv^2$ is *translational kinetic energy* because it is energy due to an object moving in a straight line. It should be remembered that rotating objects also have kinetic energy and this form of energy is known as rotational kinetic energy.

Example

Calculate the kinetic energy of a car of mass 900 kg moving at a speed of 20 m s^{-1}. State the form of energy from which the kinetic energy is derived.

$$
\begin{aligned}
\textit{kinetic energy} &= \tfrac{1}{2} mv^2 \\
&= \tfrac{1}{2} \times 900 \times 20^2 \\
&= \mathbf{1.8 \times 10^5 \; J}
\end{aligned}
$$

This energy is derived from the chemical energy of the fuel.

Now it's your turn

1 Calculate the kinetic energy of a car of mass 800 kg moving at 100 kilometres per hour.

2 A cycle and cyclist have a combined mass of 80 kg and are moving at 5.0 m s^{-1}. Calculate:
   (a) the kinetic energy of the cycle and cyclist,
   (b) the increase in kinetic energy for an increase in speed of 5.0 m s^{-1}.

Energy conversion and conservation

Newspapers sometimes refer to a 'Global Energy Crisis'. In the near future, there may well be a shortage of fossil fuels. Fossil fuels are sources of chemical energy. It would be more accurate to refer to a 'Fuel Crisis'. When chemical energy is used, the energy is transformed into other forms of energy, some of which are useful and some of which are not. Eventually, all the chemical energy is likely to end up as energy that is no longer useful to us. For example, when petrol is burned in a car engine, some of the chemical energy is converted into the kinetic energy of the car and some is wasted as heat (thermal) energy. When the car stops, its kinetic energy is converted into heat energy in the brakes. The outcome is that the chemical energy has been converted into heat energy which dissipates in the atmosphere and is of no further use. However, the total energy present in the Universe has remained constant. All energy changes are governed by the **law of conservation of energy**. This law states that

Energy cannot be created or destroyed. It can only be converted from one form to another.

There are many different forms of energy and you will meet a number of these during your A/AS Physics studies. Some of the more common forms are listed in Table 4.2.

Table 4.2 Forms of energy

energy	notes
potential energy	energy due to position
kinetic energy	energy due to motion
elastic or strain energy	energy due to stretching an object
electrical energy	energy associated with moving electric charge
sound energy	a mixture of potential and kinetic energy of the particles in the wave
wind energy	a particular type of kinetic energy
light energy	energy of electromagnetic waves
solar energy	light energy from the Sun
chemical energy	energy released during chemical reactions
nuclear energy	energy associated with particles in the nuclei of atoms
thermal energy	sometimes called heat energy

Example

Map out the energy changes taking place when a battery is connected to a lamp.

Chemical energy in battery → electrical energy in wires
→ light energy and heat energy in lamp

Now it's your turn

Map out the following energy changes:
(a) a child swinging on a swing,
(b) an aerosol can producing hair spray,
(c) a lump of clay thrown into the air which subsequently hits the ground.

Efficiency

In most energy changes some energy is 'lost' as heat (thermal) energy. For example, when a ball rolls down a slope, the total change in gravitational potential energy is not equal to the gain in kinetic energy because heat (thermal) energy has been produced as a result of frictional forces.

Efficiency gives a measure of how much of the total energy may be used and is not 'lost'.

$$efficiency = \frac{useful\ work\ done}{total\ energy\ input}$$

Efficiency may be given either as a ratio or as a percentage. Since energy cannot be created, efficiency can never be greater than 100% and a 'perpetual motion' machine is not possible (Figure 4.14).

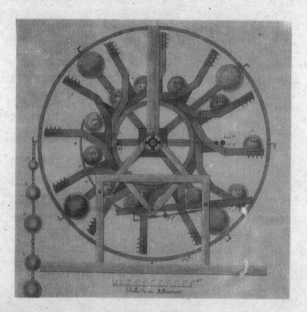

Figure 4.14 An attempt to design a machine to get something for nothing by breaking the law of conservation of energy

Example

A man lifts a weight of 480 N through a vertical distance of 3.5 m using a rope and some pulleys. The man pulls on the rope with a force of 200 N and a length of 10.5 m of rope passes through his hands. Calculate the efficiency of the pulley system.

$$
\begin{aligned}
\textit{work done by man} &= \textit{force} \times \textit{distance moved} \\
&= 200 \times 10.5 \\
&= 2100\,\text{J} \\
\textit{work done lifting load} &= 480 \times 3.5 \\
&= 1680\,\text{J} \\
\textit{efficiency} &= \text{work got out/work put in} \\
&= 1680/2100 \\
&= \textbf{0.80 or 80\%}
\end{aligned}
$$

Now it's your turn

1 An electric heater converts electrical energy into heat energy. Suggest why this process may be 100% efficient.

2 The electric motor of an elevator uses 630 kJ of electrical energy when raising the elevator and passengers, of total weight 12.500 N, through a vertical height of 29 m. Calculate the efficiency of the elevator.

Section 4.2 Summary

- Energy is needed to do work; energy is the ability to do work.
- Potential energy is the energy stored in a body due to its position or shape; examples are elastic potential energy, electric potential energy and gravitational potential energy.
- When an object of mass m moves vertically through a distance Δh, then the change in gravitational potential energy is given by:

$$\Delta E_p = mg\Delta h$$

where g is the acceleration of free fall.
- Kinetic energy is the energy stored in a body due to its motion.
- For an object of mass m moving with speed v, the kinetic energy is given by: $E_k = \frac{1}{2}mv^2$
- Energy cannot be created or destroyed. It can only be converted from one form to another.
- *Efficiency = useful work done/total energy input*

Section 4.2 Questions

1 Name each of the following types of energy:
  (a) energy used in muscles,
  (b) energy stored in the Sun,
  (c) energy of water in a mountain lake,
  (d) energy captured by a wind turbine,
  (e) energy produced when a firework explodes,
  (f) energy of a compressed gas.

2 A child of mass 35 kg moves down a sloping path on a skate board. The sloping path makes an angle of 4.5° with the horizontal. The constant speed of the child along the path is 6.5 m s^{-1}. Calculate:
  (a) the vertical distance through which the child moves in 1.0 s,
  (b) the rate at which potential energy is being lost ($g = 10$ m s^{-2}).

3 A stone of mass 120 g is dropped down a well. The surface of the water in the well is 9.5 m below ground level. The acceleration of free fall of the stone is 9.8 m s^{-2}. Calculate, for the stone falling from ground level to the water surface:
  (a) the loss of potential energy,
  (b) its speed as it hits the water, assuming all the potential energy has been converted into kinetic energy.

4 An aircraft of mass 3.2×10^5 kg accelerates along a runway. Calculate the change in kinetic energy, in MJ, when the aircraft accelerates:
  (a) from zero to 10 m s^{-1},
  (b) from 30 m s^{-1} to 40 m s^{-1},
  (c) from 60 m s^{-1} to 70 m s^{-1}.

5 In order to strengthen her legs, an athlete steps up on to a box and then down again 30 times per minute. The girl has mass 50 kg and the box is 35 cm high. The exercise lasts 4.0 minutes and as a result of the exercise, her leg muscles generate 120 kJ of heat energy. Calculate the efficiency of the leg muscles ($g = 10$ m s^{-2}).

6 By accident, the door of a refrigerator is left open. Use the law of conservation of energy to explain whether the temperature of the room will rise, stay constant or fall after the refrigerator has been working for a few hours.

4.3 The deformation of solids

At the end of Section 4.3 you should be able to:
- appreciate that deformation is caused by a force and that, in one dimension, the deformation can be tensile or compressive
- describe the behaviour of springs in terms of load, extension, elastic limit, Hooke's law and the spring constant (force per unit extension)
- distinguish between elastic and plastic deformation of a material
- deduce the strain energy in a deformed material from the area under the force–extension graph
- define and use the terms stress, strain and the Young modulus
- describe an experiment to determine the Young modulus of a metal in the form of a wire
- demonstrate knowledge of the force–extension graphs for typical ductile, brittle and polymeric materials, including an understanding of ultimate tensile stress.

When forces are applied to a solid body, its shape changes. The change may be very small, but nevertheless the forces affect the spacing of the atoms in the solid to a tiny extent and its external dimensions change. This change of shape is called **deformation**. If you think of a metal cylinder, the application of forces along the axis of the cylinder can either stretch it, making it longer, or squash it slightly, making it shorter. We call the deformation produced by these forces a **tensile** deformation for the stretching case, or a **compressive** deformation for the squashing case.

Hooke's law

A helical spring, attached to a fixed point, hangs vertically and has weights attached to its lower end, as shown in Figure 4.15. As the size of the weight is increased the spring becomes longer. The increase in length of the spring is called the **extension** of the spring and the weight attached to the spring is called the **load**.

If the load is increased greatly, the spring will permanently change its shape. However, for small loads, when the load is removed, the spring returns to its original length. The spring is said to have undergone an **elastic change**.

In an elastic change, a body returns to its original shape and size when the load on it is removed.

Figure 4.16 shows the variation with load of the extension of the spring. The section of the line from the origin to the point L is straight. In this region, the spring behaves elastically, and returns to its original length when the load is removed. The point L is referred to as the **elastic limit**. Beyond the point L, the spring is deformed permanently and the change is said to be **plastic**.

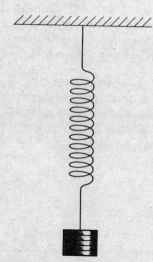

Figure 4.15 A loaded helical spring

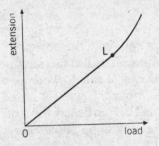

Figure 4.16 Extension of a loaded spring

The fact that there is a straight line relationship between load and extension for the elastic change is expressed in Hooke's law. It should be appreciated that, although we have used a spring as an illustration, the law applies to any object, provided the elastic limit has not been exceeded.

Hooke's law states that, provided the elastic limit is not exceeded, the extension of a body is proportional to the applied load.

The law can be expressed in the form of an equation

force $F \propto$ extension ΔL

Removing the proportionality sign gives

$F = k\Delta L$

where k is a constant, known as the **elastic constant** or **spring constant**.

The elastic constant (spring constant) is the force per unit extension.

The unit of the constant is newton per metre ($N\,m^{-1}$).

Example

An elastic cord has an unextended length of 25 cm. When the cord is extended by applying a force at each end, the length of the cord becomes 40 cm for forces of 0.75 N. Calculate the elastic constant of the cord.

extension of cord = 15 cm
elastic constant = 0.75/0.15 (extension in metres)
= **5.0 N m^{-1}**

Now it's your turn

1 Calculate the elastic constant for a spring which extends by a distance of 3.5 cm when a load of 14 N is hung from its end.

2 A steel wire extends by 1.5 mm when it is under a tension of 45 N. Calculate:
(a) the elastic constant of the wire,
(b) the tension required to produce an extension of 1.8 mm, assuming that the elastic limit is not exceeded.

Strain energy

When an object has its shape changed by forces acting on it, the object is said to be **strained**. Work has to be done by the forces to cause this strain. Provided that the elastic limit is not exceeded, the object can do work as it returns to its original shape when the forces are removed. Energy is stored in the body as potential energy when it is strained. This particular form of potential energy is called elastic potential energy or strain potential energy, or simply **strain energy**.

Strain energy is energy stored in a body due to change of shape.

Consider the spring shown in Figure 4.15. To produce an extension x, the force applied at the lower end of the spring increases linearly with extension from zero to a value F. The average force is $\frac{1}{2}F$ and the work done W by the force is therefore

$$W = average\ force \times extension \qquad \text{(see Section 4.1)}$$
$$= \tfrac{1}{2}Fx$$

However, the elastic constant k is given by the equation

$$F = k\Delta L$$

The value ΔL is equivalent to x. Therefore, substituting for F,

$$strain\ energy\ W = \tfrac{1}{2}k(\Delta L)^2$$

The energy is in joules if k is in newtons per metre and x (equal to ΔL) is in metres.

A graph of load (y-axis) against extension (x-axis) enables strain energy to be found even when the graph is not linear (see Figure 4.17). We have shown that strain energy is given by

$$strain\ energy\ W = \tfrac{1}{2}F\Delta L$$
$$= \tfrac{1}{2}Fx$$
$$= \tfrac{1}{2}kx^2$$

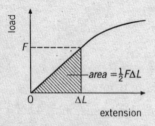

Figure 4.17 Strain energy is given by the area under the graph.

The expression $\frac{1}{2}Fx$ represents the area between the straight line on Figure 4.17 and the x-axis. This means that strain energy is represented by the area under the line on a graph of load (y-axis) plotted against extension (x-axis).

Example

A spring has an elastic constant $65\,\text{N}\,\text{m}^{-1}$ and is extended elastically by $1.2\,\text{cm}$. Calculate the strain energy stored in the spring.

$$
\begin{aligned}
\textit{strain energy } W &= \tfrac{1}{2}kx^2 \\
&= \tfrac{1}{2} \times 65 \times (1.2 \times 10^{-2})^2 \\
&= \mathbf{4.7 \times 10^{-3}\ J}
\end{aligned}
$$

Now it's your turn

1 A wire has an elastic constant of $5.5 \times 10^4\,\text{N}\,\text{m}^{-1}$. It is extended elastically by $1.4\,\text{mm}$. Calculate the strain energy stored in the wire.

2 A rubber band has an elastic constant of $180\,\text{N}\,\text{m}^{-1}$. The work done in extending the band is $0.16\,\text{J}$. Calculate the extension of the band.

The Young modulus

The difficulty with using the elastic constant is that the constant is different for each specimen of a material having a different shape. It would be far more convenient if we had a constant for a particular material which would enable us to find extensions knowing the constant and the dimensions of the specimen. This is possible using the **Young modulus**.

We have already mentioned the term **strain**. When an object of original length L is extended by an amount ΔL, the strain (ε) is defined as

$$
\textit{strain} = \frac{\textit{extension}}{\textit{original length}}
$$

$$
\varepsilon = \Delta L / L
$$

Strain is the ratio of two lengths and does not have a unit.

The strain produced within an object is caused by a **stress**. In our case, we are dealing with changes in length and so the stress is referred to as a **tensile stress**. When a tensile force F acts normally to an area A, the stress (σ) is given by

$$
\textit{stress} = \frac{\textit{force}}{\textit{area normal to the force}}
$$

$$
\sigma = F / A
$$

The unit of tensile stress is newtons per metre squared ($\text{N}\,\text{m}^{-2}$). This unit is also the unit of pressure and so an alternative unit for stress is the pascal (Pa).

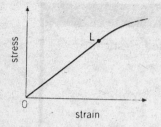

Figure 4.18 Stress–strain graph

In Figure 4.17, we plotted a graph of load against extension. Since load is related to stress and extension is related to strain, a graph of stress plotted against strain would have the same basic shape, as shown in Figure 4.18. Once again, there is a straight line region between the origin and L, the elastic limit. In this region, changes of strain with stress are elastic. In the region where the changes are elastic, it can be seen that

stress ∝ *strain*

or, removing the proportionality sign,

stress = *E* × *strain*

The constant *E* is known as the Young modulus of the material.

$$Young\ modulus\ E = \frac{stress}{strain}$$

The unit of the Young modulus is the same as that for stress because strain is a ratio and has no unit. Some values of the Young modulus for different materials are shown in Table 4.3.

Table 4.3 Young modulus for different materials

material	Young modulus E/Pa
aluminium	0.70×10^{11}
copper	1.1×10^{11}
steel	2.1×10^{11}
glass	0.41×10^{11}

The Young modulus of a metal in the form of a wire may be measured by applying loads to a wire and measuring the extensions caused. The original length and the cross-sectional area must also be measured. A suitable laboratory arrangement is shown in Figure 4.19. A copper wire is often used. This is because, for wires of the same diameter under the same load, a copper wire will give larger, more measurable, extensions than a steel wire. (Why is this?) A paper flag with a reference mark on it is attached to the wire at a distance of just less than one metre from the clamped end. The original length L is measured from the clamped end to the reference mark, using a metre rule. The diameter d of the wire is measured using a micrometer screw gauge, and the cross-sectional area A calculated from $A = \frac{1}{4}\pi d^2$. Extensions ΔL are measured as masses m are added to the mass-carrier. (Think of a suitable way of measuring these extensions.) Be careful not to exceed the linear, Hooke's law, region. The load F is calculated from $F = mg$. A graph of ΔL against F has gradient L/EA, so the Young modulus E is equal to $L(A \times \text{gradient})$.

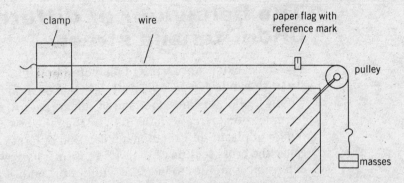

Figure 4.19 Simple experiment to measure the Young modulus of a wire

---

Example

A steel wire of diameter 1.0 mm and length 2.5 m is suspended from a fixed point and a mass of weight 45 N is suspended from its free end. The Young modulus of the material of the wire is 2.1×10^{11} Pa. Assuming that the elastic limit of the wire is not exceeded, calculate:
(a) the applied stress,
(b) the strain,
(c) the extension of the wire.

(a)   area $= \pi \times (0.5 \times 10^{-3})^2$
               $= 7.9 \times 10^{-7}$ m^2
     stress $=$ force/area
              $= 45/7.9 \times 10^{-7}$
              $= \mathbf{5.7 \times 10^7}$ **Pa**

(b)   strain $=$ stress/Young modulus
              $= 5.7 \times 10^7/2.1 \times 10^{11}$
              $= \mathbf{2.7 \times 10^{-4}}$

(c)   extension $=$ strain $\times$ length
              $= 2.7 \times 10^{-4} \times 2.5$
              $= 6.8 \times 10^{-4}$ m
              $= \mathbf{0.68}$ **mm**

Now it's your turn

1 A copper wire of diameter 1.78 mm and length 1.4 m is suspended from a fixed point and a mass of weight 32 N is suspended from its free end. The Young modulus of the material of the wire is 1.1×10^{11} Pa. Assuming that the elastic limit of the wire is not exceeded, calculate:
    (a) the applied stress,
    (b) the strain,
    (c) the extension of the wire.

2 An elastic band of area of cross-section 2.0 mm^2 has an unextended length of 8.0 cm. When stretched by a force of 0.4 N, its length becomes 8.3 cm. Calculate the Young modulus of the elastic.

The behaviour of different materials under tensile stress

The force–extension graphs of all metals in the form of wires have the general shape shown in Figure 4.20: a straight line portion through the origin (the Young modulus is measured in this linear region), and then a region in which the extension increases more rapidly than the force. Eventually, well into the region of plastic deformation, application of larger and larger forces will cause the cross-section of the wire to form a narrow neck, so that the extension continues to increase without the addition of further force. The wire eventually breaks.

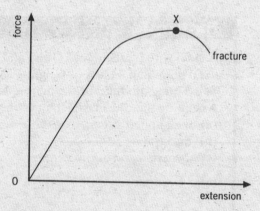

Figure 4.20 Force–extension graph for a ductile material

From the force at which the curve is a maximum (point X), a quantity known as the **ultimate tensile stress** can be calculated. This the the maximum force divided by the original cross-sectional area of the wire (not the area of the narrow neck, which would be almost impossible to specify). The ultimate tensile stress gives an idea of the maximum stress that the wire could support; note that this is *not* the same as the stress when the wire finally breaks. (Note also that the terminology 'breaking point' is not satisfactory, as it is not clear to what the 'point' refers; is it a position on the wire, the length at which the wire breaks, the force required to break the wire, the breaking stress or the breaking strain?)

Figure 4.20 is characteristic of materials which can be drawn out into wires, or **ductile** materials. Ductility is a characteristic of many metals.

Another class of material shows the behaviour sketched in Figure 4.21. A glass fibre is a typical example. The fibre extends elastically with a linear relation between force and extension over an appreciable range. Very soon after the limit of proportionality, the fibre snaps. This is called brittle fracture, and many amorphous substances like glass are classified as **brittle** materials. Their characteristic is that deformation obeys Hooke's law over practically the whole range of extensions; there is very little, if any, plastic deformation. This distinction tells us why, when a glass beaker is dropped on the floor, it shatters into many pieces. The glass is incapable of plastic deformation, and if the impact causes a deformation taking it out of the elastic region, it has to break. A metal beaker, on the other hand, is deformed plastically by the impact, ending up with a permanent deformation in the form of a dent, but not breaking.

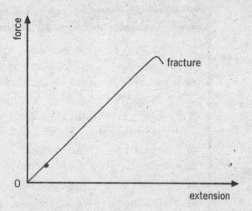

Figure 4.21 Force–extension graph for a brittle material

A typical polymeric material (such as rubber) has another entirely different force–extension graph. This is sketched in Figure 4.22 for a rubber cord. Only a very small part of this curve, near the origin, is sufficiently linear to use it to calculate the Young modulus. An important point to note is that the cord can be stretched to many times its original length before it breaks; the curve also shows a very extensive region in which the cord will return to its original length when the stretching force is removed. However, this may not be a case of simple elastic extension because, although the cord is not permanently deformed, it may not return to its original length along the same path. This is called **elastic hysteresis**. How is energy conserved? The graphs show that the strain energy required to deform the rubber (the area under the graph – see Figure 4.17) is greater than the work done by the material in returning to its original length. The excess energy, represented by the area bounded by the two curves, must be energy dissipated within the rubber, showing up as an increase in temperature. Types of rubber with large elastic hysteresis are used in vibration absorbers for mounting oscillating machinery and preventing the vibrations being transmitted to the floor. The mechanical energy of the oscillations is converted to thermal energy in the mounts.

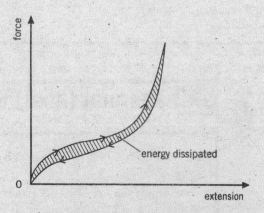

Figure 4.22 Force–extension graph for a polymeric material showing elastic hysteresis

Section 4.3 Summary

- Forces on an object can cause tensile deformation (stretching) or compressive deformation (squashing).
- An elastic change occurs when an object returns to its original shape and size when the load is removed from it.
- Hooke's law states that, for an elastic change, extension is proportional to load.
- The elastic constant (spring constant) k is the ratio of force to extension.
- *Strain energy* $= \frac{1}{2}kx^2$
- *Tensile strain = extension/original length*
- *Tensile stress = force/cross-sectional area*
- *Young modulus = stress/strain*
- Ductile and brittle materials have characteristic force–extension graphs.
- Polymeric materials such as rubber typically show elastic hysteresis.

Section 4.3 Questions

1 A spring has an unextended length of 12.4 cm. When a load of 4.5 N is suspended from the spring, its length becomes 13.3 cm. Calculate:
(a) the elastic constant of the spring,
(b) the length of the spring for a load of 3.5 N.

2 The elastic cord of a catapult has an elastic constant of 700 N m^{-1}. Calculate the energy stored in the elastic cord when it is extended by 15 cm.

3 Two wires each have length 1.8 m and diameter 1.2 mm. One wire has a Young modulus of 1.1×10^{11} Pa and the other 2.2×10^{11} Pa. One end of each wire is attached to the same fixed point and the other end of each wire is attached to the same load of 75 N so that each has the same extension. Assuming that the elastic limit of the wires is not exceeded, calculate the extension of the wires.

4 Distinguish between ductile and brittle materials, making reference to the force–extension graph for each type of material. Name an example of each type.

5 Explain what is meant by elastic hysteresis.

4.4 Thermal (heat) energy

At the end of Section 4.4 you should be able to:
- define and use the concept of specific heat capacity, and identify the main principles of its determination by electrical methods
- define and use the concept of specific latent heat, and identify the main principles of its determination by electrical methods.

In this section, we will be looking at the effect of heating on temperature.

Specific heat capacity

When a solid, a liquid or a gas is heated, its temperature rises. Plotting a graph of thermal (heat) energy supplied against temperature rise (see Figure 4.23), it is seen that the temperature rise $\Delta\theta$ is proportional to the heat energy ΔQ supplied, for a particular mass of a particular substance.

$$\Delta Q \propto \Delta\theta$$

Similarly, the heat energy required to produce a particular temperature rise is proportional to the mass m of the substance being heated (see Figure 4.24).

$$\Delta Q \propto m$$

Combining these two relations gives

$$\Delta Q \propto m\Delta\theta$$

$$\Delta Q = mc\Delta\theta$$

where c is the constant of proportionality known as the **specific heat capacity** of the substance. In this case, *specific* means *per unit mass*.

The numerical value of the specific heat capacity of a substance is the quantity of heat energy required to raise the temperature of unit mass of the substance by one degree.

The SI unit of specific heat capacity is $J\,kg^{-1}K^{-1}$. The unit of specific heat capacity is *not* the joule and this is why, in the definition of specific heat capacity, it is important to make reference to the *numerical value*. Specific heat capacity is different for different substances. Some values are given in Table 4.4.

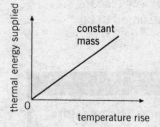

Figure 4.23

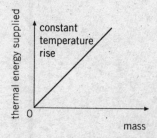

Figure 4.24

Table 4.4 Values of specific heat capacity for different materials

material	specific heat capacity/$J\,kg^{-1}\,K^{-1}$
ethanol	2500
glycerol	2420
ice	2100
mercury	140
water	4200
aluminium	913
copper	390
glass	640

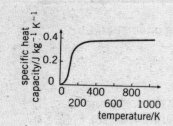

Figure 4.25

It should be noted that, for relatively small changes in temperature, specific heat capacity is approximately constant. However, over a wide range of temperature, the value for a substance may vary considerably (see Figure 4.25). Unless stated otherwise, specific heat capacity is assumed to be constant.

Example

Calculate the quantity of heat energy required to raise the temperature of a mass of 810 g of aluminium from 20 °C to 75 °C. The specific heat capacity of aluminium is $910\,\mathrm{J\,kg^{-1}K^{-1}}$.

$$\textit{heat energy required} = m \times c \times \Delta\theta$$
$$= \frac{810}{1000} \times 910 \times (75 - 20)$$
$$= \mathbf{4.1 \times 10^4\ J}$$

Now it's your turn

1 Calculate the heat energy gained or lost for the following temperature changes. Use Table 4.4 to obtain values for specific heat capacity.
   (a) 45 g of copper heated from 10 °C to 90 °C
   (b) 1.3 kg of ice at 0 °C cooled to −15 °C

2 Calculate the specific heat capacity of water given that 0.20 MJ of energy are required to raise the temperature of a mass of 600 g of water by 80 K.

Thermal (heat) capacity

Specific heat capacity may be used for a single substance, but often objects are made of several different materials. In order to find the change in heat energy, the heat capacity and mass of every substance in the object would need to be known. This can be avoided using the **thermal (heat) capacity** of the object.

> The numerical value of the thermal (heat) capacity of a body is the quantity of heat energy required to raise the temperature of the whole body by one degree.

For a body of thermal capacity C, the heat energy ΔQ supplied is related to the rise in temperature $\Delta\theta$ by the expression

$$\Delta Q = C\Delta\theta$$

Heat capacity has the SI unit $\mathrm{J\,K^{-1}}$.

Example

A room has a thermal capacity of $1.2\,\text{MJ K}^{-1}$. Calculate the heat energy required to raise the temperature of the room from $18\,°\text{C}$ to $20\,°\text{C}$.

$$\Delta Q = C \times \Delta\theta$$
$$= 1.2 \times 10^6 \times (20 - 18)$$
$$= \mathbf{2.4 \times 10^6\,J}$$

Now it's your turn

1 A kettle contains 700 g of water. The specific heat capacity of water is $4200\,\text{J kg}^{-1}\text{K}^{-1}$ and the kettle itself has heat capacity $540\,\text{J K}^{-1}$. The kettle and its contents are heated from $15\,°\text{C}$ to the boiling point of water ($100\,°\text{C}$).
   (a) Calculate the heat energy gained by:
      (i) the water,
      (ii) the kettle.
   (b) What fraction of the total energy was used in heating the water?

2 An aluminium saucepan has a copper base. Use the information below to determine the thermal capacity of the saucepan.

metal	mass/g	specific heat capacity/J kg^{-1}K^{-1}
aluminium	1160	900
copper	160	390

Specific latent heat

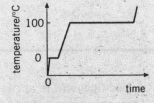

Figure 4.26

Figure 4.26 illustrates how the temperature of a mass of water varies with time when it is heated at a constant rate. At times when the substance is changing phase (ice to water or water to steam), heat energy is being supplied without any change of temperature. Because the heat energy does not change the temperature of the substance, it is said to be **latent** (i.e. hidden). The latent heat required to melt (fuse) a solid is known as **latent heat of fusion**.

The numerical value of the **specific latent heat of fusion** is the quantity of heat energy required to convert unit mass of solid to liquid without any change in temperature.

The SI unit of specific latent heat of fusion is J kg^{-1}. For a substance with latent heat of fusion L_f, the quantity of heat energy ΔQ required to fuse (melt) a mass m of solid is given by

$$\Delta Q = mL_f$$

The latent heat required to vaporise a liquid is referred to as **latent heat of vaporisation**.

The numerical value of the **specific latent heat of vaporisation** is the quantity of heat energy required to convert unit mass of liquid to vapour without any change in temperature.

The SI unit of specific latent heat of vaporisation is the same as that for fusion i.e. J kg^{-1}. For a substance with latent heat of vaporisation L_v, the quantity of heat energy ΔQ required to vaporise a mass m of liquid is given by

$$\Delta Q = mL_v$$

When a vapour condenses (vapour becomes liquid), the latent heat of vaporisation is released. Similarly, when a liquid solidifies (liquid becomes solid), the latent heat of fusion is released. Some values of specific latent heat of fusion and of vaporisation are given in Table 4.5.

Table 4.5 Values of specific latent heat

material	specific latent heat of fusion/kJ kg^{-1}	specific latent heat of vaporisation/kJ kg^{-1}
ice/water	330	2260
ethanol	108	840
copper	205	4840
sulfur	38.1	

Experiments to determine the specific heat capacity or specific latent heat of a material often involve the use of electrical heating. The energy input is determined from $\Delta Q = W = EIt$, or an equivalent expression, and then equated to $mc\Delta\theta$ or mL. The use of such an equation neglects any heat loss; experiments should be designed to minimise this error.

Example

Use the information given in Table 4.5 to determine the heat energy required to melt 50 g of sulfur at its normal melting point.

$$heat\ energy\ required = m \times L_f$$
$$= \frac{50}{1000} \times 38.1 \times 1000$$
$$= 1900\ J$$

Now it's your turn

Where appropriate, use the information given in Table 4.5.
(a) Calculate the heat energy required to:
   (i) melt 50 g of ice at 0 °C,
   (ii) evaporate 50 g of water at 100 °C.
(b) Using your answers to (a), determine how many times more energy is required to evaporate a mass of water than to melt the same mass of ice.

Exchanges of heat energy

When a hot object and a cold object come into contact, heat energy passes from the hot object to the cold one so that the two objects reach the same temperature. The law of conservation of energy applies in that the heat energy gained by the cold object is equal to the heat energy lost by the hot object. This does, of course, assume that no energy is lost to the surroundings. This simplification enables temperatures to be calculated.

Examples

1 A mass of 0.30 kg of water at 95 °C is mixed with 0.50 kg of water at 20 °C. Calculate the final temperature of the water, given that the specific heat capacity of water is 4200 J kg^{-1}K^{-1}.

Hint: always start by writing out a word equation containing all the gains and losses of heat energy.

heat energy lost by hot water = heat energy gained by cold water
$$(m \times c \times \Delta\theta_1) = (M \times c \times \Delta\theta_2)$$
$$0.30 \times 4200 \times (95 - t) = 0.50 \times 4200 \times (t - 20)$$

where t is the final temperature of the water.

$$1260 \times (95 - t) = 2100 \times (t - 20)$$
$$119\,700 - 1260t = 2100t - 42\,000$$
$$161\,700 = 3360t$$
$$t = 48\,°C$$

2 A mass of 12 g of ice at 0 °C is placed in a drink of mass 210 g at 25 °C. Calculate the final temperature of the drink, given that the specific latent heat of fusion of ice is 334 kJ kg^{-1} and that the specific heat capacity of water and the drink is 4.2 kJ kg^{-1}K^{-1}.

$$\begin{array}{ccc} \text{energy lost} & = & \text{energy gained} \\ \text{by drink} & & \text{by melting ice} \end{array} + \begin{array}{c} \text{energy gained} \\ \text{by ice-water} \end{array}$$

$$(m \times c \times \Delta\theta_1) = (M \times L_f) + (M \times c \times \Delta\theta_2)$$

$$M \times c \times \Delta\theta_2 = \frac{12}{1000} \times 4.2 \times 1000 \times (t - 0)$$

$$= 50.4t$$

$$\frac{210}{1000} \times 4.2 \times 1000 \times (25 - t) = \frac{12}{1000} \times 334 \times 1000 + 50.4t$$

where t is the final temperature of the drink. Simplifying,

$$22\,050 - 882t = 4008 + 50.4t$$
$$18\,042 = 932.4t$$
$$t = \mathbf{19\,°C}$$

Now it's your turn

Use the data in Tables 4.4 and 4.5 where appropriate.

1 A lump of copper of mass 120 g is heated in a gas flame. It is then transferred to a mass of 450 g of water, initially at 20 °C. The final temperature of the copper and the water is 31 °C. Calculate the temperature of the gas flame.

2 Steam at 100 °C is passed into a mass of 350 g of water, initially at 15 °C. The steam condenses. Calculate the mass of steam required to raise the temperature of the water to 80 °C.

Section 4.4 Summary

■ Specific heat capacity is numerically equal to the heat energy required to raise the temperature of unit mass of substance by one degree.
■ The heat energy ΔQ required to raise the temperature of a mass m of substance of specific heat capacity c by an amount $\Delta\theta$ is given by the expression: $\Delta Q = mc\Delta\theta$
■ Specific heat capacity has the SI unit of J kg^{-1}K^{-1}.
■ The heat capacity of an object is numerically equal to the heat energy required to raise the temperature of the whole body by one degree.
■ The heat energy ΔQ required to raise the temperature of an object having a heat capacity C by an amount $\Delta\theta$ is given by the expression: $\Delta Q = C\Delta\theta$
■ Heat capacity has the SI unit of J K^{-1}.

■ Specific latent heat of fusion is numerically equal to the quantity of heat energy required to convert unit mass of solid to liquid without any change in temperature.
■ Specific latent heat of vaporisation is numerically equal to the quantity of heat energy required to convert unit mass of liquid to vapour without any change in temperature.
■ When a substance of mass m changes its state the quantity of heat energy required is given by:

$$\Delta Q = mL$$

where L is the appropriate specific latent heat.
■ Specific latent heat has the SI unit of $J\,kg^{-1}$.

Section 4.4 Questions

In the following questions, use the data in Tables 4.4 and 4.5 as appropriate.

1 A piece of copper of mass 170 g is cooled in a freezer. It is then dropped into water at 0 °C, causing 4.0 g of water to freeze. Determine the temperature inside the freezer.

2 A liquid-in-glass thermometer consists of a mass of 62 g of glass and 3.5 g of mercury.
  (a) Calculate the thermal capacity of the thermometer.
  (b) The thermometer is used to measure the temperature of some glycerol of mass 90 g. Before the thermometer is inserted into the glycerol, the thermometer records 18 °C and the temperature of the glycerol is 42 °C. Calculate the temperature recorded on the thermometer when placed in the glycerol.
  (c) Using your answer to (b), suggest why such thermometers cannot be used to measure reliably the temperature of a small mass of substance.

3 (a) A jet of steam at 100 °C is directed into a hole in a large block of ice. After the steam has been switched off, the condensed steam and the melted ice are both at 0 °C. The mass of water collected in the hole is 206 g. Calculate the mass of steam condensed.
  (b) Suggest why a scald with steam is much more serious than one involving boiling water.

4 A mass of 450 g of frozen vegetables is taken from a freezer at −20 °C. The vegetables are immediately placed in a saucepan containing 1100 g of boiling water. The saucepan has a thermal capacity of 900 J K^{-1}. The final temperature of the saucepan, water and vegetables is 83 °C.
  (a) Calculate the specific heat capacity of the vegetables.
  (b) The saucepan and its contents are then heated using a heater which provides 1200 J of thermal energy each second. Determine how long it takes to bring the water back to its boiling point.

4.5 Power

At the end of Section 4.5 you should be able to:
■ define power as work done per unit time and derive power as the product of force and velocity.

We have seen that energy is the ability to do work. Consider a family car and a grand prix racing car which both contain the same amount of fuel. They are capable of doing the same amount of work, but the racing car is able to travel much faster. This is because the engine of the racing car can convert the chemical energy of the fuel into useful energy at a much faster rate. The engine is said to be more powerful. **Power** is the rate of doing work. Power is given by the formula

$$power = \frac{work\ done}{time\ taken}$$

The unit of power is the watt (symbol W) and is equal to a rate of working of 1 joule per second. This means that a light bulb of power 1 W will convert 1 J of electrical energy to other forms of energy (e.g. light and heat) every second. Some typical values of power are shown in Table 4.6.

Table 4.6 Values of power

	power/W
power to operate a small calculator	50×10^{-6}
light power from a torch	4×10^{-3}
loudspeaker output	5
manual labourer working continuously	100
water buffalo working continuously	750
hair dryer	3×10^3
motor car engine	80×10^3
electric train	5×10^6
electricity generating station output	2×10^9

Power, like energy, is a scalar quantity.

Care must be taken when referring to power. It is common in everyday language to say that a strong person is 'powerful'. In physics, strength, or force, and power are *not* the same. Large forces may be exerted without any movement and thus no work is done and the power is zero! For example, a large rock resting on the ground is not moving, yet it is exerting a large force.

Consider a force F which moves a distance x at constant speed v in the direction of the force, in time t. Referring back to Section 4.1, the work done W by the force is given by

$$W = Fx$$

Dividing both sides of this equation by time t gives

$$\frac{W}{t} = F\frac{x}{t}$$

Now, $\dfrac{W}{t}$ is the rate of doing work, i.e. the power P and $\dfrac{x}{t} = v$. Hence,

$$P = Fv$$

power = force × speed

Example

A small electric motor is used to lift a weight of 1.5 N through a vertical distance of 120 cm in 2.7 s. Calculate the useful power output of the motor.

work done = force × distance moved
    = 1.5 × 1.2 (the distance must be in metres)
    = 1.8 J

 power = work done/time taken
    = (1.8/2.7)
    = **0.67 W**

Now it's your turn

1 Calculate the electrical energy converted into thermal energy when an electric fire, rated at 2.4 kW, is left switched on for a time of 3.0 minutes.

2 The output power of the electric motors of a train is 3.6 MW when the train is travelling at 30 m s^{-1}. Calculate the total force opposing the motion of the train.

3 A boy of mass 60 kg runs up a flight of steps in a time of 1.8 s. There are 22 steps and each one is of height 20 cm. Calculate the useful power developed in the boy's legs. (The acceleration of free fall is 10 m s^{-2}.)

The kilowatt-hour

Every household has to pay the 'electricity bill'. Electricity is vital in modern living and this energy does not come free of charge. It is important to realise that what is paid for is electrical energy, not electrical power. Since many electrical appliances in the home have a power of the order of kilowatts and we use them for hours, the joule, as a unit of energy, is too small.

For example, an electric fire of power 3 kW, used for 2 hours would use $3000 \times 2 \times 60 \times 60 = 21600000$ joules, i.e. 21.6 million joules of energy! Instead, in many parts of the world, electrical energy is purchased in kilowatt-hours (kW h).

One kilowatt-hour is the energy expended when work is done at the rate of 1 kilowatt for a time of 1 hour.

$$1\,\text{kWh} = 1.0\,\text{kW} \times 1\,\text{hour}$$
$$= 1000\,\text{W} \times 3600\,\text{s}$$
$$1\,\text{kWh} = 3.6 \times 10^6\,\text{J}$$

Figure 4.27 A domestic digital electricity meter

The kilowatt-hour is sometimes referred to as the Unit of energy. Electricity meters in the home (Figure 4.27) are often shown as measuring Units, where 1 Unit = 1 kW h.

Example

Calculate the cost of using an electric fire, rated at 2.5 kW for a time of 6.0 hours, if 1 kW h of energy costs 7.0 cents.

energy used $= 2.5 \times 6.0 = 15\,\text{kWh}$
cost $= 15 \times 7.0$
$= \textbf{105 cents}$

Now it's your turn

1 A television set is rated at 280 W. Calculate the cost of watching a three-hour film if 1 kilowatt-hour of electrical energy costs 8 cents.

2 An electric kettle is rated at 2.4 kW. Electrical energy costs 8 cents per kW h. The kettle takes 1.0 minute to boil sufficient water for two mugs of coffee. Calculate the cost of making this amount of coffee on three separate occasions.

3 Electrical energy generating companies sometimes measure their output in gigawatt-years. Calculate the number of kilowatt-hours in 6.0 gigawatt-years.

Section 4.5 Summary

- Power is the rate of doing work.
- *Work done = power × time taken*
- The unit of power is the watt (W).
- 1 watt = 1 joule per second
- *Power = force × speed*
- Electrical energy may be measured in kilowatt-hours (kW h).
- 1 kW h is the energy expended when work is done at the rate of 1000 watts for a time of 1 hour.

Section 4.5 Questions

1 The lights in a school laboratory have a total power of 600 W and are left on for 7.0 hours each day. In order to reduce fuel bills, it is decided to have the lights switched on only when there are people in the laboratory. This amounts to a total time of 4.5 hours per day. Assuming that the laboratory is used for 200 days each year, calculate the saving, if 1 kW h of energy costs 7.0 cents.

2 A car travelling at speed v along a horizontal road moves against a resistive force F given by the equation

$$F = 400 + kv^2$$

where F is in newtons, v in m s^{-1} and k is a constant.

At speed $v = 15$ m s^{-1}, the resistive force F is 1100 N.
(a) Calculate, for this car:
   (i)   the power necessary to maintain the speed of 15 m s^{-1},
   (ii) the total resistive force at a speed of 30 m s^{-1},
   (iii) the power required to maintain the speed of 30 m s^{-1}.
(b) Determine the energy expended in travelling 1.2 km at a constant speed of:
   (i) 15 m s^{-1},
   (ii) 30 m s^{-1}.
(c) Using your answers to (b), suggest why, during a fuel shortage, the maximum permitted speed of cars may be reduced.

4.6 Moment of a force

At the end of Section 4.6 you should be able to:
- define and apply the moment of a force
- show an understanding that a couple is a pair of forces which tends to produce rotation only
- define and apply the torque of a couple
- show an understanding that, when there is no resultant force and no resultant torque, a system is in equilibrium
- apply the principle of moments
- show an understanding that the weight of a body may be taken as acting at a single point known as its centre of gravity.

When a force acts on an object, the force may cause the object to move in a straight line. It could also cause the object to spin (rotate).

Think about a metre rule held in the hand at one end so that the rule is horizontal (Figure 4.28). If a weight is hung from the ruler we can feel a turning effect on the ruler. The turning effect increases if the weight is increased or it is moved further from the hand along the ruler. The turning

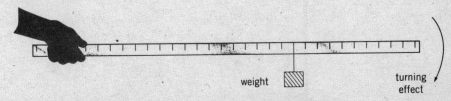

weight   turning effect

Figure 4.28 Turning effect on a metre rule

effect acts at the hand where the metre rule is pivoted. Keeping the weight and its distance along the rule constant, the turning effect can be changed by holding the ruler at an angle to the horizontal. The turning effect becomes smaller as the rule approaches the vertical position.

The turning effect of a force is called the **moment** of the force.

The moment of a force depends on the magnitude of the force and also on the distance of the force from the pivot or fulcrum. This distance must be defined precisely. In the simple experiment above, we saw that the moment of the force depended on the angle of the ruler to the horizontal. Varying this angle meant that the line of action of the force from the pivot varied (see Figure 4.29). The distance required when finding the moment of a force is the perpendicular distance d of the line of action of the force from the pivot.

The moment of a force is defined as the product of the force and the perpendicular distance of the line of action of the force from the pivot.

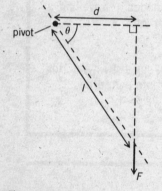

Figure 4.29 Finding the moment of a force

Referring to Figure 4.29, the force has magnitude F and acts at a point distance l from the pivot. Then, when the ruler is at angle θ to the horizontal,

$$moment\ of\ force = F \times d$$
$$= F \times l \cos \theta$$

Since force is measured in newtons and distance is measured in metres, the unit of the moment of a force is newton-metre (N m).

Example

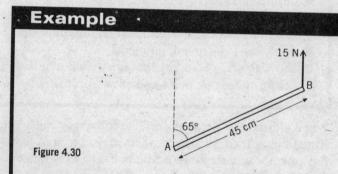

Figure 4.30

In Figure 4.30, a light rod AB of length 45 cm is held at A so that the rod makes an angle of 65° to the vertical. A vertical force of 15 N acts on the rod at B. Calculate the moment of the force about the end A.

$$moment\ of\ force = force \times perpendicular\ distance\ from\ pivot$$
$$= 15 \times 0.45 \sin 65$$

(Remember that the distance must be in metres.)
$$= \textbf{6.1 N m}$$

Now it's your turn

Referring to Figure 4.30, calculate the moment of the force about A for a vertical force of 25 N with the rod at an angle of 30° to the vertical.

Couples

When a screwdriver is used, we apply a turning effect to the handle. We do not apply one force to the handle because this would mean the screwdriver would move sideways. Rather, we apply two forces of equal size but opposite direction on opposite sides of the handle (see Figure 4.31).

> A **couple** consists of two forces, equal in magnitude but opposite in direction whose lines of action do not coincide.

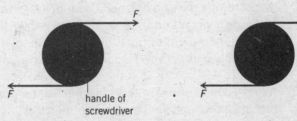

Figure 4.31 Two forces acting as a couple

Figure 4.32 Torque of a couple

Consider the two parallel forces, each of magnitude F acting as shown in Figure 4.32 on opposite ends of a diameter of a disc of radius r. Each force produces a moment about the centre of the disc of magnitude Fr in a clockwise direction. The total moment about the centre is $F \times 2r$ or $F \times$ *perpendicular distance between the forces.*

Although a turning effect is produced, this turning effect is not called a moment because it is produced by two forces, not one. Instead, this turning effect if referred to as a **torque**. The unit of torque is the same as that of the moment of a force i.e. newton-metre.

> The torque of a couple is the product of one of the forces and the perpendicular distance between the forces.

It is interesting to note that, in engineering, the tightness of nuts and bolts is often stated as the maximum torque to be used when screwing up the nut on the bolt. Spanners used for this purpose are called torque wrenches because they have a scale on them to indicate the torque that is being applied.

Figure 4.33 Tightening a wheel nut requires the application of a torque.

Example

Calculate the torque produced by two forces, each of magnitude 30 N, acting in opposite directions with their lines of action separated by a distance of 25 cm.

$$torque = force \times separation\ of\ forces$$
$$= 30 \times 0.25 \quad \text{(distance in metres)}$$
$$= 7.5\,\text{N m}$$

Now it's your turn

The torque produced by a person using a screwdriver is 0.18 N m. This torque is applied to the handle of diameter 4.0 cm. Calculate the force applied to the handle.

The principle of moments

A metre rule may be balanced on a pivot so that the rule is horizontal. Hanging a weight on the rule will make the rule rotate about the pivot. Moving the weight to the other side of the pivot will make the rule rotate in the opposite direction. Hanging weights on both sides of the pivot as shown in Figure 4.34 means that the ruler may rotate clockwise, or anticlockwise or it may remain horizontal. In this horizontal position, there is no resultant turning effect and so the total turning effect of the forces in the clockwise direction equals the total turning effect in the anticlockwise direction. You can check this very easily with the apparatus of Figure 4.34.

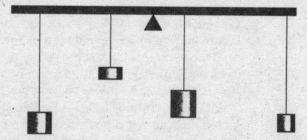

Figure 4.34

When a body has no tendency to change its speed of rotation, it is said to be in **rotational equilibrium**.

The **principle of moments** states that, for a body to be in rotational equilibrium, the sum of the clockwise moments about any point must equal the sum of the anticlockwise moments about that same point.

Example

Some weights are hung from a light rod AB as shown in Figure 4.35. The rod is pivoted. Calculate the magnitude of the force F required to balance the rod horizontally.

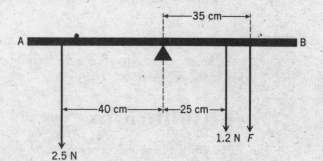

Figure 4.35

Sum of clockwise moments = $(0.25 \times 1.2) + 0.35F$
Sum of anticlockwise moments = 0.40×2.5
By the principle of moments
$$(0.25 \times 1.2) + 0.35F = 0.40 \times 2.5$$
$$0.35F = 1.0 - 0.3$$
$$F = \textbf{2.0 N}$$

Now it's your turn

Some weights are hung from a light rod AB as shown in Figure 4.36. The rod is pivoted. Calculate the magnitude of the force F required to balance the rod horizontally.

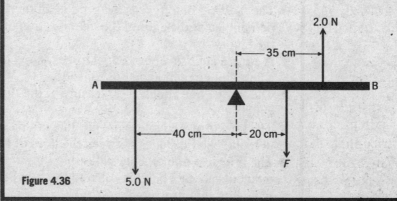

Figure 4.36

Centre of gravity

An object may be made to balance at a particular point. When it is balanced at this point, the object does not turn and so all the weight on one side of the pivot is balanced by the weight on the other side. Supporting the object at the pivot means that the only force which has to be applied at the pivot is one to stop the object falling – that is, a force equal to the weight of the object.

Although all parts of the object have weight, the whole weight of the object appears to act at this balance point. This point is called the **centre of gravity (C.G.)** of the object.

> The centre of gravity of an object is the point at which the whole weight of the object may be considered to act.

The weight of a body can be shown as a force acting vertically downwards at the centre of gravity. For a uniform body such as a ruler, the centre of gravity is at the geometrical centre.

Equilibrium

The principle of moments gives the condition necessary for a body to be in rotational equilibrium. However, the body could still have a resultant force acting on it which would cause it to accelerate linearly. Thus, for complete equilibrium, there cannot be any resultant force in any direction.

For a body to be **in equilibrium**:

1 The sum of the forces in any direction must be zero,
2 The sum of the moments of the forces about any point must be zero.

Section 4.6 Summary

- The moment of a force is a measure of the turning effect of the force.
- The moment of a force is the product of the force and the perpendicular distance of the line of action of the force from the pivot.
- A couple consists of two equal forces acting in opposite directions whose lines of action do not coincide.
- The torque of a couple is a measure of the turning effect of the couple.
- The torque of a couple is the product of one of the forces and the perpendicular distance between the lines of action of the forces.
- The principle of moments states that the sum of the clockwise moments about a point is equal to the sum of the anticlockwise moments about the point.
- The centre of gravity of a body is the point at which the whole weight of the body may be considered to act.
- For a body to be in equilibrium,

  1 the sum of the forces in any direction must be zero,
  2 the sum of the moments of the forces about any point must be zero.

Section 4.6 Questions

1 A uniform rod has a weight of 14 N. It is pivoted at one end and held in a horizontal position by a thread tied to its other end, as shown in Figure 4.37. The thread makes an angle of 50° with the horizontal. Calculate:
   (a) the moment of the weight of the rod about the pivot,
   (b) the tension T in the thread required to hold the rod horizontally.

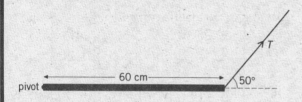

Figure 4.37

2 A ruler is pivoted at its centre of gravity and weights are hung from the ruler as shown in Figure 4.38. Calculate:
   (a) the total anticlockwise moment about the pivot,
   (b) the magnitude of the force F.

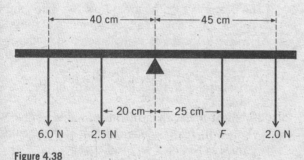

Figure 4.38

3 A uniform plank of weight 120 N rests on two stools as shown in Figure 4.39. A weight of 80 N is placed on the plank, midway between the stools. Calculate:
   (a) the force acting on the stool at A,
   (b) the force acting on the stool at B.

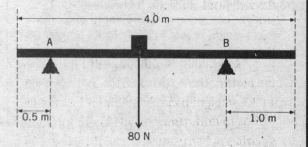

Figure 4.39

4 A nut is to be tightened to a torque of 16 N m. Calculate the force which must be applied to the end of a spanner of length 24 cm in order to produce this torque.

Exam-style Questions

1 (a) State what is meant by:
   (i) work done,
   (ii) power.
   (b) (i) Show that, when a gas expands by an amount ΔV against a constant pressure p, the work done W is given by

   $$W = p\Delta V$$

   (ii) Explain why, when the gas in (i) expands suddenly, it is likely to cool.

   (c) The air in a car tyre has a volume of $7.8 \times 10^3 \, cm^3$. The tyre suddenly bursts and the air expands to 3.5 times its original volume in a time of 2.5 ms. Atmospheric pressure is $1.0 \times 10^5 \, Pa$. Calculate:
   (i) the work done by the air in the tyre during the expansion,
   (ii) the mean power dissipated in the tyre burst.

2 (a) Explain, by reference to two examples from different areas of physics, what is meant by *potential* energy.

(b) (i) State the law of conservation of energy.
(ii) Explain how the law may be applied to a ball dropped from a height h on to a horizontal surface and which finally comes to rest on the surface after several bounces.

(c) A car of mass 950 kg is travelling along a straight horizontal road at a speed of 32 m s^{-1}. It is brought to rest by applying its brakes. All of the kinetic energy of the car may be assumed to be converted into thermal energy of the discs of the brakes. There are four discs, each of mass 1.9 kg and made of metal of specific heat capacity 420 J kg^{-1} K^{-1}. Calculate the temperature rise of the brakes when the car stops.

3 (a) Define:
(i) specific heat capacity,
(ii) specific latent heat of vaporisation.

(b) (i) Explain what is meant by the *thermal capacity* of a body.
(ii) A metal has mass m and specific heat capacity c. Derive an expression relating m and c to the thermal capacity C of the metal.
(iii) In what circumstances is thermal capacity a more useful concept than specific heat capacity?

(c) An electric kettle has a power output of 2.3 kW. The kettle is used to heat a mass of 950 g of water of specific heat capacity 4200 J kg^{-1} K^{-1}.
(i) Calculate the time taken to heat the water from 15 °C to 100 °C.
(ii) The kettle is left switched on when the water temperature reaches 100 °C. Calculate the rate at which the water is boiled away. (The specific latent heat of vaporisation of water is 2.26 MJ kg^{-1}.)

(d) The kettle in (c) is used six times each day to boil water. On each occasion, approximately 200 g excess water is heated. Given that 1 kW h of energy costs 6.5 cents, estimate the extra cost involved each year.

4 An electric heater is placed in some water of mass 700 g. The heater is switched on and the variation with time t of the temperature θ of the water is shown in Figure 4.40.

(a) (i) Given that the specific heat capacity of water is 4200 J kg^{-1} K^{-1}, use Figure 4.40 to determine the power of the heater.
(ii) State two assumptions made in your calculation in (i).

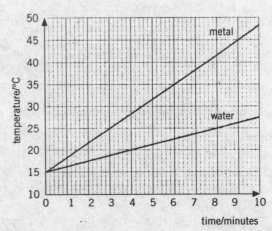

Figure 4.40

(b) The heater is then removed from the water and is placed in a hole in a block of metal of mass 1.2 kg. The heater is switched on and readings of temperature θ and time t are taken. These are also shown on Figure 4.40.
(i) Determine the specific capacity of the metal.
(ii) Suggest why the experiment would not be suitable for the determination of the specific heat capacity of an insulator such as concrete.

5. Force and collisions

In Chapter 3 we described motion in terms of displacement, velocity and acceleration. Now we will try to explain *why* bodies move. We shall realise that a force is required to make a body accelerate. This will introduce Newton's three laws of motion, the fundamental laws which state and define what forces do in relation to motion. We shall then see how Newton's laws help us to explain what happens in collisions, which are important in many areas of physics. The total momentum of a system of colliding bodies remains constant. This leads us to the principle of conservation of momentum, one of the important conservation laws of physics.

5.1 Force

At the end of Sections 5.1–5.4 you should be able to:
- show a qualitative understanding of frictional forces and viscous forces, including air resistance
- state each of Newton's laws of motion
- recall and solve problems using the relationship $F = ma$ appreciating that acceleration and force are always in the same direction
- show an understanding that mass is the property of a body which resists changes in motion
- describe and use the concept of weight as the effect of a gravitational field on a mass
- show an understanding of the effects of fields of force, as exemplified by the forces on mass and charge in uniform gravitational and electric fields respectively.

When you push a trolley in a supermarket or pull a case behind you at an airport, you are exerting a force. When you hammer in a nail, a force is being exerted. When you drop a book and it falls to the floor, the book is falling because of the force of gravity. When you lean against a wall or sit on a chair, you are exerting a force. Forces can change the shape or dimensions of bodies. You can crush a drinks can by squeezing it and applying a force; you can stretch a rubber band by pulling it. In everyday life, we have a good understanding of what is meant by force and the situations in which forces are involved. In physics the idea of force is used to add detail to the descriptions of moving objects.

As with all physical quantities, a method of measuring force must be established. One way of doing this is to make use of the fact that forces can change the dimensions of bodies in a reproducible way. It takes the same force to stretch a spring by the same amount (provided the spring is not overstretched by applying a very large force). This principle is used in the spring balance. A scale shows how much the spring has been extended, and the scale can be calibrated in terms of force. Laboratory spring balances are often called newton balances, because the newton is the SI unit of force.

Forces are vector quantities: they have magnitude as well as direction. Forces acting on a body are often shown by means of a force diagram drawn to scale, in which the forces are represented by lines of length proportional to the magnitude of the force, and in the appropriate direction (see Chapter 1).

5.2 Force and motion

The Greek philosopher Aristotle believed that the natural state of a body was a state of rest, and that a force was necessary to make it move and to keep it moving. In this argument, the greater the force, the greater the speed of the body.

Nearly two thousand years later, Galileo questioned this idea. He suggested that motion at a constant speed could be just as natural a state as the state of rest. He introduced an understanding of the effect of **friction** on motion.

Figure 5.1

Imagine a heavy box being pushed along a rough floor at constant speed (Figure 5.1). This may take a considerable force. The force required can be reduced if the floor is made smooth and polished, and reduced even more if a lubricant, for example grease, is applied between the box and the floor. We can imagine a situation where, when friction is reduced to a vanishingly small value, the force required to push the box at constant speed is also vanishingly small.

Galileo realised that the force of friction was a force that opposed the pushing force. When the box is moving at constant speed, the pushing force is exactly equal to the frictional force, but in the opposite direction, so that there is a net force of zero acting on the box. In the situation of vanishingly small friction, the box will continue to move with constant speed, because there is no force to slow it down.

Frictional forces are also important in considering the motion of a body through a liquid or gas. We use the term **viscous force** to describe the frictional force in a fluid (such as air resistance). The property of the fluid determining this force is the **viscosity** of the fluid. In Section 3.5 we considered the fact that parachutists eventually fall with a constant, terminal velocity because of air resistance.

Newton's laws of motion

Isaac Newton (1642–1727) used Galileo's ideas to produce a theory of motion, expressed in his three laws of motion. The **first law of motion** re-states Galileo's deduction about the natural state of a body.

> Every body continues in its state of rest, or of uniform motion in a straight line, unless compelled to change that state by a net force.

This law tells us what a force does: it disturbs the state of rest or uniform motion of a body. The property of a body to stay in a state of rest or uniform motion is called **inertia**.

Newton's second law tells us what happens if a force is exerted on a body. It causes the velocity to change. A force exerted on a body at rest makes it move – it gives it a velocity. A force exerted on a moving body may make its speed increase or decrease, or change its direction of motion. A change in speed or velocity is an acceleration. Newton's second law defines the magnitude of this acceleration. It also introduces the idea of the mass of a body. Mass is a measure of the inertia of a body. The bigger the mass, the more difficult it is to change its state of rest or uniform motion. A simplified form of Newton's **second law** is

> For a body of constant mass, its acceleration is directly proportional to the net force applied to it.

The direction of the acceleration is in the direction of the net force. In a word equation, the law is

> *force = mass × acceleration*

and in symbols

> $F = ma$

where F is the force, m is the mass and a is the acceleration. Here we have made the constant of proportionality equal to unity (that is, we use an equals sign rather than a proportionality sign) by choosing quantities with units which will give us this simple relation. In SI units, the force F is in newtons, the mass m in kilograms and the acceleration a in metres per second per second.

Figure 5.2 Isaac Newton

One newton is defined as the force which will give a mass of one kilogram an acceleration of one metre per second per second in the direction of the force.

When a force is applied to a body, that force is always applied by another body. When you push a supermarket trolley, the trolley experiences a force. That force has been applied by another body – you. Newton understood that the body on which the force is exerted applies another force back on the body which is applying the force. The supermarket trolley exerts a force on you as well. Newton's **third law** relates these two forces.

Figure 5.3

Whenever one body exerts a force on another, the second exerts an equal and opposite force on the first.

Very often this law is stated as

To every action, there is an equal and opposite reaction.

But this statement does not highlight the very important point that the action force and the reaction force act on *different objects*. To take the example of the supermarket trolley, the action force exerted by you on the trolley is equal and opposite to the reaction force exerted by the trolley on you.

Newton's third law has applications in every branch of everyday life. We walk because of this law. When you take a step forward, your foot presses against the ground. The ground then exerts an equal and opposite force on you. This is the force, on you, which propels you in your path. Space rockets work because of the law (Figure 5.4). To expel the exhaust gases from the rocket, the rocket exerts a force on the gases. The gases exert an equal and opposite force on the rocket, sending it forward.

Figure 5.4 Space shuttle launch

Examples

1 An object of mass 1.5 kg is to be accelerated at $2.2 \, \text{m s}^{-2}$. What force is required?

By Newton's second law, $F = ma = 1.5 \times 2.2 = \mathbf{3.3 \, N}$.

2 A car of mass 1.5 tonnes ($1.5 \times 10^3 \, \text{kg}$), travelling at $80 \, \text{km h}^{-1}$, is to be stopped in 11 s. What force is required?

The acceleration of the car can be obtained from $v = u + at$ (see Chapter 3). The initial speed u is $80 \, \text{km h}^{-1}$, or $22 \, \text{m s}^{-1}$. The final speed v is 0. Then $a = -22/11 = -2.0 \, \text{m s}^{-2}$. This is negative because the car is decelerating.

By Newton's second law, $F = ma = 1.5 \times 10^3 \times 2.0 = \mathbf{3.0 \times 10^3 \, N}$.

Now it's your turn

1 A force of 5.0 N is applied to a body of mass 3.0 kg. What is the acceleration of the body?

2 A stone of mass 50 g is accelerated from a catapult to a speed of 8.0 m s^{-1} from rest over a distance of 30 cm. What average force is applied by the rubber of the catapult?

5.3 Weight

We saw in Chapter 3 that all objects released near the surface of the Earth fall with the same acceleration (the acceleration of free fall) if air resistance can be neglected. The force causing this acceleration is the gravitational attraction of the Earth on the object, or the force of gravity. The force of gravity which acts on an object is called the **weight** of the object. We can apply Newton's second law to the weight. For a body of mass m falling with the acceleration of free fall g, the weight W is given by

$$W = mg$$

The SI unit of force is the newton (N). This is also the unit of weight. The weight of an object is obtained from its mass in kilograms by multiplying the mass by the acceleration of free fall, 9.8 m s^{-2}. Thus a mass of one kilogram has a weight of 9.8 N. Because weight is a force and force is a vector, we ought to be aware of the direction of the weight of an object. It is towards the centre of the Earth. Because weight always has this direction, we do not need to specify direction every time we give the magnitude of the weight of objects.

The weight of a body is an example of the force acting on a mass in what is called a **field of force**. Near the surface of the Earth, this field is approximately constant and uniform. This means that in calculations we can take the same value of g, whatever the position on the surface of the Earth or for a short distance (compared with the Earth's radius) above it.

There are other sorts of fields of force. An important example is an electric field. An electric charge experiences a force in an electric field. There are significant similarities between the behaviour of a mass in a gravitational field and an electric charge in an electric field. We shall explore these similarities in Chapter 11.

How do we measure mass and weight? If you hang an object from a newton balance, you are measuring its weight (Figure 5.5). The unknown weight of the object is balanced by a force provided by the spring in the balance. From a previous calibration, this force is related to the extension of the spring. There is the possibility of confusion here. Laboratory newton balances may, indeed, be calibrated in newtons. But all commercial spring

Figure 5.5 A newton balance

balances – for example, the balances at fruit and vegetable counters in supermarkets – are calibrated in kilograms. Such balances are really measuring the weight of the fruit and vegetables, but the scale reading is in mass units, because there is no distinction between mass and weight in everyday life. The average shopper thinks of 5 kg of mangoes as having a weight of 5 kg. In fact, the mass of 5 kg has a weight of 49 N.

The word 'balance' in the spring balance and in the laboratory top-pan balance relates to the balance of forces. In each case, the unknown force (the weight) is equalled by a force which is known through calibration.

A way of comparing masses is to use a beam balance, or lever balance (see Figure 2.17). Here the weight of the object is balanced against the weight of some masses, which have previously been calibrated in mass units. The word 'balance' here refers to the equilibrium of the beam: when the beam is horizontal the moment of the weight on one side of the pivot is equal and opposite to the moment on the other side of the pivot. Because weight is given by the product of mass and the acceleration of free fall, the equality of the weights means that the masses are also equal.

We have introduced the idea of weight by thinking about an object in free fall. But objects at rest also have weight: the gravitational attraction on a book is the same whether it is falling or whether it is resting on a table. The fact that the book is at rest tells us, by Newton's first law, that the resultant force acting on it is zero. So there must be another force acting on the book which exactly balances its weight. Here the table exerts an upwards force on the book. This force is equal in magnitude to the weight but opposite in direction. It is a **normal contact force**: 'contact' because it occurs due to the contact between book and table, and 'normal' because it acts perpendicularly to the plane of contact. The weight and normal contact forces are shown in Figure 5.6.

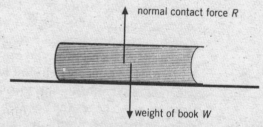

Figure 5.6 A book resting on a table: forces on the book. (The forces act in the same vertical line, but are separated slightly for clarity.)

The book remains at rest on the table because the weight W of the book downwards is exactly balanced by the normal contact force R exerted by the table on the book. The vector sum of these forces is zero, so the book is in equilibrium. A very common mistake is to state that 'By Newton's third law, W is equal to R'. But these two forces are both acting on the book, and cannot be related by the third law. Third-law forces always act on *different* bodies. To see the application of the third law, think about the normal contact force R. This is an upwards force exerted by the *table*. The reaction to this is a downwards force R' exerted by the *book*. By Newton's third law, these forces are equal and opposite. This situation is illustrated in Figure 5.7.

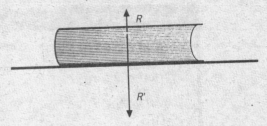

Figure 5.7 A book resting on a table: action and reaction forces, acting at the point of contact

Having considered the action and reaction forces between book and table, we ought to think about the reaction force to the weight of the book, regarded as an action force, even when the book is not on the table. This is not so easy, because there does not seem to be an obvious second force. But remember that the weight is due to the gravitational attraction of the Earth on the book. If the Earth attracts the book, the book also attracts the Earth. This gravitational force of the book on the Earth is the reaction force. We can test whether the two forces do, indeed, act on different bodies. The action force (the weight of the book) acts on the book. The reaction force (the attraction of the Earth to the book) acts on the Earth. Thus, the condition that action and reaction forces should act on different bodies is satisfied.

5.4 Problem solving

In dealing with problems involving Newton's laws, start by drawing a general sketch of the situation. Then consider each body in your sketch. Show all the forces acting on that body, both known forces and unknown forces you may be trying to find. Here it is a real help to try to draw the arrows which represent the forces in approximately the correct direction and approximately to scale. Label each force with its magnitude, or with a symbol if you do not know the magnitude. For each force, you must know the cause of the force (gravity, friction, and so on), and you must also know *on* what object that force acts and *by* what object it is exerted. This labelled diagram is referred to as a **free-body diagram**, because it detaches the body from the others in the situation. Having established all the forces acting on the body, you can use Newton's second law to find unknown quantities. This procedure is illustrated in the example which follows.

Newton's second law equates the resultant force acting on a body to the product of its mass and its acceleration. In some problems, the system of bodies is in equilibrium. They are at rest, or are moving in a straight line with uniform speed. In this case, the acceleration is zero, so the resultant force is also zero. In other cases, the resultant force is not zero and the objects in the system are accelerating.

For whichever case applies, you should remember that forces are vectors. You will probably have to resolve the forces into two components at right-angles, and then apply the second law to each set of components separately. Problems can often be simplified by making a good choice of directions for resolution. You will end up with a set of equations, based on the application of Newton's second law, which must be solved to determine the unknown quantity.

Example

A box of mass 5.0 kg is pulled along a horizontal floor by a force P of 25 N, applied at an angle of 20° to the horizontal (Figure 5.8). A frictional force F of 20 N acts parallel to the floor.

Calculate the acceleration of the box.

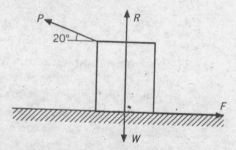

Figure 5.8

The free-body diagram is shown in Figure 5.8. Resolving the forces parallel to the floor, the component of the pulling force, acting to the left, is 25 cos 20 = 23.5 N. The frictional force, acting to the right, is 20 N. The resultant force to the left is thus 23.5 − 20.0 = 3.5 N.

By Newton's second law, $a = F/m = 3.5/5.0 = \mathbf{0.70\,m\,s^{-2}}$.

Now it's your turn

Figure 5.9

A gardener pushes a lawnmower of mass 18 kg at constant speed. To do this requires a force P of 80 N directed along the handle, which is at an angle of 40° to the horizontal (Figure 5.9).
(a) Calculate the horizontal retarding force F on the mower.
(b) If this retarding force were constant, what force, applied along the handle, would accelerate the mower from rest to 1.2 m s⁻¹ in 2.0 s?

Sections 5.1–5.4 Summary

- The force of friction opposes motion.
- Newton's laws of motion define *force*. The laws are:

First law: Every body continues in its state of rest, or of uniform motion in a straight line, unless acted upon by a force.

Second law: Force is proportional to mass × acceleration or $F = ma$, where force F is in newtons, mass m is in kilograms and acceleration a is in metres per second per second.

Third law: Whenever one body exerts a force on another, the second exerts an equal and opposite force on the first.

- Weight W and mass m are related by:

$$W = mg$$

where g is the acceleration of free fall.

Sections 5.1–5.4 Questions

1 A net force of 95 N accelerates an object at 1.9 m s^{-2}. Calculate the mass of the object.

2 A parachute trainee jumps from a platform 3.0 m high. When he reaches the ground, he bends his knees to cushion the fall. His torso decelerates over a distance of 0.65 m. Calculate:
(a) the speed of the trainee just before he reaches the ground,
(b) the deceleration of his torso,
(c) the average force exerted on his torso (of mass 45 kg) by his legs during the deceleration.

3 If the acceleration of a body is zero, does this mean that no forces act on it?

4 A railway engine pulls two carriages of equal mass with uniform acceleration. The tension in the coupling between the engine and the first carriage is T. Deduce the tension in the coupling between the first and second carriages.

5 What is your mass? What is your weight?

5.5 Momentum

At the end of Sections 5.5–5.8 you should be able to:
- define linear momentum as the product of mass and velocity
- define force as the rate of change of momentum
- state the principle of conservation of momentum
- apply the principle of conservation of momentum to solve simple problems including elastic and inelastic interactions between two bodies in one dimension
- recognise that, for a perfectly elastic collision, the relative speed of approach is equal to the relative speed of separation
- show an understanding that, whilst momentum of a system is always conserved in interactions between bodies, some change in kinetic energy usually takes place.

We shall now introduce a quantity called **momentum,** and see how Newton's laws may be related to it.

The momentum of a particle is defined as the product of its mass and its velocity.

In words

momentum = mass × velocity

and in symbols

$p = mv$

The unit of momentum is the unit of mass times the unit of velocity; that is, kg m s^{-1}. An alternative unit is the newton second (N s). Momentum, like velocity, is a vector quantity. Its complete name is **linear momentum**, to distinguish it from angular momentum, which does not concern us here.

Newton's first law states that every body continues in a state of rest, or of uniform motion in a straight line, unless acted upon by a force. We can express this law in terms of momentum. If a body maintains its uniform motion in a straight line, its momentum is unchanged. If a body remains at rest, again its momentum (zero) does not change. Thus, an alternative statement of the first law is that the momentum of a particle remains constant unless an external force acts on the particle. As an equation

$p = $ constant

This is a special case, for a single particle, of a very important conservation law: the principle of conservation of momentum. This word 'conservation' here means that the quantity remains constant.

Newton's second law can also be expressed in terms of momentum. We already have it in a form which relates the force acting on a body to the product of the mass and the acceleration of the body. Remember that the acceleration of a body is the rate of change of its velocity. The product of mass and acceleration then is just the mass times the rate of change of velocity. For a body of constant mass, this is just the same as the rate of change of (mass times velocity). But (mass times velocity) is momentum, so the product of mass and acceleration is identical with the rate of change of momentum. Thus, the second law may be stated as

The resultant force acting on a body is equal to the rate of change of its momentum.

This is the formal definition of the second law. Expressed in terms of symbols

$$F = ma = m(\Delta v/\Delta t) = \Delta(mv)/\Delta t = \Delta p/\Delta t$$

Continuing with the idea of force being equal to rate of change of momentum, the third law relating to action and reaction forces becomes: the rate of change of momentum due to the action force on one body is equal and opposite to the rate of change of momentum due to the reaction force on the other body. Simplifying this by considering a unit time interval, this becomes: when two bodies exert action and reaction forces on each other, their changes of momentum are equal and opposite.

5.6 The principle of conservation of momentum

Figure 5.10 System of two particles

We have already seen that Newton's first law states that the momentum of a single particle is constant, if no external force acts on the particle. Now think about a system of two particles (Figure 5.10). We allow these particles to exert some sort of force on each other: it could be gravitational attraction or, if the particles were charged, it could be electrostatic attraction or repulsion.

These two particles are isolated from the rest of the Universe, and experience no outside forces at all. If the first particle exerts a force F on the second, Newton's third law tells us that the second exerts a force $-F$ on the first. The minus sign indicates that the forces are in opposite directions. As we saw in the last section, we can express this law in terms of change of momentum. The change of momentum of the second particle as a result of the force exerted on it by the first is equal and opposite to the change of momentum of the first particle as a result of the force exerted on it by the second. Thus, the changes of momentum of the individual particles cancel out, and the momentum of the system of two particles remains constant. The particles have merely exchanged some momentum. The situation is expressed by the equation

$$p = p_1 + p_2 = \text{constant}$$

where p is the total momentum, and p_1 and p_2 are the individual momenta.

We could extend this idea to a system of three, four, or finally any number n of particles.

If no external force acts on a system, the total momentum of the system remains constant, or is conserved.

A system on which no external force acts is often called an *isolated system*. The fact that the total momentum of an isolated system is constant is the **principle of conservation of momentum**. It is a direct consequence of Newton's third law of motion.

5.7 Collisions

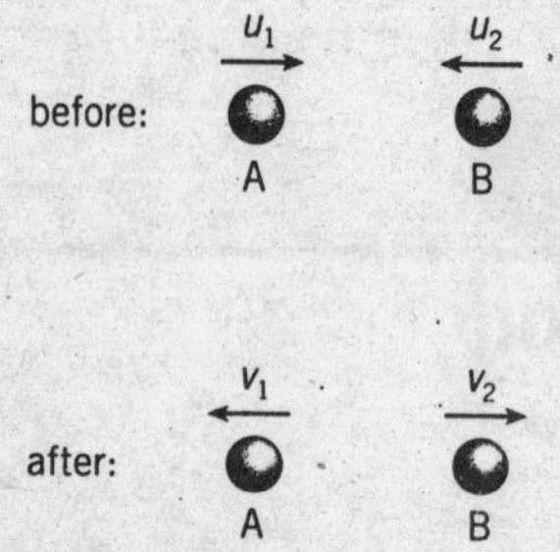

Figure 5.11 Collision between two particles

We now use the principle of conservation of momentum to analyse a system consisting of two colliding particles. (If you want a real example to think about, try billiard balls.)

Consider two particles A and B making a direct, head-on collision. Particle A has mass m_1 and is moving with velocity u_1 in the direction from left to right; B has mass m_2 and has velocity u_2 in the direction from right to left (Figure 5.11). As velocity is a vector quantity, this is the same as saying that the velocity is $-u_2$ from left to right. The particles collide. After the collision they have velocities $-v_1$ and v_2 respectively in the direction from left to right. That is, both particles are moving back along their directions of approach.

According to the principle of conservation of momentum, the total momentum of this isolated system remains constant, whatever happens as a result of the interaction of the particles. Thus, the total momentum before the collision must be equal to the total momentum after the collision. The momentum before the collision is

$$m_1u_1 - m_2u_2$$

and the momentum after is

$$-m_1v_1 + m_2v_2$$

Because total momentum is conserved

$$m_1u_1 - m_2u_2 = -m_1v_1 + m_2v_2$$

Knowing the masses of the particles and the velocities before collision, this equation would allow us to calculate the relation between the velocities after the collision.

The way to approach collision problems is as follows.

- Draw a labelled diagram showing the colliding bodies before collision. Draw a separate diagram showing the situation after the collision. Take care to define the directions of all the velocities.
- Obtain an expression for the total momentum before the collision, remembering that momentum is a vector quantity. Similarly, find the total momentum after the collision, taking the same reference direction.
- Then equate the momentum before the collision to the momentum afterwards.

Examples

1 What is the magnitude of the momentum of an α-particle of mass 6.6×10^{-27} kg travelling with a speed of 2.0×10^7 m s^{-1}?

$$p = m\dot{v} = 6.6 \times 10^{-27} \times 2.0 \times 10^7 = \mathbf{1.3 \times 10^{-19}\,kg\,m\,s^{-1}}$$

2 A cannon of mass 1.5 tonnes (1.5×10^3 kg) fires a cannon-ball of mass 5.0 kg (Figure 5.12). The speed with which the ball leaves the cannon is 70 m s^{-1} relative to the Earth. What is the initial speed of recoil of the cannon?

Figure 5.12

The system under consideration is the cannon and the cannon-ball. The total momentum of the system before firing is zero. Because the total momentum of an isolated system is constant, the total momentum after firing must also be zero. That is, the momentum of the cannon-ball, which is $5.0 \times 70 = 350$ kg m s^{-1} to the right, must be exactly balanced by the momentum of the cannon. If the initial speed of recoil is v, the momentum of the cannon is $1500v$ to the left. Thus, $1500v = 350$ and $v = \mathbf{0.23\,m\,s^{-1}}$.

Now it's your turn

1 What is the magnitude of the momentum of an electron of mass 9.1×10^{-31} kg travelling with a speed of 7.5×10^6 m s^{-1}?

2 An ice-skater of mass 80 kg, initially at rest, pushes his partner, of mass 65 kg, away from him so that she moves with an initial speed of 1.5 m s^{-1}. What is the initial speed of the first skater after this manoeuvre?

Momentum and impulse

It is now useful to introduce a quantity called **impulse** and relate it to a change in momentum.

If a constant force F acts on a body for a time Δt, the impulse of the force is given by $F\Delta t$.

The unit of impulse is given by the unit of force, the newton, multiplied by the unit of time, the second: it is the newton second (N s).

We know from Newton's second law that the force acting on a body is equal to the rate of change of momentum of the body. We have already expressed this as the equation

$$F = \Delta p / \Delta t$$

If we multiply both sides of this equation by Δt, we obtain

$$F \Delta t = \Delta p$$

We have already defined $F \Delta t$ as the impulse of the force. The right-hand side of the equation (Δp) is the change in the momentum of the body. So, Newton's second law tells us that the **impulse of a force is equal to the change in momentum**. It is useful for dealing with forces that act over a short interval of time, as in a collision. The forces between colliding bodies are seldom constant throughout the collision, but the equation can be applied to obtain information about the average force acting.

Note that the idea of impulse explains why there is an alternative unit for momentum. In Section 5.5 we introduced the kg m s^{-1} and the N s as possible units for momentum. The kg m s^{-1} is the logical unit, the one you arrive at if you take momentum as being the product of mass and velocity. The N s comes from the impulse–momentum equation: it is the unit of impulse, and because impulse is equal to change of momentum, it is also a unit for momentum.

Example

Some tennis players can serve the ball at a speed of 55 m s^{-1}. The tennis ball has a mass of 60 g. In an experiment it is determined that the ball is in contact with the racket for 25 ms during the serve (Figure 5.13). Calculate the average force exerted by the racket on the ball.

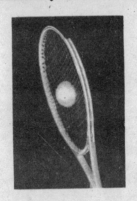

Figure 5.13

The change in momentum of the ball as a result of the serve is $0.060 \times 55 = 3.3$ kg m s^{-1}. By the impulse–momentum equation, the change in momentum is equal to the impulse of the force. Since impulse is the product of force and time, $Ft = 3.3$ N s.

Here t is 0.025 s; thus $F = 3.3/0.025 =$ **132 N**.

Now it's your turn

Figure 5.14

A golfer hits a ball of mass 45 g at a speed of 40 m s^{-1} (Figure 5.14). The golf club is in contact with the ball for 3.0 ms. Calculate the average force exerted by the club on the ball.

5.8 Elastic and inelastic collisions

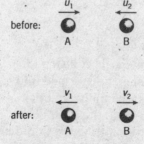

Figure 5.15 Collision between two particles

In some collisions, kinetic energy is conserved as well as momentum. By the **conservation of kinetic energy**, we mean that the total kinetic energy of the colliding bodies before collision is the same as the total kinetic energy afterwards. This means that no energy is lost in the permanent deformation of the colliding bodies, or as heat and sound. There is a transformation of energy during the collision: while the colliding bodies are in contact, some of the kinetic energy is transformed into elastic potential energy, but as the bodies separate, it is transformed into kinetic energy again.

Using the same notation for the masses and speeds of the colliding particles as in Section 5.7 (see Figure 5.15), the total kinetic energy of the particles before collision is

$$\tfrac{1}{2}m_1u_1^2 + \tfrac{1}{2}m_2u_2^2$$

The total kinetic energy afterwards is

$$\tfrac{1}{2}m_1v_1^2 + \tfrac{1}{2}m_2v_2^2$$

If the collision is **elastic**, the kinetic energy before collision is equal to the kinetic energy after collision.

$$\tfrac{1}{2}m_1u_1^2 + \tfrac{1}{2}m_2u_2^2 = \tfrac{1}{2}m_1v_1^2 + \tfrac{1}{2}m_2v_2^2$$

Note that because energy is a scalar, the directions of motion of the particles are not indicated by the signs of the various terms.

In solving problems about elastic collisions, this equation is useful because it gives another relation between masses and velocities, in addition to that obtained from the principle of conservation of momentum.

When the velocity directions are as defined in Figure 5.15, application of the two conservation conditions shows that

$$u_1 + u_2 = v_1 + v_2$$

for a perfectly elastic collision. That is, the relative speed of approach $(u_1 + u_2)$ is equal to the relative speed of separation $(v_1 + v_2)$. Note that this useful relation applies *only for a perfectly elastic collision.*

Elastic collisions occur in the collisions of atoms and molecules. We shall see in Chapter 6 that one of the most important assumptions in the kinetic theory of gases is that the collisions of the gas molecules with the walls of the container are perfectly elastic. However, in larger scale collisions, such as those of billiard balls, collisions cannot be perfectly elastic. (The 'click' of billiard balls on impact indicates that a very small fraction of the total energy of the system has been transformed into sound.) Nevertheless, we often make the assumption that such a collision is perfectly elastic.

Collisions in which the total kinetic energy is not the same before and after the event are called **inelastic**.

Total energy must, of course, be conserved. But in an inelastic collision the kinetic energy that does not re-appear in the same form is transformed into heat, sound and other forms of energy. In an extreme case, all the kinetic energy may be lost. A lump of modelling clay dropped on to the floor does not bounce. All the kinetic energy it possessed just before hitting the floor has been transformed into the work done in flattening the lump, and (a much smaller amount) into the sound energy emitted as a 'squelch'.

Although kinetic energy may or may not be conserved in a collision, momentum is always conserved, and so is total energy.

The truth of this statement may not be entirely obvious, especially when considering examples such as the lump of modelling clay which was dropped on to the floor. Surely the clay had momentum just before the collision with the floor, but had no momentum afterwards? True! But for the system of the lump of clay alone, external forces (the attraction of the Earth on the clay, and the force exerted by the floor on the clay on impact) were acting. When external forces act, the conservation principle does not apply. We need to consider a system in which no external forces act. Such a system is the lump of modelling clay and the Earth. While the clay falls towards the floor, gravitational attraction will also pull the Earth towards the clay. Conservation of momentum can be applied in that the total momentum of clay and Earth remains constant throughout the process: before the collision, and after it. The effects of the transfer of the clay's momentum to the Earth are not noticeable due to the difference in mass of the two objects.

Example

A billiard ball A moves with speed u_A directly towards a similar ball B which is at rest (Figure 5.16). The collision is perfectly elastic. What are the speeds v_A and v_B after the collision?

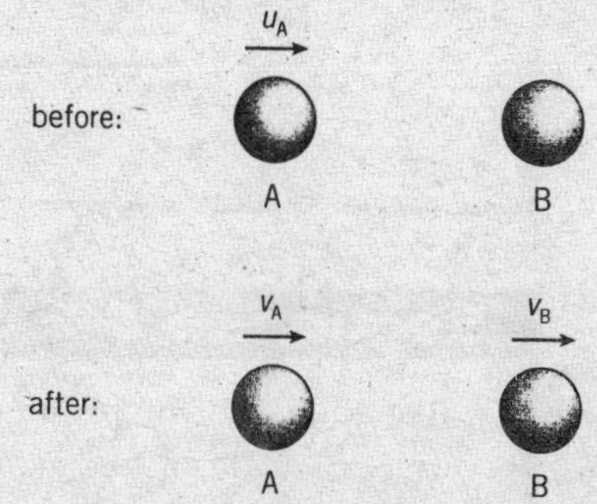

before:

u_A

A   B

after:

v_A v_B

Figure 5.16   A   B

It is convenient to take the direction from left to right as the direction of positive momentum. If the mass of a billiard ball is m, the total momentum of the system before the collision is mu_A. By the principle of conservation of momentum, the total momentum after collision is the same as that before, or

$$mu_A = mv_A + mv_B$$

The collision is perfectly elastic, so the total kinetic energy before the collision is the same as that afterwards, or

$$\tfrac{1}{2}mu_A^2 = \tfrac{1}{2}mv_A^2 + \tfrac{1}{2}mv_B^2$$

Solving these equations gives $v_A = 0$ and $v_B = u_A$. That is, ball A comes to a complete standstill, and ball B moves off with the same speed as that with which ball A struck it. (Another solution is possible algebraically: $v_A = u_A$ and $v_B = 0$. This corresponds to a non-collision.

Ball A is still moving with its initial speed, and ball B is still at rest. In cases where algebra gives us two possible solutions, we need to decide which is the one that is physically appropriate.)

Now it's your turn

A trolley A moves with speed u_A towards a trolley B of equal mass which is at rest (Figure 5.17). The trolleys stick together and move off as one with speed $v_{A,B}$.
(a) Find $v_{A,B}$.
(b) What fraction of the initial kinetic energy of trolley A is converted into other forms in this inelastic collision?

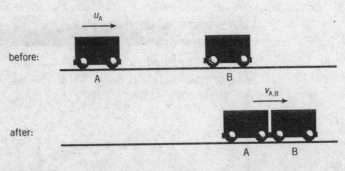

Figure 5.17

Sections 5.5–5.8 Summary

- The linear momentum p of a body is the product of its mass m and its velocity v, described as $p = mv$. Momentum has units kg m s^{-1} or N s. It is a vector quantity.
- Newton's laws of motion can be stated in terms of momentum.
  First law: The momentum of a particle remains constant unless an external force acts on the particle: p = constant
  Second law: The resultant force acting on a body is equal to the rate of change of its momentum: $F = \Delta p/\Delta t$
  Third law: When two bodies exert action and reaction forces on each other, their changes of momentum are equal and opposite.
- The principle of conservation of momentum states that the total momentum of an isolated system is constant. An isolated system is one on which no external force acts.
- In collisions between bodies, application of the principle of conservation of momentum shows that the total momentum of the system before the collision is equal to the total momentum after the collision.
- An elastic collision is one in which the total kinetic energy remains the same. In this situation, the relative speed of approach is equal to the relative speed of separation.

▶▶

- An inelastic collision is one in which the total kinetic energy is not the same before and after the event.
- Although kinetic energy may or may not be conserved in a collision, momentum is always conserved, and so is total energy.
- The impulse of a force F is the product of the force and the time Δt for which it acts:

  impulse $= F\Delta t$

- The impulse of a force acting on a body is equal to the change of momentum of the body:

  $F\Delta t = \Delta p$

- The unit of impulse is N s.

Sections 5.5–5.8 Questions

1 Calculate the magnitude of the momentum of a car of mass 1.5 tonnes (1.5×10^3 kg) travelling at a speed of 22 m s^{-1}.

2 When a certain space rocket is taking off, the propellant gases are expelled at a rate of 900 kg s^{-1} and speed of 40 km s^{-1}. Calculate the thrust on the rocket.

3 An insect of mass 4.5 mg, flying with a speed of 0.12 m s^{-1}, encounters a spider's web, which brings it to rest in 2.0 ms. Calculate the average force exerted by the insect on the web.

4 An atomic nucleus at rest emits an α-particle of mass 4 u. The speed of the α-particle is found to be 5.6×10^6 m s^{-1}. Calculate the speed with which the daughter nucleus, of mass 218 u, recoils.

5 A heavy particle of mass m_1, moving with speed u, makes a head-on collision with a light particle of mass m_2, which is initially at rest. The collision is perfectly elastic, and m_2 is very much less than m_1. Describe the motion of the particles after the collision.

6 A light body and a heavy body have the same momentum. Which has the greater kinetic energy?

Exam-style Questions

1 A bullet of mass 12 g is fired horizontally from a gun with a velocity of 180 m s^{-1}. It hits, and becomes embedded in, a block of wood of mass 2000 g, which is freely suspended by long strings, as shown in Figure 5.18.

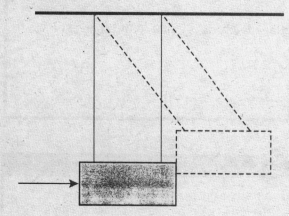

Figure 5.18

Calculate:
(a) (i) the magnitude of the momentum of the bullet just before it enters the block,
(ii) the magnitude of the initial velocity of the block and bullet after impact,
(iii) the kinetic energy of the block and embedded bullet immediately after the impact.
(b) Deduce the maximum height above the equilibrium position to which the block and embedded bullet rise after impact.

2 A nucleus A of mass 222 u is moving at a speed of 350 m s^{-1}. While moving, it emits an α-particle of mass 4 u. After the emission, it is determined that the daughter nucleus, of mass 218 u, is moving with speed 300 m s^{-1} in the original direction of the parent nucleus. Calculate the speed of the α-particle.

3 A safety feature of modern cars is the air-bag, which, in the event of a collision, inflates and is intended to decrease the risk of serious injury. Use the concept of impulse to explain why an air-bag might have this effect.

4 Two frictionless trolleys A and B, of mass m and $3m$ respectively, are on a horizontal track (Figure 5.19). Initially they are clipped together by a device which incorporates a spring, compressed between the trolleys. At time $t = 0$ the clip is released. The velocity of trolley B is u to the right.

Figure 5.19

(a) Calculate the velocity of trolley A as the trolleys move apart.
(b) At time $t = t_1$, trolley A collides elastically with a fixed spring and rebounds. (Compression and expansion of the spring take a negligibly short time.) Trolley A catches up with trolley B at time $t = t_2$.
  (i) Calculate the velocity of trolley A between $t = t_1$ and $t = t_2$.
  (ii) Find an expression for t_2 in terms of t_1.
(c) When trolley A catches up with trolley B at time t_2 the clip operates so as to link them again, this time without the spring between them, so that they move together with velocity v. Calculate the common velocity v in terms of u.
(d) Initially the trolleys were at rest and the total momentum of the system was zero. However, the answer to (c) shows that the total momentum after $t = t_2$ is not zero. Discuss this result with reference to the principle of conservation of momentum.

5 A ball of mass m makes a perfectly elastic head-on collision with a second ball, of mass M, initially at rest. The second ball moves off with half the original speed of the first ball.
(a) Express M in terms of m.
(b) Determine the fraction of the original kinetic energy retained by the ball of mass m after the collision.

6. Thermal physics

The aim of this chapter is to describe a number of phenomena associated with temperature and thermal energy. We shall start by looking at the physical basis of thermometers, which are instruments to measure temperature, and then see how scales of temperature are defined. The fundamental scale of temperature, the thermodynamic scale, is based on the properties of an ideal gas. To develop the idea of the temperature of a gas, we shall need to think of the gas as a collection of very small particles in continuous random motion. This is the basis of the kinetic theory of gases. Finally, we shall consider the energy of the particles of a gas and link this internal energy to the quantities of work and heat through the first law of thermodynamics.

6.1 Temperature

At the end of Sections 6.1–6.2 you should be able to:
- show an appreciation that thermal energy is transferred from a region of higher temperature to a region of lower temperature
- show an understanding that regions of equal temperature are in thermal equilibrium
- show an understanding that there is an absolute scale of temperature which does not depend on the property of any particular substance (the thermodynamic scale), and the concept of absolute zero
- convert temperatures measured in kelvin to degrees Celsius:
  $T/K = \theta/°C + 273.15$
- show an understanding that a physical property which varies with temperature may be used for the measurement of temperature, and state examples of such properties
- sketch the temperature characteristic of a thermistor
- compare the relative advantages and disadvantages of resistance and thermocouple thermometers as previously calibrated instruments.

Our everyday idea of temperature is based on our sense of touch. Putting your hand into a bowl containing ice immediately gives a sense of cold; putting the other hand into a bowl of warm water gives the sensation of heat (Figure 6.1). Intuitively, we would say that the water is at a higher temperature than the ice.

Think about a thick metal bar with one end at a higher temperature than the other. For example, one end can be heated by pouring hot water over it, and the other end cooled by holding it under a cold-water tap. The effect of the temperature difference between the ends is that thermal energy is

Figure 6.1

transferred along the bar from the high temperature end to the low temperature end. We can think of this in terms of the vibrations of the atoms of the metal. One atom passes on some of its vibrational energy to its neighbour, which originally had less. If the bar is removed from the arrangement for keeping the ends at different temperatures, eventually the whole of the bar will end up in equilibrium at the same temperature. When different regions in thermal contact are at the same temperature, they are said to be in **thermal equilibrium**.

In physics, we look for ways of defining and measuring quantities. In the case of temperature, we will first look at ways of measuring this quantity, and then think about the definition.

Many physical properties change with temperature. Most materials (solids, liquids and gases) expand as their temperature is increased. The electrical resistance of a metal wire increases as the temperature of the wire is increased. If two wires of different metals are twisted together at one end, and the other ends are connected to a voltmeter, the reading on the voltmeter depends on the temperature of the junction of the wires. All these properties may be made use of in different types of thermometer.

A thermometer is an instrument for measuring temperature. The physical property on which a particular thermometer is based is called the thermometric property, and the working material of the thermometer, the property of which varies with temperature, is called the thermometric substance. Thus, in the familiar mercury-in-glass thermometer, the thermometric substance is mercury and the thermometric property is the length of the mercury thread in the capillary tube of the thermometer.

Remember that temperature measures the degree of 'hotness' of a body. It does not measure the *amount* of heat energy.

Temperature scales

Each type of thermometer can be used to establish its own temperature scale. To do this, the fact that substances change state (from solid to liquid, or from liquid to gas) at fixed temperatures is used to define reference temperatures, which are called fixed points. By taking the value of the thermometric property at two fixed points, and dividing the range of values into a number of equal steps (or degrees), we can set up what is called an empirical scale of temperature for that thermometer. ('Empirical' means 'derived by experiment'.) If the fixed points are the melting point of ice and the boiling point of water, and if we choose to have one hundred equal degrees between the temperatures corresponding to these fixed points, taken as 0 degrees and 100 degrees respectively, we arrive at the empirical centigrade scale of temperature for that thermometer. If the values of the thermometric property P are P_i and P_s at the ice- and steam-points respectively, and if the property has the value P_θ at an unknown temperature θ, the unknown temperature is given by

$$\theta = \frac{100(P_\theta - P_i)}{(P_s - P_i)}$$

on the empirical centigrade scale of this particular thermometer. This equation is illustrated in graphical form in Figure 6.2. It is important to realise that the choice of a different thermometric substance and thermometric property would lead to a different centigrade scale. Agreement between scales occurs only at the two fixed points. This happens because the property may not vary linearly with temperature.

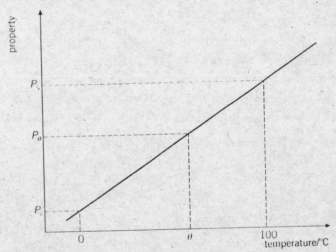

Figure 6.2 Empirical centigrade scale

This situation, with temperature values depending on the type of thermometer on which they are measured, is clearly unsatisfactory for scientific purposes. It is found that the differences between empirical scales are small in the case of thermometers based on gases as thermometric substances. In the constant-volume gas thermometer (Figure 6.3), the pressure of a fixed volume of gas (measured by the height difference h) is used as the thermometric property. The differences between the scales of different gas thermometers become even less as the pressures used are reduced. This is because the lower the pressure of a real gas, the more linear is the variation of pressure or volume. If we set up an empirical centigrade scale for a real gas in a constant-volume thermometer by obtaining the pressures of the gas at the ice-point ($0\,^\circ$C) and the steam-point ($100\,^\circ$C) we can extrapolate the graph of pressure p against the centigrade temperature θ

to find the temperature at which the pressure of the gas would become zero. This is shown in Figure 6.4. The extrapolated temperature will be found to be close to −273 degrees on the empirical centigrade constant-volume gas thermometer scale. If the experiment is repeated with lower and lower pressures of gas in the thermometer, the extrapolated temperature tends to a value of −273.15 degrees. This temperature is known as **absolute zero** on the thermodynamic scale of temperature. It does not depend on the properties of any particular substance.

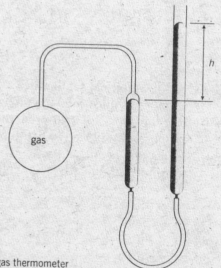

Figure 6.3 Constant-volume gas thermometer

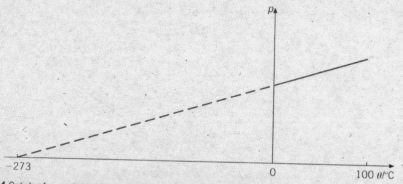

Figure 6.4 Graph of p against θ for constant-volume gas thermometer

Example

The length of the mercury column in a mercury-in-glass thermometer is 25 mm at the ice-point and 180 mm at the steam-point. What is the temperature when the length of the column is 55 mm?

Using the equation $\theta = 100(P_\theta - P_i)/(P_s - P_i)$, the thermometric property P is the length of the mercury column.

Thus, $\theta = 100(55 - 25)/(180 - 25) = \textbf{19.4° C}$ on the centigrade scale of this mercury-in-glass thermometer.

Thermodynamic temperature

As mentioned above, decreasing the pressure of a real gas makes it behave more and more like an ideal gas. For an ideal gas, the relation between pressure p, volume V and temperature T is

$$\frac{pV}{T} = \text{constant}$$

Here T is the **thermodynamic temperature**. We now use the property of an ideal gas to define the thermodynamic scale of temperature. We have already seen from the idea of extrapolating the p, θ graph for a constant-volume gas thermometer that there seems to be a natural zero of temperature, absolute zero. This is used as one of the fixed points of the thermodynamic scale. The upper fixed point is taken as the triple point of water – the temperature at which ice, water and water vapour are in equilibrium. This is found to be less dependent on environmental conditions, such as pressure, than the ice-point. The thermodynamic temperature of the triple point of water is taken as 273.16 units, by international agreement. This defines the kelvin (symbol K), the unit of thermodynamic temperature.

One kelvin is the fraction 1/273.16 of the thermodynamic temperature of the triple point of water.

Thus, if a constant-volume gas thermometer gives a pressure reading of p_{tr} at the triple point, and a pressure reading of p at an unknown temperature T, the unknown temperature (in K) is given by

$$T = 273.16(p/p_{tr})$$

The Celsius scale

Why choose 273.16 as the number of units between the two fixed points of this scale? The reason is that this number will give 100 K between the ice- and steam-points, allowing agreement between the thermodynamic temperature scale and a centigrade scale based on the pressure of an ideal gas. The ideal-gas centigrade scale is based on experiments with real gases at decreasing pressures. This agreement is based on a slightly awkward linking up of the theoretical thermodynamic scale and the empirical constant-volume gas thermometer scale. To avoid this complication, a new scale, the Celsius scale, was defined by international agreement.

The unit of temperature on the Celsius scale is the degree Celsius (°C), which is exactly equal to the kelvin.

The equation linking temperature θ on the Celsius scale and thermodynamic temperature T is

$$\theta/°C = T/K - 273.15 \quad \text{or} \quad T/K = \theta/°C + 273.15$$

In this equation, θ is measured in °C and T in K. Note that the degree sign ° always appears with the Celsius symbol, but it is never used with the kelvin symbol K.

Be careful to avoid confusion between the numbers 273.15 and 273.16. Absolute zero on the ideal gas constant-volume scale is –273.15 degrees. The fact that 273.16 occurs in the definition of the kelvin means that the temperature of the triple point of water is 0.01°C. For many purposes, in particular in calculations involving conversions of temperatures from the Celsius to the thermodynamic scale or the other way, it is sufficient to work with the number 273.

Example

The pressure reading of a constant-volume gas thermometer is 1.50×10^4 Pa at the triple point of water. When the bulb is placed in a certain liquid, the pressure is 4.28×10^3 Pa. Find the temperature of the liquid.

We use the equation $T = 273.16(p/p_{tr})$. Because the pressure readings are given to three significant figures only, we can approximate the 273.16 to 273. Thus, $T = 273 \times (4.28 \times 10^3/1.50 \times 10^4) = \mathbf{77.9\,K}$. (The liquid is liquid nitrogen at its boiling-point.)

Now it's your turn

The pressure reading of a constant-volume gas thermometer is 1.367×10^3 Pa at the steam-point. What is the pressure reading at the triple point of water?

6.2 Thermometers

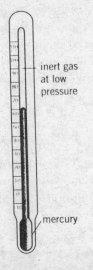

inert gas at low pressure

mercury

Figure 6.5 Mercury-in-glass thermometer

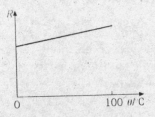

Figure 6.6 Resistance of a metal wire over a small range of temperatures

Figure 6.7 Thermistors are useful temperature measuring devices.

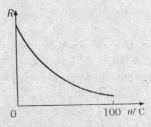

Figure 6.8 Resistance of a thermistor over a small range of temperatures

We have already mentioned some of the many different types of thermometer. In choosing a thermometer for a particular application, we need to consider a number of aspects. These include accuracy, sensitivity (the distance between divisions on its scale), and the range of temperatures it is able to measure. In some cases the thermometer has to measure rapidly varying temperatures, in which case its speed of response will be important, and whether it can be directly read, or requires time-consuming adjustments by the operator. For other applications it is important that the sensitive part of the thermometer is small and does not absorb much heat, so that it does not change the temperature of the object during measurement.

Gas thermometers (Figure 6.3) are important because they provide a link with the thermodynamic scale. However, they are bulky and inconvenient. Because of the adjustments necessary, they are certainly not suitable for the measurement of rapidly varying temperatures. The bulb containing the gas may have a volume of as much as a litre ($1000\,cm^3$), so a gas thermometer could not be used for measuring the temperature of a small body.

The most familiar type of thermometer is probably the liquid-in-glass type (Figure 6.5), based on the expansion of the liquid. Such thermometers are convenient, sensitive and moderately quick-acting. The range of the thermometer is limited by the liquid used and the glass containing it. Mercury-in-glass thermometers cover the range from about $-40\,°C$ to $350\,°C$; ethanol-in-glass (the ethanol is normally coloured with a red dye) from about $-120\,°C$ to $80\,°C$. An advantage of liquid-in-glass thermometers is that, over their temperature range, their empirical centigrade scale is very close to the thermodynamic scale.

An important class of thermometers is based on the temperature variation of electrical resistance. The resistance of a metal wire increases with increasing temperature. The range of metal resistance thermometers is very wide, from about $-260\,°C$ to $1700\,°C$. The temperature sensor is a coil of fine wire, often platinum. The variation with temperature of the resistance of a metal wire is not exactly linear with thermodynamic temperature, and for accurate work the resistance thermometer needs to be calibrated at a number of standard temperatures. However, over a small range of temperature, the variation of resistance is linear, as shown in Figure 6.6. Semiconducting materials are also used as resistance thermometers: these are called **thermistors** (see Figure 6.7). The electrical resistance of such devices decreases very rapidly with increasing temperature, as shown in Figure 6.8. (Note that this is in the opposite sense to the behaviour of a metal wire.) Because of the rapid change of resistance, thermistors are very sensitive devices, but their scales are very non-linear compared with thermodynamic temperature, and calibration is essential. By selection of suitable semiconducting materials, a wide range of temperatures can be covered. Their very small size (some smaller than a pin-head) can be used to measure the temperature of small objects, and also varying temperatures. A simple circuit allows the resistance of the thermistor, or a quantity related to the resistance, to be displayed on a meter, which can then be calibrated in terms of temperature. A common use of a thermistor thermometer is as a temperature sensor in car radiators.

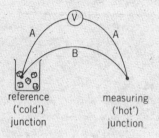

Figure 6.9 Thermocouple

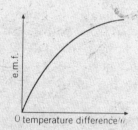

Figure 6.10 Graph of thermocouple e.m.f. against temperature

Another type of electrical thermometer makes use of the **thermoelectric effect**. When the junctions of two different conductors (Figure 6.9) are at different temperatures, an e.m.f. (voltage) is developed. The device is known as a thermocouple. The relation between the e.m.f. and the temperature difference θ is not linear with thermodynamic temperature (see Figure 6.10), and calibration is essential. The e.m.f. generated is rather small (for a thermocouple made of a copper wire and a constantan wire, it is about 5 mV for a temperature difference of 100 °C), but it is easy to arrange a circuit to amplify this e.m.f. and provide a direct reading, which can then be calibrated. Thermocouples cover a large temperature range. Because the sensing part (the junction between the two wires) is very small the thermocouple can be used to measure rapidly varying temperatures, and the temperatures of small objects. One is shown in Figure 2.30 (page 35).

Sections 6.1–6.2 Summary

- Two regions that are at the same temperature are said to be in thermal equilibrium.
- Empirical centigrade scale of temperature: $\theta = 100(P_\theta - P_i)/(P_s - P_i)$, where θ is the centigrade temperature on the scale of a thermometer based on a property P. P_s, P_i and P_θ are the values of the property at the steam-point, ice-point and at temperature θ respectively.
- The kelvin (K), unit of thermodynamic temperature, is the fraction $1/273.16$ of the thermodynamic temperature of the triple point of water.
- Thermodynamic scale of temperature: for a constant-volume gas thermometer, $T = 273.16(p/p_{tr})$, where T is the thermodynamic temperature, p is the pressure reading at temperature T and p_{tr} is the reading at the triple point of water.
- Celsius scale: $\theta = T - 273.15$, where θ is the Celsius temperature (in °C) and T is the thermodynamic temperature (in K).
- The thermometric properties of common thermometers include the length of a column of liquid (in liquid-in-glass thermometers), the pressure of a gas (in constant-volume gas thermometers), the resistance of a coil of wire or of a sample of semiconductor (in thermistors), and the thermoelectric e.m.f. (in thermocouples).

Sections 6.1–6.2 Questions

1 The e.m.f. of a certain thermocouple is 5.60 mV when one junction is placed in melting ice and the other in steam. With one junction in melting ice and the other in a boiling liquid, the e.m.f. is −2.46 mV. Calculate the boiling point of the liquid on the centigrade scale of this thermocouple.

2 The gain of an operational amplifier (op-amp) is found to depend on temperature. At 5 °C, the gain is 110; at 18 °C, it is 125. Assuming that the gain depends linearly on Celsius temperature, estimate the gain when the op-amp is used on a hot day, at a temperature of 25 °C.

Figure 6.11 Robert Boyle

6.3 The gas laws

At the end of Section 6.3 you should be able to:

■ recall and solve problems using the equation of state for an ideal gas expressed as $pV = nRT$, where n is the number of moles
■ show an understanding of the Avogadro constant as the number of atoms in 0.012 kg of carbon-12
■ use molar quantities, where one mole of any substance is the amount containing a number of particles equal to the Avogadro constant.

Experiments in the seventeenth and eighteenth centuries showed that the volume, pressure and temperature of a given sample of gas are all related. The precise relation between these quantities and the mass of gas in the sample is called the **equation of state** of the gas. For a given mass of gas, it was found that

The volume V of the gas is inversely proportional to its pressure p, provided that the temperature is held constant.

Expressed mathematically, this is

$$V \propto 1/p$$

or

$$pV = \text{constant}$$

Another way of writing this equation is

$$p_1V_1 = p_2V_2$$

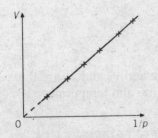

Figure 6.12 Graph of V against $1/p$

where p_1 and V_1 are the initial pressure and volume of the gas, and p_2 and V_2 are the final values after a change of pressure and volume carried out at constant temperature.

A graph of V against $1/p$ is a straight line through the origin (Figure 6.12). This relation is known as **Boyle's law**, after Robert Boyle (1627–91), who first stated this relation as a result of his own experiments.

The effect of temperature on the volume of a gas was investigated by the French scientist Jacques Charles (1746–1823). Charles found that the graph of volume V against temperature θ is a straight line (Figure 6.13). Because gases

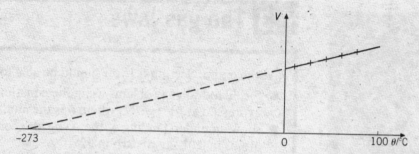

Figure 6.13 Graph of V against θ

liquefy when the temperature is reduced, experimental points could not be obtained below the liquefaction temperature. But if the graph was projected backwards, it was found that it cut the temperature axis at about –273 °C.

In the description of the constant-volume gas thermometer, we have already met the effect of temperature on the pressure of a gas. This effect was investigated by another Frenchman, Joseph Gay-Lussac (1778–1850).

Remember that the graph of pressure p against temperature θ is a straight line, which, if projected like the volume–temperature graph, also meets the temperature axis at about –273 °C (Figure 6.4). We used this fact to introduce the thermodynamic scale of temperature and the idea of the absolute zero of temperature.

If Celsius temperatures are converted to thermodynamic temperatures, Charles' results are expressed as

For a given mass of gas maintained at constant pressure, the volume V of the gas is directly proportional to its thermodynamic temperature T.

Expressed mathematically, this is

$$V \propto T$$

This is **Charles' law**. Another way of writing the equation is

$$\frac{V_1}{T_1} = \frac{V_2}{T_2}$$

where V_1 and T_1 are the initial volume and temperature and V_2 and T_2 are the final values. The corresponding relation between pressure and temperature is

For a given mass of gas maintained at constant volume, the pressure p of the gas is directly proportional to its thermodynamic temperature T.

Expressed mathematically, this is

$$p \propto T$$

This is **Gay-Lussac's law**, or the **law of pressures**. The equation can also be written as

$$\frac{p_1}{T_1} = \frac{p_2}{T_2}$$

where the subscripts 1 and 2 again refer to initial and final values.

We can combine the three gas laws into a single relation between pressure p, and volume V and temperature T for a fixed mass of gas. This relation is

$$pV \propto T$$

or

$$\frac{p_1 V_1}{T_1} = \frac{p_2 V_2}{T_2}$$

All the above laws relate to a fixed mass of gas. Another series of experiments could be carried out to find out how the volume of a gas, held at constant pressure and temperature, depends on the mass of gas present. It would be found that the volume is proportional to the mass. This would give the combined relation

$$pV \propto mT$$

where m is the mass of gas, or

$$pV = AmT$$

where A is a constant of proportionality. However, this is not a very useful way of expressing the relation, as the constant A has different values for different gases. We need to find a way of including the quantity of gas. The way to do this is to express the fixed mass of gas in Boyle's, Charles' and Gay-Lussac's laws in terms of the number of moles of gas present.

We have met the mole in Chapter 1, as one of the base SI units.

The **mole** (abbreviated mol) is the amount of substance which contains as many elementary entities as there are atom in 0.012 kg of carbon-12.

The **Avogadro constant N_A** is the number of atoms in 0.012 kg of carbon-12. That is, it is the number of atoms in ° mole of atoms. It has the value 6.02×10^{23} mol^{-1}.

The **relative atomic mass A_r** is the ratio of the mass of an atom to one-twelfth of the mass of an atom of carbon-12. Since the definitions of the mole and the Avogadro constant relate to 0.012 kg (or 12 g) of carbon-12, the relative atomic mass is numerically equal to the mass in grams of a mole of atoms.

Similarly, the **relative molecular mass** M_r is the ratio of the mass of a molecule to one-twelfth of the mass of an atom of carbon-12, and is numerically equal to the mass in grams of a mole of molecules.

> The **atomic mass unit** u is one-twelfth of the mass of an atom of carbon-12, and has the value 1.66×10^{-27} kg. (This quantity is also called the unified atomic mass constant.)

Thus, a simpler way of finding the mass of one mole of a particular element is to take its nucleon number (see Chapter 13) expressed in grams. For example, the nucleon number of argon (^{40}Ar) is 40. One mole of argon then has mass 40 g.

Returning to the combination of the three gas laws and using the number n of moles of the gas, we have

$$pV \propto nT$$

or, putting in a new constant of proportionality R,

$$pV = nRT$$

R is called the **molar gas constant** (sometimes the **universal gas constant**, because it has the same value for all gases). It has the value $8.3 \, \mathrm{J\,K^{-1}mol^{-1}}$.

Sometimes the equation $pV = nRT$ is expressed in the form

$$pV_m = RT$$

where the molar volume V_m is the volume occupied by one mole of the gas. Another version is

$$pV = NkT$$

where N is the number of molecules in the gas and k is a constant called the **Boltzmann constant**. It has the value $1.38 \times 10^{-23} \, \mathrm{J\,K^{-1}}$. The molar gas constant R and the Boltzmann constant k are connected through the Avogadro constant N_A

$$R = kN_A$$

Strictly speaking, the laws of Boyle, Charles and Gay-Lussac are not really laws, as their validity is restricted. They are accurate for real gases only if the pressure of the sample is not too great, and if the gas is well above its liquefaction temperature. But they can be used to define an 'ideal' gas.

An **ideal gas** is one which obeys the equation of state

$$pV = nRT$$

at all pressures, volumes and temperatures.

We have already used the ideal gas equation to introduce the idea of thermodynamic temperature. For approximate calculations, the ideal gas equation can be used with real gases.

Examples

1 Find the volume occupied by 1 mole of air at standard temperature and pressure (273 K and 1.01×10^5 Pa), taking R as $8.3 \, \text{J K}^{-1} \, \text{mol}^{-1}$ for air.

Since $pV = nRT$, $V = nRT/p$. Substituting the values $n = 1$ mol, $R = 8.3 \, \text{J K}^{-1} \, \text{mol}^{-1}$, $T = 273$ K and $p = 1.01 \times 10^5$ Pa, $V = (1 \times 8.3 \times 273)/1.01 \times 10^5 = \mathbf{2.24 \times 10^{-2} \, m^3}$.

2 Find the number of molecules per cubic metre of air at standard temperature and pressure.

We have just shown that the volume occupied by one mole of air at standard temperature and pressure is $2.24 \times 10^{-2} \, \text{m}^3$. One mole of air contains N_A molecules, where N_A is the Avogadro constant ($6.02 \times 10^{23} \, \text{mol}^{-1}$). Thus the number of molecules per cubic metre of air is $6.02 \times 10^{23}/2.24 \times 10^{-2} = \mathbf{2.69 \times 10^{25} \, m^{-3}}$.

Note: It is useful to remember these two quantities, the *molar volume* of a gas and the *number density* of molecules in it. They give an idea of the relatively small volume occupied by a mole of gas at standard temperature and pressure (a cube of side about 28 cm), and the enormous number of molecules in every cubic metre of a gas under these conditions.

3 A syringe contains $25 \times 10^{-6} \, \text{m}^3$ of helium gas at a temperature of 20 °C and a pressure of 5.0×10^4 Pa. The temperature is increased to 400 °C and the pressure on the syringe is increased to 2.4×10^5 Pa. Find the new volume of gas in the syringe.

Boyle's law, pV = constant, and Charles' law, V/T = constant, may be combined to give pV/T = constant. This is written in the form $p_1V_1/T_1 = p_2V_2/T_2$ and re-arranged as $V_2 = p_1V_1T_2/p_2T_1$. Substituting the values $p_1 = 5.0 \times 10^4$ Pa, $V_1 = 25 \times 10^{-6} \, \text{m}^3$, $T_2 = 673$ K, $p_2 = 2.4 \times 10^5$ Pa, $T_1 = 293$ K (again note that temperatures are converted from °C to K),

$$V_2 = 5.0 \times 10^4 \times 25 \times 10^{-6} \times 673/2.4 \times 10^5 \times 293 = \mathbf{12 \times 10^{-6} \, m^3}.$$

Now it's your turn

1 The number of molecules per cubic metre of air at standard temperature and pressure is about $2.7 \times 10^{25} \, \text{m}^{-3}$. What is the average separation of these molecules?

2 The mean mass of one mole of air (which is made up mainly of nitrogen, oxygen and argon) is 0.029 kg. What is the density of air at standard temperature and pressure (273 K and $1.01 \times 10^5 \, \text{Pa}$)?

3 The volume of a sample of gas is $3.2 \times 10^{-2} \, \text{m}^3$ when the pressure is $8.6 \times 10^4 \, \text{Pa}$ and the temperature is 27 °C. How many moles of gas are there in the sample? How many molecules? What is the number of molecules per cubic metre?

4 A sample of air has volume $15 \times 10^{-6} \, \text{m}^3$ when the pressure is $5.0 \times 10^5 \, \text{Pa}$. The pressure is reduced to $3.0 \times 10^5 \, \text{Pa}$, without changing the temperature. What is the new volume?

5 A sample of gas, originally at standard temperature and pressure (273 K and $1.01 \times 10^5 \, \text{Pa}$), has volume $3.0 \times 10^{-5} \, \text{m}^3$ under these conditions. When the pressure is increased to $4.0 \times 10^5 \, \text{Pa}$, the temperature rises to 40 °C. What is the new volume?

Section 6.3 Summary

- Boyle's law states that for a fixed mass of gas at constant temperature, the product of the pressure of a gas and its volume is a constant: pV = constant or $p_1V_1 = p_2V_2$
- Charles' law states that if the pressure of a fixed mass of gas is kept constant, the ratio of volume to thermodynamic temperature is a constant: V/T = constant or $V_1/T_1 = V_2/T_2$
- Gay-Lussac's law (the law of pressures) states that if the volume of a fixed mass of gas is kept constant, the ratio of pressure to thermodynamic temperature is a constant: p/T = constant or $p_1/T_1 = p_2/T_2$
- The equation of state for an ideal gas relates the pressure p, volume V and thermodynamic temperature T of n moles of gas: $pV = nRT$ where $R = 8.3 \, \text{J K}^{-1} \, \text{mol}^{-1}$, the molar gas constant.
- For N molecules of gas: $pV = NkT$ where $k = 1.38 \times 10^{-23} \, \text{J K}^{-1}$, the Boltzmann constant.
- The relation between R and k is: $R = N_A k$ where $N_A = 6.02 \times 10^{23} \, \text{mol}^{-1}$, the Avogadro constant (the number of atoms in 0.012 kg of carbon-12).
- Another expression of the ideal gas equation (or of Boyle's law, Charles' law and Gay-Lussac's law combined) is: pV/T = constant or $p_1V_1/T_1 = p_2V_2/T_2$

Section 6.3 Questions

1 On a day when the atmospheric pressure is 102 kPa and the temperature is 8 °C, the pressure in a car tyre is 190 kPa above atmospheric pressure. After a long journey the temperature of the air in the tyre rises to 29 °C. Calculate the pressure above atmospheric of the air in the tyre at 29 °C. Assume that the volume of the tyre remains constant.

2 A helium-filled balloon is released at ground level, where the temperature is 17 °C and the pressure is 1.0 atmosphere. The balloon rises to a height of 2.5 km, where the pressure is 0.75 atmospheres and the temperature is 5 °C. Calculate the ratio of the volume of the balloon at 2.5 km to that at ground level.

3 A metal box in the form of a cube of side 30 cm is filled with air at atmospheric pressure (1.01×10^5 Pa) at 15 °C. The box is sealed and heated in an oven to 200 °C. Calculate the net force on each side of the box.

6.4 A microscopic model of a gas

At the end of Section 6.4 you should be able to:
- describe a simple kinetic model for solids, liquids and gases
- infer from a Brownian motion experiment the evidence for the movement of molecules
- state the basic assumptions of the kinetic theory of gases
- define the term pressure and use the kinetic theory of a gas to explain the pressure exerted by a gas
- use the kinetic theory of a gas to deduce the relationship $p = \frac{1}{3}Nm\langle c^2\rangle/V$, where N = number of molecules
- compare $pV = \frac{1}{3}Nm\langle c^2\rangle$ with $pV = NkT$ and hence deduce that the average translational kinetic energy of a molecule is proportional to T.

One of the aims of physics is to describe and explain the behaviour of various systems. For mechanical systems, this involves calculating the motion of the parts of the system in detail. For example, we have already seen how to predict the motion of a stone thrown in a uniform gravitational field. Using the equations of uniformly accelerated motion, it is not too difficult to calculate the position and velocity of the stone at any time (Chapter 3). However, there are some cases in which it is quite impossible to describe what happens to each component of the system.

This sort of problem arises if we try to describe the properties of a gas in terms of the motion of each of its molecules. The difficulty is that the numbers are so large. One cubic metre of atmospheric air contains about 3×10^{25} molecules. There is no practical method of determining the position and velocity of every single molecule at a given time. Even the most advanced computer would be unable to handle the calculation of the motions of such a very large number of molecules.

In some ways, the fact that the gas is made up of such an enormous number of molecules is an advantage. It means that we can give a large-scale description of the gas in terms of only a few variables. These variables are quantities such as pressure p, volume V and temperature T. They tell us about average conditions in the gas, instead of describing the behaviour of each molecule. We have already met the experimental laws relating to these quantities. Our aim now is to relate the ideal gas equation, which deals with the large-scale (macroscopic) quantities p, V and T, to the small-scale (microscopic) behaviour of the particles of the gas. We shall do this by taking averages over the very large numbers of molecules involved. We shall find that we can derive the equation for Boyle's law when we make very simple assumptions about the atoms or molecules which make up the gas. This is the **kinetic theory of an ideal gas**. We shall also see that temperature can be related to the kinetic energy of the molecules of the gas.

Before we go too far with this model of a gas, is there any experimental evidence for believing that its molecules are moving around all the time? In 1827 the biologist Robert Brown was observing, under a microscope, tiny pollen grains suspended in water. He saw that the grains were always in a jerky, haphazard motion, even though the water was completely still. This movement is now called **Brownian motion**. If we make the assumption that the water molecules are in rapid, random motion, then it is easy to see how the pollen grains could move so jerkily under the random bombardment, from all sides, of the water molecules. We can also reproduce Brownian motion by observing the motion of tiny soot particles in smoke. These particles, too, move in the jerky manner of Brown's pollen grains. Thus it seems that the molecules of both gases and liquids are in the rapid, random motion as required by the molecular model. Nearly a century later, Einstein made a theoretical analysis of Brownian motion and, from the masses and distances moved by the suspended particles, was able to estimate that the diameter of a typical atom is of the order of 10^{-10} m.

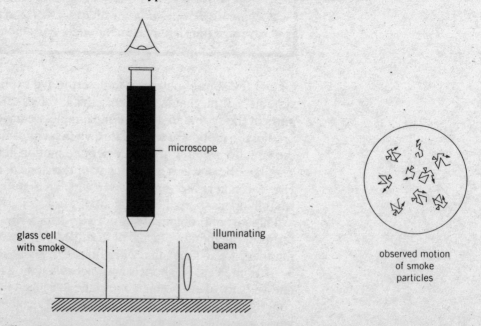

microscope

glass cell
with smoke

illuminating
beam

observed motion
of smoke
particles

Figure 6.14 Observing Brownian motion

If we are able to think of gases and liquids as consisting of large numbers of atoms or molecules in rapid, random motion, what about solids? The obvious difference between a solid and a liquid or gas is that the solid has rigidity. However, the microscopic model explains this by suggesting that the atoms are held in more or less fixed positions by much stronger interatomic forces than exist in liquids or gases. This is because the atoms are arranged quite close together in a solid, whereas in a gas there is a great deal of empty space. Atoms in a solid can still move, but their motion is restricted to vibration about their equilibrium positions. In the change of state of a solid to a liquid, work must be done to break the rigid interatomic forces so that the atoms can move freely. This work is the latent heat of fusion of the solid, which we met in Section 4.3.

The kinetic theory

An explanation of how a gas exerts a pressure was developed by Robert Boyle in the seventeenth century, and in greater detail by Daniel Bernoulli in the eighteenth century. The basic idea was that the gas consists of atoms or molecules moving about at great speed (later visualised by Robert Brown). The gas exerts a pressure on the walls of its container because of the continued impacts of the molecules with the walls. We shall make some simplifying assumptions about the molecules of the gas. The assumptions of the kinetic theory of an ideal gas are:

1 All molecules behave as identical, hard, perfectly elastic spheres.
2 The volume of the molecules is negligible compared with the volume of the containing vessel.
3 There are no forces of attraction or repulsion between molecules.
4 There are many molecules, all moving randomly.

The number of molecules must be very large, so that average behaviour can be considered.

Suppose that the container is a cube of side L (Figure 6.15). The motion of each molecule can be resolved into x-, y- and z-components. For convenience we shall take the x-, y- and z-directions to be parallel to the edges of the cube.

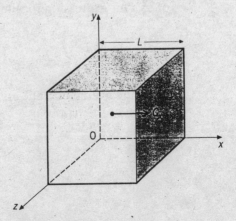

Figure 6.15 A gas molecule in a cubic container

Consider one molecule. Let the x-component of its velocity be c_x. When this molecule collides with the wall of the container perpendicular to the x-axis, the x-component of velocity will be exactly reversed, because (from our assumptions) the collision of the molecule with the wall is perfectly elastic. The time taken for the molecule to move between the two walls perpendicular to the x-axis is L/c_x. To make the round trip from one wall to the opposite one and back again takes $2L/c_x$. This is the time between one collision of the molecule with a wall and its next collision with the same wall.

When the molecule strikes a wall, the component of velocity is reversed in direction, from c_x to $-c_x$. Thus, during each collision with a wall, the x-component of momentum changes by

$$\Delta p_x = 2mc_x$$

where m is the mass of the molecule. The rate at which this molecule transfers momentum to the wall is

$$\text{change of momentum/time between collisions} = 2mc_x/(2L/c_x) = mc_x^2/L$$

From Newton's second law, this rate of change of momentum is the average force exerted by this particular molecule on the wall through its collisions with the wall. If there are N molecules in the container, the total force is Nmc_x^2/L. Pressure is force divided by area, and the area of the wall is L^2, so the pressure exerted by all N molecules on this wall is Nmc_x^2/L^3. The volume V of the container is L^3, giving us

$$p = Nmc_x^2/V$$

as an expression for the pressure. This expression relates only to the x-component of velocity of the molecules. For a molecule moving with velocity c, an extension of Pythagoras' theorem to three dimensions gives the relation between c and the three components of velocity c_x, c_y and c_z: $c^2 = c_x^2 + c_y^2 + c_z^2$. Because we are dealing with a large number of molecules in random motion, the average value of the components in the x-direction will be the same as for those in the y-direction or the z-direction. Therefore, taking the averages, $\langle c_x^2 \rangle = \langle c_y^2 \rangle = \langle c_z^2 \rangle$ and $\langle c_x^2 \rangle = \frac{1}{3}\langle c^2 \rangle$. The notation $\langle c_x^2 \rangle$ means the average value of c_x^2. Our expression for the pressure can now be written

$$pV = \tfrac{1}{3} Nm\langle c^2 \rangle$$

Since the average kinetic energy of a molecule is

$$\langle E_k \rangle = \tfrac{1}{2} m\langle c^2 \rangle$$

there seems to be a link between our kinetic theory equation for pV and energy. We can find this relation by re-writing the pV equation as

$$pV = \tfrac{2}{3}N(\tfrac{1}{2}m\langle c^2\rangle) = \tfrac{2}{3}N\langle E_k\rangle$$

We have already stated the ideal gas law in the forms $pV = nRT = NkT$. Indeed, even real gases obey this law reasonably well under normal conditions. If our kinetic theory model and the subsequent theoretical derivation of the equation are correct, we can bring together the two equations for pV. This will allow us to relate the temperature of a gas to the average kinetic energy of its molecules.

$$pV = \tfrac{2}{3}N\langle E_k\rangle = NkT$$

and

$$\langle E_k\rangle = \tfrac{3}{2}kT$$

This is an important result. We have obtained a relation between the average kinetic energy of a molecule in a gas and the temperature of the gas. This will allow us to obtain an idea of the average speed of the molecules. Since

$$\langle E_k\rangle = \tfrac{1}{2}m\langle c^2\rangle = \tfrac{3}{2}kT$$

we have

$$\langle c^2\rangle = 3kT/m$$

and

$$\sqrt{\langle c^2\rangle} = \sqrt{3kT/m}$$

The quantity $\sqrt{\langle c^2\rangle}$ is called the **root-mean-square speed** or **r.m.s. speed (c_{rms})** of the molecules. It is not exactly equal to the average speed of the molecules, but is often taken as being so. The average speed is about 0.92 of the root-mean-square speed. The difference between the r.m.s. speed and the average speed is highlighted in Example 1 on page 152.

Note that the r.m.s. speed is proportional to the square root of the thermodynamic temperature of the gas, and inversely proportional to the square root of the mass of the molecule. Thus, at a given temperature, less massive molecules move faster, on average, than more massive molecules. For a given gas, the higher the temperature, the faster the molecules move.

Examples

1 The speeds of seven molecules in a gas are numerically equal to 2, 4, 6, 8, 10, 12 and 14 units. Find the numerical values of **(a)** the mean speed $\langle c \rangle$, **(b)** the mean speed squared $\langle c \rangle^2$, **(c)** the mean square speed $\langle c^2 \rangle$, **(d)** the r.m.s. speed.

(a) $\langle c \rangle = (2 + 4 + 6 + 8 + 10 + 12 + 14)/7 = $ **8 units**
(b) $\langle c \rangle^2 = 8^2 = $ **64 units²**
(c) $\langle c^2 \rangle = (4 + 16 + 36 + 64 + 100 + 144 + 196)/7 = $ **80 units²**
(d) r.m.s. speed $= \sqrt{80} = $ **8.94 units**.

2 Find the total kinetic energy of the molecules in one mole of an ideal gas at standard temperature (273 K).

We know that the average kinetic energy of one molecule is $\frac{3}{2}kT$.

For one mole of molecules, that is N_A molecules, the energy is

$$\frac{3}{2}N_A kT \text{ or } \frac{3}{2}RT$$

Substituting the values $R = 8.3 \, \mathrm{J \, K^{-1} \, mol^{-1}}$ and $T = 273$ K, we have

$$E_k = \frac{3}{2} \times 8.3 \times 273 = \textbf{3400 J mol}^{-1}.$$

3 Find the root-mean-square speed of the molecules in nitrogen gas at 27 °C. The mass of a nitrogen molecule is 4.6×10^{-26} kg.

Since $c_{rms} = \sqrt{3kT/m}$, we have

$$c_{rms} = \sqrt{3 \times 1.38 \times 10^{-23} \times 300/4.6 \times 10^{-26}} = \textbf{520 m s}^{-1}.$$

Now it's your turn

1 Find the average kinetic energy of a molecule in an ideal gas at a temperature of 500 K.

2 Find the root-mean-square speed of the molecules in hydrogen gas at 500 °C. The relative molecular mass of hydrogen is 2.

Section 6.4 Summary

■ Assumptions of the kinetic theory of gases:

1 Molecules behave as identical, hard, perfectly elastic spheres.
2 Volume of the molecules is negligible compared with the volume of the containing vessel.
3 There are no forces of attraction or repulsion between atoms.
4 There are many molecules, all moving randomly.

■ Kinetic theory pV equation: $pV = \frac{1}{3}Nm\langle c^2 \rangle$
■ Average kinetic energy of a molecule: $\langle E_k \rangle = \frac{1}{2}m\langle c^2 \rangle = \frac{3}{2}kT$
■ Root-mean-square speed of molecules: $c_{rms} = \sqrt{3kT/m}$

1 Estimate the root-mean-square speed of helium atoms near the surface of the Sun, where the temperature is about 6000 K. (mass of helium atom = 6.6×10^{-27} kg)

2 In the kinetic theory of the pressure of a gas, the impulse delivered by a molecule to the wall of the container is proportional to the velocity of the molecule. Why then does the kinetic theory equation for pV include the square of the velocity?

6.5 Internal energy

At the end of Sections 6.5–6.6 you should be able to:
- show an understanding of the concept of internal energy
- relate a rise in temperature of a body to an increase in its internal energy
- show an understanding that internal energy is determined by the state of the system and that it can be expressed as the sum of a random distribution of kinetic and potential energies associated with the molecules of a system
- recall and use the first law of thermodynamics expressed in terms of the change in internal energy, the heating of the system and the work done on the system.

We have seen that the molecules of an ideal gas possess kinetic energy, and that this kinetic energy is proportional to the thermodynamic temperature of the gas. The sum of the kinetic energies of all the molecules, due to their random motion, is called the **internal energy** of the ideal gas. Not all molecules have the same kinetic energy, because they are moving with different speeds, but the sum of all the kinetic energies will be a constant if the gas is kept at a constant temperature.

For a real gas, the situation is a little more complicated. Because the molecules of a real gas exert forces on each other, at any instant there will be a certain potential energy associated with the positions the molecules occupy in space. Because the molecules are moving randomly, the potential energy of a given molecule will also vary randomly. But at a given temperature the total potential energy of all the molecules will remain constant. If the temperature changes, the total potential energy will also change. Furthermore, the molecules of a real gas collide with each other, and will interchange kinetic energy during the collisions. For this real gas, the internal energy is given by the sum of the potential energies and the kinetic energies of all the molecules. It is important to realise that looking at a single molecule will give us very little information. Its kinetic energy will be changing all the time as it collides with other gas molecules, and its potential energy is also changing as its position

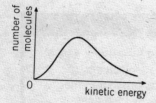

Figure 6.16 Distribution of molecular kinetic energies

relative to the other molecules in the gas changes. This single molecule has a kinetic energy which is part of the very wide range of kinetic energies of the molecules of the gas. We say that there is a *distribution* of molecular kinetic energies. The distribution is illustrated in Figure 6.16. Similarly, there is a distribution of molecular potential energies. But by adding up the kinetic and potential energies of all the molecules in the gas, the random nature of the kinetic and potential energies of the single molecule is removed.

The idea of internal energy can be extended to all states of matter. In a liquid, intermolecular forces are stronger as the molecules are closer together, so the potential energy contribution to internal energy becomes more significant. The kinetic energy contribution is still due to the random motion of the molecules in the liquid. We can think of a solid as being made up of atoms or molecules which oscillate (vibrate) about equilibrium positions. Here, the potential energy contribution is caused by the strong binding forces between atoms, and the kinetic energy contribution is due to the motion of the vibrating atoms.

The concept of internal energy is particularly useful as it helps us to distinguish between temperature and heat. Using an ideal gas as an example, temperature is a measure of the *average* kinetic energy of the molecules. It therefore does not depend on how many molecules are present in the gas. Internal energy (again for an ideal gas), however, is the *total* kinetic energy of the molecules, and clearly does depend on how many molecules there are. In general heat refers to a *transfer* of energy from one substance to another, often as a result of a temperature difference. If two objects at different temperatures are placed in contact, there will be a flow of heat from the object at the higher temperature to the one at the lower. The direction of heat flow is determined by the difference in temperature, not by any difference in internal energies. If 10 g of a liquid at 30 °C is placed in contact, or mixed into, 100 g of the same liquid at 20 °C, the direction of heat flow is from the liquid at 30 °C to the liquid at 20 °C, even though the liquid at 20 °C has a greater internal energy than the smaller mass of liquid at 30 °C.

6.6 The first law of thermodynamics

We have already met the law of conservation of energy (Chapter 4). There, it was stated in the following form: Energy can neither be created nor destroyed, it can only be transformed from one form to another. In this section we shall see how this conservation law may be re-stated in relation to terms like work, heat and internal energy. This will lead to an understanding of the **first law of thermodynamics**.

Thermodynamics is the study of processes involving the transfer of heat and the doing of work. In thermodynamics it is necessary to define the **system** under consideration. For example, the system may be an ideal gas in a cylinder fitted with a piston, or an electric heating coil in a container of liquid.

In Chapter 4 we established the scientific meaning of work. Work is done when energy is transferred by mechanical means. We have just seen that heat is a transfer of energy due to a difference in temperature. Work and heat both involve a transfer of energy, but by different means.

We also know that the internal energy of a system is the total energy, kinetic and potential, of the various parts of the system. For a system consisting of an ideal gas, the internal energy is simply the total kinetic energy of all the atoms or molecules of the gas. For such a system, we would expect the internal energy to increase if heat were added to the gas, or if work were done on it. In both cases we are adding energy to the system. By the law of conservation of energy, this energy cannot just disappear; it must be transformed to another type of energy. It appears as an increase in the internal energy of the gas; that is, the total kinetic energy of the molecules is increased. But if the *total* kinetic energy of the molecules increases, their *average* kinetic energy is also increased. Because the average kinetic energy is a measure of temperature, the addition of energy to the ideal gas shows up as an increase in its temperature. We can express this transformation of energy as an equation.

The increase in internal energy of a system is equal to the sum of the heat added to the system and the work done on it.

The increase in internal energy is given the symbol ΔU, heat added is represented by Q and work done on the system by W. The equation is then

$$\Delta U = Q + W$$

Note the sign convention which has been adopted. A positive value of Q means that heat has been *added to the system*. A positive value of W means that work is done *on the system*. A positive value for ΔU means an increase in internal energy.

If the system does work, then we show this by writing $-W$. If heat leaves the system, we show this by writing $-Q$. Take care! There is an alternative sign convention that takes the work done *by* the system as a positive quantity. To avoid confusion, write down the sign convention you are using every time you quote the first law.

Let's see how the first law of thermodynamics applies to some simple processes. First, think about a change in the pressure and volume of a gas in a cylinder fitted with a piston. The cylinder and piston are insulated, so that no heat can enter or leave the gas. (The thermodynamic name for such a change is an **adiabatic** change. That is, no thermal energy is allowed to enter or leave the system.) If no heat enters or leaves the gas, Q in the first law equation is zero. Thus

$$\Delta U = 0 + W$$

If work is done on the gas by pushing the piston in, W is positive (remember the sign convention!) and ΔU will also be positive. That is, the internal energy increases, and because temperature is proportional to internal energy, the temperature of the gas rises. An adiabatic change can be achieved even if the cylinder and piston are not well insulated. Moving the piston rapidly, so that the heat has no time to flow in or out, is just as effective. You will have noticed that a cycle pump gets hot as a result of brisk pumping: this is because the gas in the pump is being compressed adiabatically. Work is being done on the gas, the pump strokes are too rapid for the heat to escape, and the internal energy, and hence the temperature, increases. Another example is the diesel engine, where air in the cylinder is compressed so rapidly that the temperature rises to a point that, when fuel is injected into the cylinder, it is above its ignition temperature.

Now think about an electric kettle containing water. Here the element provides heat to the system. The quantity Q in the first law equation is positive (the sign convention is 'heat added, Q positive'). No mechanical work is done on or by the water, so W in the first law equation is zero. Thus

$$\Delta U = Q + 0$$

The fact that Q is positive means that ΔU is also positive. Internal energy, and hence temperature, increases.

When a substance changes from solid to liquid, intermolecular bonds are broken, thus increasing the potential energy component of the internal energy. During the melting process, the temperature does not change, and therefore the kinetic energy of the molecules does not change. Most substances expand on melting, and thus external work is done. By the first law, thermal energy must be supplied to the system, and this thermal energy is the latent heat.

Volume changes associated with evaporation are much greater than those associated with melting. The external work done is much greater during vaporisation, and thus latent heat of vaporisation is much greater than latent heat of fusion.

Example

200 J of heat is added to a system, which does 150 J of work. Find the change in internal energy of the system.

We use the first law of thermodynamics in the form $\Delta U = Q + W$ with the sign convention that Q is positive if heat is supplied to the system and W is positive if work is done on the system. Here $Q = 200$ J and $W = -150$ J (the system is doing the work, hence the minus sign). Therefore $\Delta U = 200 - 150 = \mathbf{50\,J}$. This change in internal energy is an increase.

Now it's your turn

An **isothermal** change is one which takes place at constant temperature. Explain why, in any isothermal change, the change in internal energy is zero. In such a change, 200 J of heat is added to a system. How much work is done on or by the system?

Sections 6.5–6.6 Summary

- The internal energy of a system is the sum of the random kinetic and potential energies of the various parts of the system. For an ideal gas, the internal energy is the total kinetic energy of the molecules. Internal energy is a measure of the temperature of the system.
- The first law of thermodynamics expresses the law of conservation of energy. The increase in internal energy ΔU of a system is equal to the sum of the heat Q added to the system and the work W done on it:

$$\Delta U = Q + W \qquad \bullet$$

(Sign convention: positive Q, heat is added to the system; positive W, work is done on the system; positive ΔU, increase in internal energy.)

Sections 6.5–6.6 Questions

1 An ideal gas expands isothermally, doing 250 J of work. What is the change in internal energy? How much heat is absorbed in the process?

2 50 J of heat energy is supplied to a fixed mass of gas in a cylinder. The gas expands, doing 20 J of work. Calculate the change in internal energy of the gas.

6.7 Melting, boiling and evaporation

At the end of Section 6.7 you should be able to:
- distinguish between the processes of melting, boiling and evaporation
- relate the difference in the structures and densities of solids, liquids and gases to simple ideas of the spacing, ordering and motion of molecules
- distinguish between the structure of crystalline and non-crystalline solids with particular reference to metals, polymers and amorphous materials.

We have already used a simple kinetic model of matter to describe solids, liquids and gases in terms of the particles (ions, atoms or molecules) making up the particular sample we are thinking of.

Let us now consider heating a solid. We think of the molecules, atoms or ions having definite equilibrium positions, but vibrating about these positions with kinetic energy that depends on temperature. As we come to the melting point, the energy supplied does not increase the kinetic energy, and hence the temperature, of the solid, but instead is used to overcome the forces between the atoms. This means that the potential energy of the molecules is increased. The increase in potential energy is the latent heat of fusion of the solid. As heating continues, the atoms can now move about relatively freely in the liquid phase, but they are still close enough to experience interatomic forces, and still have potential energy associated with these forces as well as kinetic energy corresponding to the temperature.

Eventually the temperature is increased to the boiling point. Here, if atoms are to escape into the gaseous phase, the interatomic forces in the liquid must be overcome. The energy input here is the latent heat of vaporisation. This goes to moving the molecules far enough apart for the interatomic forces, and hence the associated potential energy, to be negligible. (Remember that one of the assumptions of the kinetic theory is that, for an ideal gas, there are no interatomic forces.) As we have seen, the latent heat of vaporisation is much greater than the latent heat of fusion.

Note that boiling occurs only at a particular temperature for a given atmospheric pressure. We have associated boiling with a definite input of energy, the latent heat. However, molecules are lost from the surface of a liquid at any temperature by the process of evaporation. The molecules in a liquid do not all move with the same speed, as we know from our work on the kinetic theory. A molecule with a high enough speed may escape from the attractive forces of the molecules in the surface region of the liquid and leave the liquid entirely. This process can take place at any temperature. However, the higher the liquid temperature, the greater the number of very fast molecules, and the greater the rate of loss of molecules from the surface. This fits with the everyday observation that the evaporation rate increases with an increase in the temperature of the surroundings.

The loss of the fastest molecules means that the average speed of those remaining falls, with a corresponding fall in average kinetic energy and hence in temperature. This means that evaporation tends to cool the remaining liquid. This, too, is an everyday observation.

So, although both represent a change of state from liquid to vapour, we distinguish between evaporation and boiling by the fact that evaporation can take place at any temperature, whereas boiling takes place at a fixed temperature for a given pressure of the surroundings. Furthermore, evaporation takes place at the surface of the liquid, whereas boiling occurs in the body of the liquid.

In describing this simple model, the words 'atoms' and 'molecules' are often used interchangeably. More precisely, we ought to consider whether we are describing an element, for which the word 'atom' is usually appropriate (though some elements, such as oxygen O_2, exist as molecules), or a simple molecular compound, for which 'molecule' should be used. For example, in talking about the boiling of water, we are concerned with the forces between water molecules and the loss of water molecules from the surface.

The structure of solids, liquids and gases

Let us now try to get an idea of the spacing of the atoms or molecules in our simple model of matter. We can estimate this quantity for a solid by using known values of its density and the mass containing one mole (the molar mass). You will find that an answer of the order of 10^{-10}m will come up each time in calculations such as the Example on page 159, and we can take this as being the order of magnitude of the spacing of atoms in a solid. We can do similar calculations for liquids and gases, although of course the assumption about the cubic arrangement of atoms now does not make sense. In general, the average spacing of atoms in liquids is up to about twice that in solids, or still of the order of 10^{-10} m. For gases, the average spacing is of the order of 10^{-9} m.

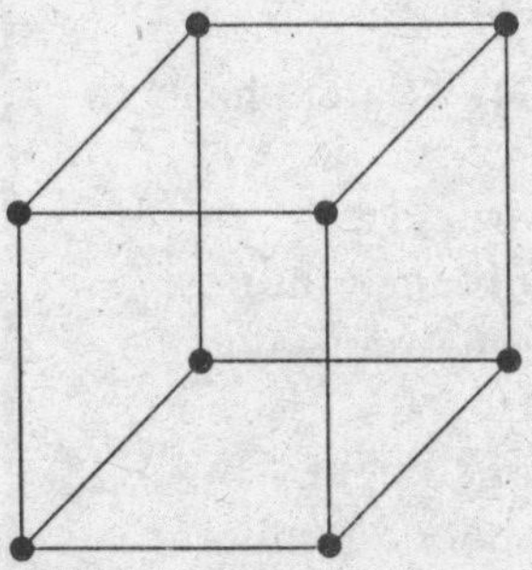

Example

Assuming that the atoms in iron are arranged in a simple cubic lattice (Figure 6.17), estimate the spacing between atoms in iron. The density of iron is 7900 kg m^{-3} and its molar mass is 56×10^{-3} kg.

One mole contains $N_A = 6.0 \times 10^{23}$ atoms. From the density value, we know that 1.0 m^3 has mass 7900 kg, which is 7900/56 $\times 10^{-3}$ = 1.4×10^5 mol. This contains $1.4 \times 10^5 \times 6.0 \times 10^{23} = 8.5 \times 10^{28}$ atoms. If these atoms are arranged regularly in a cubic lattice, along one side of the 1.0 m^3 cube there will be $\sqrt[3]{8.5 \times 10^{28}} = 4.4 \times 10^9$ atoms. But the side of the cube has length 1.0 m, so the interatomic spacing is $1/4.4 \times 10^9 = 2.3 \times 10^{-10}$ m. Because we have made an assumption about the simple cubic arrangement of atoms in iron (which is not, in fact, true), we should not give this estimate to more than one significant figure: interatomic spacing = $\mathbf{2 \times 10^{-10}}$ **m**. You can carry out this calculation for other metals.

Now it's your turn

Estimate the interatomic spacing in aluminium, for which density = 2700 kg m^{-3} and molar mass = 27×10^{-3} kg.

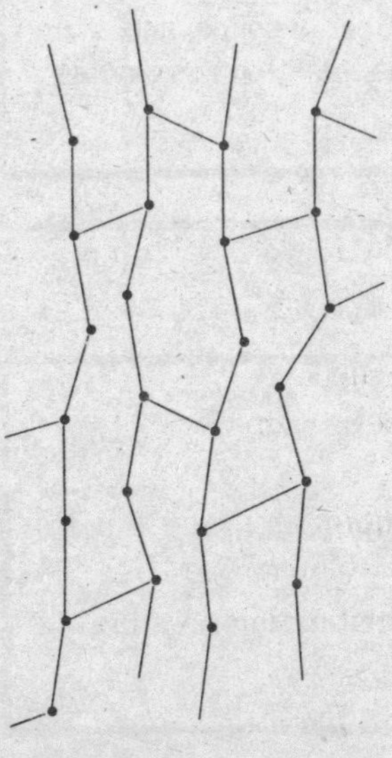

Figure 6.17 Unit cell of a simple cubic lattice – this cell is repeated regularly throughout the crystal

Figure 6.18 Linear polymer, showing long molecules with some cross-linking

Crystalline and non-crystalline solids

We have already made use of an assumption that, in metals like iron and aluminium, the atoms are arranged in a simple cubic lattice. The regular arrangement of atoms or ions in a lattice is a characteristic of **crystalline** solids. Crystalline solids include materials that we would immediately recognise as having a crystalline form, such as common salt or sugar, and also metals. The crystals of which a metal such as iron is made are not immediately visible. The normal form of a metal is polycrystalline; that is, consisting of a large number of small crystals. The atoms within each of these small crystals are arranged in a regular lattice, but the small crystals are arranged randomly with respect to one another. It is possible to grow large single crystals of metals in which a single lattice continues right through the solid in just the same way as in large crystals of common salt.

A **polymer** is a solid made up of very large molecules, usually a compound of carbon. These molecules often take the form of long chains, sometimes with cross-links to neighbouring chains (Figure 6.18). If the polymer shows a certain amount of order, for example if the chains are parallel, the material is sometimes referred to as a crystalline polymer, but the order is never as perfect as in a true crystalline material. If the chains are mixed up and tangled, the polymer is said to be **amorphous**.

An amorphous solid is one of which the main feature is a disordered arrangement of molecules. If the material shows any crystalline order, it is over a short range only. Amorphous polymers have already been mentioned, but the typical example of an amorphous material is glass. Glass is sometimes thought of as being a supercooled liquid, so that the arrangement of the silicate molecules of the glass can be thought of as being the positions in which the molecules of a liquid might be found at a particular instant. However, there is no mobility in the amorphous arrangement.

Section 6.7 Summary

- Melting, boiling and evaporation are all examples of changes of phase (solid to liquid, and liquid to vapour).
- All these changes of phase require an input of energy (latent heat) to overcome the interatomic forces.
- Boiling takes place at a fixed temperature for a particular atmospheric pressure, but evaporation occurs at all temperatures. Boiling occurs in the body of the liquid; evaporation occurs at its surface.
- In a solid, the interatomic spacing is of the order of 10^{-10} m.
- Crystalline solids show a regular, lattice arrangement of atoms.
- Polymers consist of long chains of molecules, which may be arranged with some regularity, or randomly.
- Amorphous substances have a disordered structure; any order is over a short range only.

Section 6.7 Question

Research values of densities and molar volumes for a number of solids, and also for liquids and gases, to find the order of magnitude of the interatomic spacing in these phases of matter.

6.8 Density and pressure

At the end of Section 6.8 you should be able to:
- define the term density
- derive, from definitions of pressure and density, the equation $p = \rho g h$
- use the equation $p = \rho g h$
- show an understanding of the origin of the upthrust acting on a body in a liquid.

We have already used the terms density and pressure in this chapter, and you will be familiar with them from your earlier work. Here we will bring them together to see an important link between them.

The density of a substance is defined as its mass per unit volume.

$\rho = m/V$

The symbol for density is ρ (Greek rho) and its SI unit is kg m^{-3}.

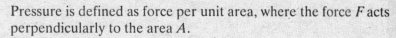

Example

An iron sphere of radius 0.18 m has mass 190 kg. Calculate the density of iron.

First calculate the volume of the sphere from $V = \frac{4}{3}\pi r^3$. This works out at 0.024 m^3. Application of the formula for density gives $\rho = \mathbf{7800\ kg\,m^{-3}}$.

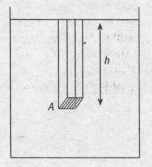

Figure 6.19 Column of liquid above the area A

Pressure is defined as force per unit area, where the force F acts perpendicularly to the area A.

$p = F/A$

The symbol for pressure is p and its SI unit is the pascal (Pa), which is equal to one newton per square metre (N m^{-2}).

The link between pressure and density comes when we deal with liquids, or with fluids in general. Consider a point at a depth h below the surface of a liquid in a container. What is the pressure due to the liquid? Very simply, the pressure is caused by the weight of the column of liquid above a small area at that depth, as shown in Figure 6.19. The weight of the column is $W = mg = \rho Ahg$, and the pressure is $W/A = \rho gh$.

$p = \rho gh$

The pressure is proportional to the depth below the surface of the liquid. If an external pressure, such as atmospheric pressure, acts on the surface of the liquid, this must be taken into account in calculating the absolute pressure. The absolute pressure is the sum of the external pressure and the pressure due to the depth below the surface of the liquid.

Example

Calculate the excess of pressure over atmospheric at a point 1.2 m below the surface of the water in a swimming pool. The density of water is 1.0×10^3 kg m^{-3}.

This is a straightforward calculation from $p = \rho gh$. Substituting, $p = 1.0 \times 10^3 \times 9.8 \times 1.2 = \mathbf{1.2 \times 10^4\ Pa}$.

If the total pressure had been required, this value would be added to atmospheric pressure p_A. Taking p_A to be 1.01×10^5 Pa, the total pressure is 1.13×10^5 Pa.

▶▶

Now it's your turn

Calculate the difference in blood pressure between the top of the head and the soles of the feet of a student 1.3 m tall, standing upright. Take the density of blood as $1.1 \times 10^3 \, \text{kg m}^{-3}$.

Upthrust

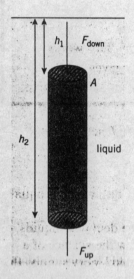

Figure 6.20 Origin of the buoyancy force (upthrust)

When an object is immersed in a fluid (a liquid or a gas), it appears to weigh less than when in a vacuum. It is easier to lift large stones under water than when they are out of the water. The reason for this is that immersion in the fluid provides an **upthrust** or buoyancy force.

We can see the reason for the upthrust when we think about an object, such as the cylinder in Figure 6.20, in water. Remember that the pressure in a liquid increases with depth. Thus, the pressure at the bottom of the cylinder is greater than the pressure at the top of the cylinder. This means that there is a bigger force acting upwards on the base of the cylinder, than there is acting downwards on the top. The difference in these forces is the upthrust or buoyancy force F_b. Looking at Figure 6.20, we can see that

$$F_b = F_{up} - F_{down}$$

and, since

$$p = \rho g H = F/A,$$
$$F_b = \rho g A (h_2 - h_1) = \rho g A h$$
$$= \rho g V$$

where h is the length of the cylinder, and V is its volume. The upthrust is simply the **weight of the liquid displaced** by the immersed object. This relation has been derived for a cylinder, but it will also apply to objects of any shape.

The rule that the upthrust acting on a body immersed in a fluid is equal to the weight of the fluid displaced is known as **Archimedes' principle**.

Section 6.8 Summary

- Density ρ is defined by the equation $\rho = m/V$, where m is the mass of an object and V is its volume.
- Pressure p is defined by the equation $p = F/A$, where F is the force acting perpendicularly to an area A.
- The total pressure p at a point at a depth h below the surface of a fluid of density ρ is $p = p_A + \rho g h$, p_A being the atmospheric pressure; the difference in pressure between the surface and a point at a depth h is $\rho g h$.
- The upthrust on a body immersed in a fluid is equal to the weight of the fluid displaced.

Section 6.8 Question

The water in a storage tank is 15 m above a water tap in the kitchen of a house. Calculate the pressure of the water leaving the tap. Density of water = $1.0 \times 10^3 \, kg \, m^{-3}$.

Exam-style Questions

1. Over the range of temperature between about 250 K and 450 K, the resistance R of a certain platinum wire is given by

   $$R = a + bT + cT^2$$

   where T is the thermodynamic temperature and

   $a = -1.32 \, \Omega$
   $b = +4.31 \times 10^{-2} \, \Omega \, K^{-1}$
   $c = -6.04 \times 10^{-6} \, \Omega \, K^{-2}$

   (a) A certain solid melts at 150.0 °C (degrees Celsius). Calculate the resistance of the wire at this temperature.
   (b) If the wire were used as a resistance thermometer, calculate the melting point of the solid on the empirical centigrade scale of this thermometer.

2. Consider one mole of gas at standard temperature and pressure (273 K and $1.01 \times 10^5 \, Pa$).
   (a) Calculate:
   (i) the number of molecules there are per cubic metre of this gas (this is known as the Loschmidt number N_L),
   (ii) the average distance apart of the molecules.
   (b) If the diameter of a molecule of this gas is about $2.5 \times 10^{-10} \, m$, estimate the fraction of each cubic metre occupied by matter.

3. In the ideal gas equation $pV = nRT$, could one use values of Celsius temperature instead of thermodynamic temperature by using an appropriate value R^* of the molar gas constant?

4. A volume of $1.50 \times 10^3 \, m^3$ of hydrogen, at a pressure of $2.00 \times 10^5 \, Pa$ and a temperature of 30 °C, is heated until both volume and pressure are doubled. Calculate:
   (a) the final temperature,
   (b) the mass of hydrogen. (mass of one mole of hydrogen = $2.00 \times 10^{-3} \, kg$)

5. Two containers, each of equal, constant volume, are connected by a tube of negligible volume. When the temperature of both containers is 27 °C, the pressure of the gas in the containers is $2.1 \times 10^5 \, Pa$. One container is heated to a temperature of 127 °C, whilst the other is maintained at 27 °C. Calculate the final pressure in the containers.

6. The kinetic theory leads to the expression

   $$\langle E \rangle = \frac{3}{2} kT$$

   for the average kinetic energy of a molecule of a gas. The constant k does not depend on the type of molecule. Can this result really be true both for hydrogen and chlorine? (The mass of a chlorine molecule is about 35 times that of a hydrogen molecule.) Explain why.

7. The mass of an oxygen molecule is 16 times the mass of a hydrogen molecule. Equal masses of hydrogen and oxygen gases are mixed, and are allowed to come to equilibrium. Calculate:
   (a) the ratio of numbers of molecules,
   (b) the ratio of the average kinetic energy per molecule,
   (c) the ratio of pressures exerted on the walls of the container.

8. An ideal gas is compressed adiabatically to one-third of its volume. The work done on the gas is 500 J.
   (a) State how much heat enters or leaves the gas.
   (b) State the change in internal energy of the gas.
   (c) What happens to the temperature of the gas?

7. Electricity

Electricity is so vital to modern life that physicists and engineers need to understand it fully so that we can enjoy all its benefits. The aim of this chapter is to consider fundamental ideas about electric current and electric circuits in order to provide a basis for further study of electricity. We start by thinking about current as the flow of charge. We link the idea of potential difference to energy changes in a circuit. Resistance comes in as a measure of opposition to the flow of charge. The idea of resistance introduces Ohm's law. Kirchhoff's first and second laws are seen as the laws of conservation of charge and of energy, applied to simple electric circuits. Moving on to circuit theory, we shall derive the rules for combining resistors in series and in parallel, and also analyse complete circuits.

7.1 Charge and current

At the end of Section 7.1 you should be able to:

- show an understanding that electric current is the rate of flow of charged particles
- define charge and the coulomb
- recall and solve problems using the equation $Q = It$
- define potential difference and the volt
- recall and solve problems using $V = W/Q$
- define resistance and the ohm
- recall and solve problems using $V = IR$
- recall and solve problems using $P = VI$, $P = I^2R$ and $P = V^2/R$.

All matter is made up of tiny particles called atoms, each consisting of a positively charged nucleus with negatively charged electrons moving around it.

Charge is measured in units called **coulombs** (symbol C). The charge on an electron is -1.6×10^{-19} C. Normally atoms have equal numbers of positive and negative charges, so that their overall charge is zero. But for some atoms it is relatively easy to remove an electron, leaving an atom with an unbalanced number of positive charges. This is called a **positive ion**.

Atoms in metals have one or more electrons which are not held tightly to the nucleus. These **free** (or mobile) **electrons** wander at random throughout the metal. But when a battery is connected across the ends of the metal, the free electrons drift towards the positive terminal of the battery, producing an **electric current**.

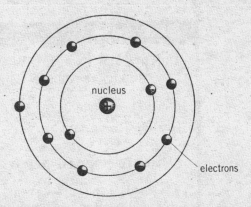

Figure 7.1 Atoms consist of a positively charged nucleus with negative electrons outside.

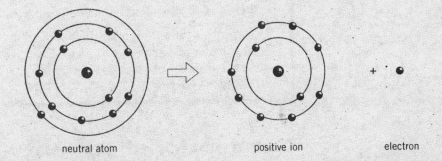

neutral atom   positive ion   electron

Figure 7.2 An atom with one or more electrons missing is a positive ion.

The size of the electric current is given by the rate of flow of charge, and is measured in units called amperes (or amps for short), with symbol A.
A current of 3 amperes means that 3 coulombs pass a point in the circuit every second. In 5 seconds, a total charge of 15 coulombs will have passed the point. So,

charge = current × time

or

$Q = It$

where the charge Q is in coulombs when the current I is in amperes and the time t is in seconds. This gives a definition of the coulomb as

The **coulomb** is that charge passing a point in a circuit when there is a current of one ampere for one second.

Example

The current in the filament of a torch bulb is 0.03 A. How much charge flows through the bulb in 1 minute?

Using $Q = It$, $Q = 0.03 \times 60$ (remember the time must be in seconds), so $Q = \mathbf{1.8\ C}$.

Now it's your turn

1 Calculate the current when a charge of 240 C passes a point in a circuit in a time of 2 minutes.

2 In a silver-plating experiment, 9.65×10^4 C of charge is needed to deposit a certain mass of silver. Calculate the time taken to deposit this mass of silver when the current is 0.20 A.

3 The current in a wire is 200 mA. Calculate:
   (a) the charge which passes a point in the wire in 5 minutes,
   (b) the number of electrons needed to carry this charge.
   (electron charge $e = -1.6 \times 10^{-19}$ C)

Conventional current

Early studies of the effects of electricity led scientists to believe that an electric current is the flow of 'something'. In order to develop a further understanding of electricity, they needed to know the direction of flow. It was decided that this flow in the circuit should be from the positive terminal of the battery to the negative. This current is called the **conventional current**, and is in the direction of flow of positive charge. We now know, in a metal, that the electric current is the flow of electrons in the opposite direction, from the negative terminal to the positive terminal. However, laws of electricity had become so firmly fixed in the minds of people that the idea of conventional current has persisted. But be warned! Occasionally we need to take into account the fact that electron flow is in the opposite direction to conventional current.

Potential difference

A cell makes one end of the circuit positive and the other negative. The cell is said to set up a **potential difference** across the circuit. Potential difference (p.d. for short) is measured in volts (symbol V), and is often called the voltage. You should never talk about the potential difference or voltage *through* a device, because it is in fact a difference *across* the ends of the device. The potential difference provides the energy to move charge through the device.

The potential difference between any two points in a circuit is a measure of the electrical energy transferred, or the work done, by each coulomb of charge as it moves from one point to the other. We already know that the unit of potential difference is the volt (V). Energy W is measured in joules, and charge Q in coulombs.

$$potential\ difference = \frac{energy\ transferred\ (or\ work\ done)}{charge}$$

or

$$V = \frac{W}{Q}$$

We can turn this relation round to get an expression for the electrical energy transferred or converted when a charge Q is moved through a potential difference V

$$energy\ transferred\ (work\ done) = potential\ difference \times charge$$

$$W = VQ$$

The relation gives a definition of the volt as

One **volt** is the potential difference between two points when one joule of energy is transferred by one coulomb passing from one point to the other.

In, Figure 7.3, one lamp is connected to a 240 V mains supply and the other to a 12 V car battery. Both lamps have the same current yet the 240 V lamp glows more brightly. This is because the energy supplied to each coulomb of charge in the 240 V lamp is 20 times greater than for the 12 V lamp.

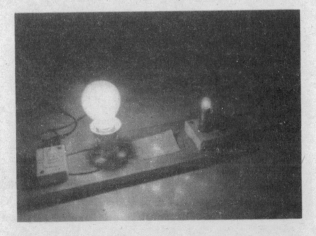

Figure 7.3 A 240 V, 100 W mains lamp is much brighter than a 12 V, 5 W car light, but both have the same current. (**Do not try this experiment yourself** as it involves a large voltage.)

Example

Electrons in a particular television tube are accelerated by a potential difference of 20 kV between the filament and the screen. The charge of the electron is -1.6×10^{-19} C. Calculate the gain in kinetic energy of each electron.

Since $V = W/Q$, then $W = VQ$. The electrical energy transferred to the electron shows itself as the kinetic energy of the electron. Thus,

kinetic energy $= VQ = 20 \times 10^3 \times 1.6 \times 10^{-19}$
$$= 3.2 \times 10^{-15} \text{ J}.$$

(Don't forget to turn the 20 kV into volts.)

Now it's your turn

1 An electron in a particle accelerator is said to have 1 MeV (1 mega electronvolt) of energy when it has been accelerated through a potential difference of 1 million volts. Calculate the energy, in joules, gained by the electron.

2 A torch bulb is rated 2.2 V, 0.25 A. Calculate:
(a) the charge passing through the bulb in one second,
(b) the energy transferred by the passage of each coulomb of charge.

Electrical power

Remember that power P is the rate of doing work, or of transferring energy. Remember also that $V = W/Q$. Divide each term on the right-hand side of this equation by time t, so that $V = (W/t)/(Q/t)$. W/t is power P, and Q/t is current I, so

$$potential\ difference = \frac{power}{current}$$

or

$$V = \frac{P}{I}$$

(W/Q is a quotient, so dividing W by t and dividing Q by t at the same time does not affect the final value because the t terms cancel each other out.)

The power is measured in watts (W) when the potential difference is in volts (V) and the current is in amperes (A). A voltmeter can measure the p.d. across a device and an ammeter the current through it; the equation above can then be used to calculate the power in the device.

Resistance

Connecting wires in circuits are often made from copper, because copper offers little opposition to the movement of electrons. The copper wire is said to have a low electrical resistance. In other words, copper is a good conductor.

Some materials, such as plastics, are poor conductors. These materials are said to be insulators, because under normal circumstances they conduct little or no current.

The **resistance** R of a wire is defined as the ratio of the potential difference V across the wire to the current I in it.

or

$$R = \frac{V}{I}$$

where the resistance is in **ohms** when the potential difference is in volts and the current in amperes. The symbol for ohms is the Greek capital letter omega, Ω. We have defined resistance in terms of a wire, but many devices have resistance. The general term for such a device is a **resistor**. (Note that the resistance of a resistor is measured in ohms, just as the volume of a tank is measured in m^3. We do not refer to the 'm^3' of a tank, nor to the 'ohms' of a resistor.)

The relation between resistance, potential difference and current means that, for a given potential difference, a high resistance means a small current, while a low resistance means a large current.

Example

The current in an electric immersion heater in a school experiment is 6.3 A when the p.d. across it is 12 V. Calculate the resistance of the heater.

Since $R = V/I$, the resistance $R = 12/6.3 = $ **1.9 Ω**.

Now it's your turn

The current in a light-emitting diode is 20 mA when it has a potential difference of 2.0 V across it. Calculate its resistance.

Electrical heating

When an electric current passes through a resistor, it gets hot. This heating effect is sometimes called **Joule heating**. The electrical power P produced (dissipated) is given by $V = P/I$, which can be rearranged to give $P = VI$.

We can obtain alternative expressions for power in terms of the resistance R of the resistor. Since $V = IR$, then

$$P = I^2 R$$

and

$$P = \frac{V^2}{R}$$

For a given resistor, the power dissipated depends on the square of the current. This means that if the current is doubled, the power will be four times as great. Similarly, a doubling of voltage will increase the power by a factor of four.

Example

An electric immersion heater used in a school experiment has a current of 6.3 A when the p.d. across it is 12 V. Calculate the power of the heater.

Since $P = VI$, power $= 12 \times 6.3 = 76\,\text{W}$.

Now it's your turn

1 Show that a 100 W lamp connected to a mains supply of 240 V will have the same current as a 5 W car lamp connected to a 12 V battery. (See Figure 7.3 on page 167.)

2 An electric kettle has a power of 2.2 kW at 240 V. Calculate:
   (a) the current in the kettle,
   (b) the resistance of the kettle element.

Section 7.1 Summary

- Electric current is the rate of flow of charge: $I = Q/t$
- Conventional current is a flow of positive charge from positive to negative. In metals, current is carried by electrons, which travel from negative to positive.
- Potential difference (or voltage) measures the electrical energy transferred by each coulomb of charge: $V = W/Q$
- Resistance R of a resistor is defined as: $R = V/I$
- Electrical power $P = VI = I^2 R = V^2/R$

Section 7.1 Questions

1 A 240 V heater takes a current of 4.2 A. Calculate:
  (a) the charge that passes through the heater in 3 minutes,
  (b) the rate at which heat energy is produced by the heater,
  (c) the resistance of the heater.

2 A small torch has a 3.0 V battery connected to a bulb of resistance 15 Ω.
  (a) Calculate:
    (i) the current in the bulb,
    (ii) the power delivered to the bulb.

(b) The battery supplies a constant current to the bulb for 2.5 hours. Calculate the total energy delivered to the bulb.

3 The capacity of storage batteries is rated in ampere-hours (A h). An 80 A h battery can supply a current of 80 A for 1 hour, or 40 A for 2 hours, and so on. Calculate the total energy, in J, stored in a 12 V, 80 A h car battery.

4 An electric kettle is rated at 2.2 kW, 240 V. The supply voltage is reduced from 240 V to 230 V. Calculate the new power of the kettle.

7.2 More about resistance

At the end of Section 7.2 you should be able to:
  ■ sketch and explain the I–V characteristics of a metallic conductor at constant temperature, a semiconductor diode and a filament lamp
  ■ state Ohm's law
  ■ recall and solve problems using $R = \rho l/A$.

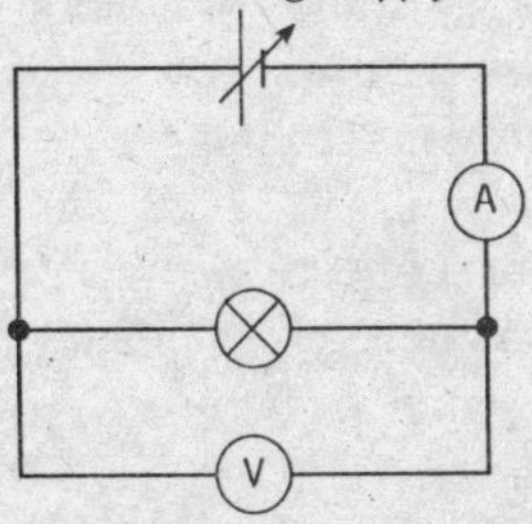

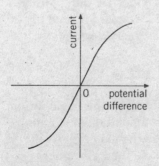

Figure 7.4 Circuit for plotting graphs of current against voltage

Figure 7.5 Current against voltage for a filament lamp

Ohm's law

The relation between the potential difference across an electrical component and the current through it may be investigated using the circuit of Figure 7.4. For example, if a filament lamp is to be investigated, adjust the voltage across the lamp and measure the corresponding currents and voltages. The variation of current with potential difference is shown in Figure 7.5.

The resistance R of the lamp can be calculated from $R = V/I$. At first the resistance is constant (where the graph is a straight line), but then the resistance increases with current (where the graph curves).

If the lamp is replaced by a length of constantan wire, the graph of the results is as shown in Figure 7.6. It is a straight line through the origin. This shows that for constantan wire, the current is proportional to the voltage. The resistance of the wire is found to stay the same as the current increases. The difference between Figures 7.5 and 7.6 is that the temperature of the constantan wire was constant for all currents used in the experiment, whereas the temperature of the filament of the lamp increased to about 1500 °C.

Graphs like Figure 7.6. would be obtained for wires of any metal, provided that the temperature of the wires did not change during the experiment. The graph illustrates a law discovered by the German scientist Georg Simon Ohm (Figure 7.7). (Ohm's name is now used for the unit of resistance.)

Figure 7.6 Current against voltage for a constantan wire

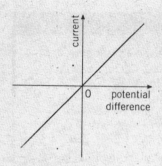

Figure 7.7 Georg Ohm

Ohm's law states that, for a conductor at constant temperature, the current in the conductor is proportional to the potential difference across it.

Conductors whose current against voltage graphs are a straight line through the origin, like that in Figure 7.6, are said to obey Ohm's law. It is found that Ohm's law applies to metal wires, provided that the current is not too large. What does 'too large' mean here? It means that the current must not be so great that there is a pronounced heating effect, causing an increase in temperature of the wire.

A lamp filament consists of a thin metal wire. Why does it not obey Ohm's law? (Figure 7.5 shows that the current against voltage graph is not a straight line.) This is because, as stated previously, the temperature of the filament does not remain constant. The increase in current causes the temperature to increase so much that the filament glows.

Current–voltage graphs

When other devices are tested in the same way, current–voltage graphs like those in Figures 7.8 to 7.10 are obtained.

We have already met the thermistor as a type of electrical resistance thermometer (Chapter 6). Thermistors are made from semiconducting material. If a thermistor is held at constant temperature in a water bath, its current–voltage graph is a straight line. At a higher temperature, the gradient is steeper, showing that the resistance is lower. Figure 7.8 illustrates this. The resistance of a sample of semiconductor decreases as the temperature increases. This is the opposite behaviour to a metal: the resistance of a metal wire increases as the temperature increases. Thermistors are used as temperature sensors, making use of the fact that their resistance changes considerably with temperature.

Light-dependent resistors (LDRs) are also made from semiconducting material. As their name suggests, their resistance can be altered by a change in light intensity: the brighter the light, the lower the resistance (Figure 7.9). LDRs are used in light sensors and camera exposure meters.

Diodes are also made from semiconducting material. The diode conducts when the current is in the direction of the arrowhead on the symbol. This condition is called **forward bias**. When the voltage is reversed, there is negative bias. This is called **reverse bias**. Figure 7.10 shows this important difference in the current–voltage graph when the voltage is reversed. Diodes are used to change alternating current into direct current in devices called **rectifiers**.

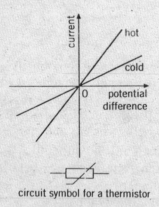

circuit symbol for a thermistor

Figure 7.8 Current against voltage graphs for a thermistor

Resistance and temperature

All solids are made up of atoms that constantly vibrate about their equilibrium positions. The higher the temperature, the greater the amplitude of vibration.

Electric current is the flow of free electrons through the material. As the electrons move, they collide with the vibrating atoms, so their movement is impeded. The more the atoms vibrate, the greater is the chance of collision. This means that the current is less and the resistance is greater.

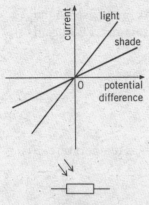

circuit symbol for a light-dependent resistor

Figure 7.9 Current against voltage graphs for a light-dependent resistor

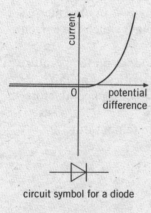

circuit symbol for a diode

Figure 7.10 Current against voltage graph for a diode

A temperature rise can cause an increase in the number of free electrons. If there are more electrons free to move, this may outweigh the effect due to the vibrating atoms, and thus the flow of electrons, or the current, will increase. The resistance is therefore reduced. This is the case in semiconductors. Insulators, too, show a reduction in resistance with temperature rise.

For metals there is no increase in the number of free electrons. The increased amplitude of vibration of the atoms makes the resistance of metals increase with temperature.

Resistivity

All materials have some resistance to a flow of charge. A potential difference across the material causes free charges inside to accelerate. As the charges move through the material, they collide with the atoms of the material which get in their way. They transfer some or all of their kinetic energy, and then accelerate again. It is this transfer of energy on collision that causes electrical heating.

As you might guess, the longer a wire, the greater its resistance. This is because the charges have further to go through the material; there is more chance of collisions with the atoms of the material. In fact the resistance is proportional to the length of the wire, or $R \propto l$.

Also, the thicker a wire is, the smaller its resistance will be. This is because there is a bigger area for the charges to travel through, with less chance of collision. In fact the resistance is inversely proportional to the cross-sectional area of the wire, or $R \propto 1/A$. These relations are illustrated in Figures 7.11 and 7.12.

Figure 7.11 The longer the room, the greater the resistance the waiter meets.

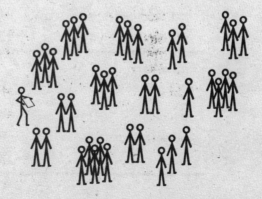

Figure 7.12 The wider the room, the easier it is for the waiter to pass through.

Finally, the resistance depends on the type of material. As previously stated, copper is a good conductor, whereas plastics are good insulators. Putting all of this together gives

$$R = \frac{\rho l}{A}$$

where ρ is a constant for a particular material at a particular temperature. ρ is called the **resistivity** of the material at that temperature.

> The resistivity ρ of a material is numerically equal to the resistance between opposite faces of a cube of the material, of unit length and unit cross-sectional area.

So if $\rho = RA/l$ and if the resistance is in ohms, the cross-sectional area in metres squared and the length in metres, then the resistivity is in ohm metres (Ω m). Remember that A is the cross-sectional area through which the current is passing, not the surface area.

We have already seen that the resistance of a wire depends on temperature. Thus, resistivity also depends on temperature. The resistivities of metals increase with increasing temperature, and the resistivities of semiconductors decrease very rapidly with increasing temperature.

The values of the resistivity of some materials at room temperature are given in Table 7.1.

Table 7.1 Resistivities at room temperature

material	resistivity/Ω m
metals:	
copper	1.7×10^{-8}
gold	2.4×10^{-8}
aluminium	2.6×10^{-8}
semiconductors:	
germanium (pure)	0.6
silicon (pure)	2.3×10^{3}
insulators:	
glass	about 10^{12}
perspex	about 10^{13}
polyethylene	about 10^{14}
sulfur	about 10^{15}

Note the enormous range of resistivities spanned by the materials in this list – a range of 23 orders of magnitude, from $10^{-8}\,\Omega\,\mathrm{m}$ to $10^{15}\,\Omega\,\mathrm{m}$.

Note, too, that the resistivity is a property of a material, while the resistance is a property of a particular wire or device.

Example

Calculate the resistance per metre at room temperature of a constantan wire of diameter 1.25 mm. The resistivity of constantan at room temperature is $5.0 \times 10^{-7}\,\Omega\,\mathrm{m}$.

The cross-sectional area of the wire is calculated using πr^2.

Area $= \pi\,(1.25 \times 10^{-3}/2)^2$

(Don't forget to change the units from mm to m.)

The resistance per metre is given by R/l, and $R/l = \rho/A$. So
resistance per metre $= 5.0 \times 10^{-7}/\pi\,(1.25 \times 10^{-3}/2)^2$
$= \mathbf{0.41\ \Omega\ m^{-1}}$

Now it's your turn

1 Find the length of copper wire, of diameter 0.63 mm, which has a resistance of $1.00\,\Omega$. The resistivity of copper at room temperature is $1.7 \times 10^{-8}\,\Omega\,\mathrm{m}$.

2 Find the diameter of a copper wire which has the same resistance as an aluminium wire of equal length and diameter 1.20 mm. The resistivities of copper and aluminium at room temperature are $1.7 \times 10^{-8}\,\Omega\,\mathrm{m}$ and $2.6 \times 10^{-8}\,\Omega\,\mathrm{m}$ respectively.

Section 7.2 Summary

- Ohm's law: for a conductor at constant temperature, the current in the conductor is proportional to the potential difference across it.
- The resistance of a metallic conductor increases with increasing temperature; the resistance of a semiconductor decreases with increasing temperature.
- A diode has a low resistance when connected in forward bias, and a very high resistance in reverse bias.
- Resistivity ρ of a material is the resistance between opposite faces of a unit cube of the material: $R = \rho l/A$

Section 7.2 Questions

1 The element of an electric kettle has resistance $26\,\Omega$ at room temperature. The element is made of nichrome wire of diameter $0.60\,\text{mm}$ and resistivity $1.1 \times 10^{-6}\,\Omega\,\text{m}$ at room temperature. Calculate the length of the wire.

2 These are values of the current I through an electrical component for different potential differences V across it:

V/V 0 0.19 0.48 1.47 2.92 4.56 6.56 8.70
I/A 0 0.20 0.40 0.60 0.80 1.00 1.20 1.40

(a) Draw a diagram of the circuit that could be used to obtain these values.

(b) Calculate the resistance of the component at each value of current.

(c) Plot a graph to show the variation with current of the resistance of the component.

(d) Suggest what the component is likely to be, giving a reason for your answer.

3 The current in a $2.50\,\text{m}$ length of wire of diameter $1.5\,\text{mm}$ is $0.65\,\text{A}$ when a potential difference of $0.40\,\text{V}$ is applied between its ends. Calculate:

(a) the resistance of the wire,

(b) the resistivity of the material of the wire.

7.3 Electrical circuits

At the end of Section 7.3 you should be able to:

- recall and use appropriate circuit symbols to interpret and draw circuit diagrams
- recall Kirchhoff's first law and appreciate the link to conservation of charge
- define e.m.f. in terms of the energy transferred by a source in driving unit charge round a complete circuit
- distinguish between e.m.f. and p.d. in terms of energy considerations
- recall Kirchhoff's second law and appreciate the link to conservation of energy
- show an understanding of the effects of the internal resistance of a source of e.m.f. on the terminal potential difference and output power.

When reporting an electrical experiment, or describing a circuit, it is essential to know exactly how the components are connected. This could be done by taking a photograph, but this has disadvantages; the photograph on page 167, for example, is not clear and does not show all the components. Besides, you may not always have a camera with you. You could sketch a block diagram, in which the components are indicated as rectangular boxes labelled 'cell', 'ammeter', 'resistor', etc. The blocks would then be connected with lines to indicate the wiring. This is also unsatisfactory; it takes a lot of time to label all the boxes. It is much better to draw the circuit diagram using a set of symbols recognised by everyone, which do not need to be labelled.

Figure 7.13 shows the symbols that you are likely to need in school and college work, and which you will meet in examination questions. (You will have met many of them already.) It is important that you learn these so that

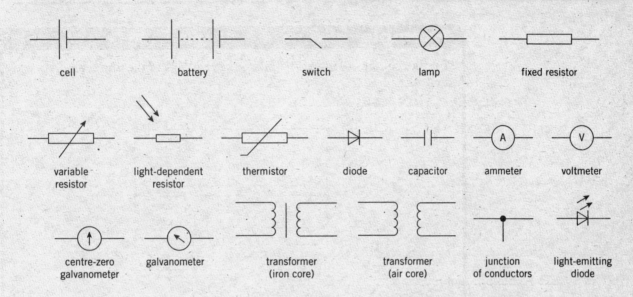

Figure 7.13 Circuit symbols

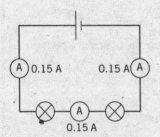

Figure 7.14 The current at each point in a series circuit is the same.

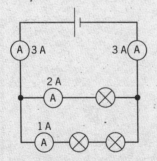

Figure 7.15 The current divides in a parallel circuit.

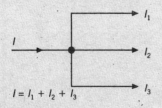

Figure 7.16

you can recognise them straight away. The only labels you are likely to see on them will be the values of the components, for example 1.5 V for a cell or 22 Ω for a resistor.

Series and parallel circuits: Kirchhoff's first law

A **series circuit** is one in which the components are connected one after another, forming one complete loop. You have probably connected an ammeter at different points in a series circuit to show that it reads the same current at each point (see Figure 7.14).

A **parallel circuit** is one where the current can take alternative routes in different loops. In a parallel circuit, the current divides at a junction, but the current entering the junction is the same as the current leaving it (see Figure 7.15). The fact that the current does not get 'used up' at a junction is because current is the rate of flow of charge, and charges cannot accumulate or get 'used up' at a junction. The consequence of this conservation of electric charge is known as **Kirchhoff's first law**. This law is usually stated as

The sum of the currents entering a junction in a circuit is always equal to the sum of the currents leaving it.

At the junction shown in Figure 7.16,

$$I = I_1 + I_2 + I_3$$

Example

For the circuit of Figure 7.17, state the readings of the ammeters A_1, A_2 and A_3.

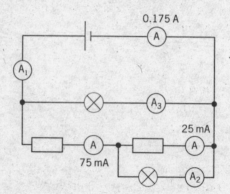

Figure 7.17

A_1 would read **175 mA**, as the current entering the power supply must be the same as the current leaving it.
A_2 would read $75 - 25 =$ **50 mA**, as the total current entering a junction is the same as the total current leaving it.
A_3 would read $175 - 75 =$ **100 mA**.

Now it's your turn

1 The lamps in Figure 7.18 are identical. There is a current of 0.50 A through the battery. What is the current in each lamp?

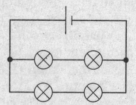

Figure 7.18

2 Figure 7.19 shows one junction in a circuit. Calculate the ammeter reading.

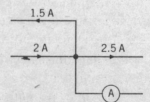

Figure 7.19

Electromotive force and potential difference

When charge passes through a power supply such as a battery, it **gains** electrical energy. The power supply is said to have an **electromotive force**, or **e.m.f.** for short. The electromotive force measures, in volts, the electrical energy gained by each coulomb of charge that passes through the power supply. Note that, in spite of its name, the e.m.f. is *not* a force. The energy gained by the charge comes from the chemical energy of the battery.

$$e.m.f. = \frac{(energy\ converted\ from\ other\ forms\ to\ electrical)}{charge}$$

When charge passes through a resistor, its electrical energy is converted to heat energy in the resistor. The resistor has a **potential difference (p.d.)** across it. The potential difference measures, in volts, the electrical energy lost by each coulomb of charge that passes through it.

$$p.d. = \frac{(energy\ converted\ from\ electrical\ to\ other\ forms)}{charge}$$

Conservation of energy: Kirchhoff's second law

Charge flowing round a circuit gains electrical energy on passing through the battery and loses electrical energy on passing through the rest of the circuit. From the law of conservation of energy, we know that the total energy must remain the same. The consequence of this conservation of energy is known as **Kirchhoff's second law**. This law may be stated as

The sum of the electromotive forces in a closed circuit is equal to the sum of the potential differences.

Figure 7.20 shows a circuit containing a battery, lamp and resistor in series. Applying Kirchhoff's second law, the electromotive force in the circuit is the e.m.f. E of the battery. The sum of the potential differences is the p.d. V_1 across the lamp plus the p.d. V_2 across the resistor. Thus, $E = V_1 + V_2$. If the current in the circuit is I and the resistances of the lamp and resistor are R_1 and R_2 respectively, the p.d.s can be written as $V_1 = IR_1$ and $V_2 = IR_2$, so $E = IR_1 + IR_2$.

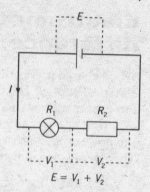

Figure 7.20

$$E = V_1 + V_2$$

It should be remembered that both electromotive force and potential difference have direction. This must be considered when working out the equation for Kirchhoff's second law. For example, in the circuit of Figure 7.21 two cells have been connected in opposition. Here the total electromotive force in the circuit is $E_1 - E_2$, and by Kirchhoff's second law $E_1 - E_2 = V_1 + V_2 = IR_1 + IR_2$.

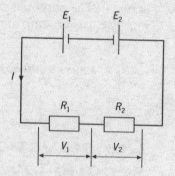

Figure 7.21

Internal resistance

When a car engine is started while the headlights are switched on, the headlights sometimes dim. This is because the car battery has resistance.

All power supplies have some resistance between their terminals, called **internal resistance**. This causes the charge circulating in the circuit to dissipate some electrical energy in the power supply itself. The power supply becomes warm when it delivers a current.

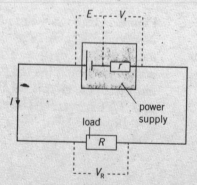

Figure 7.22

Figure 7.22 shows a power supply which has e.m.f. E and internal resistance r. It delivers a current I when connected to an external resistor of resistance R. (called the load). V_R is the potential difference across the load, and V_r is the potential difference across the internal resistance. By Kirchhoff's second law,

$$E = V_R + V_r$$

The potential difference V_R across the load is thus given by

$$V_R = E - V_r$$

V_R is called the **terminal potential difference**.

> The terminal potential difference is the p.d. between the terminals of a cell when a current is being delivered.

The terminal potential difference is always less than the electromotive force when the power supply delivers a current. This is because of the potential difference across the internal resistance. The potential difference across the internal resistance is sometimes called the **lost volts**.

lost volts = e.m.f. − terminal p.d.

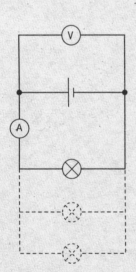

Figure 7.23 Effect of circuit current on terminal potential difference

In contrast, the electromotive force is the terminal potential difference when the cell is on open circuit (when it is delivering no current). This e.m.f. may be measured by connecting a very high resistance voltmeter across the terminals of the cell.

You can use the circuit in Figure 7.23 to show that the greater the current delivered by the power supply, the lower its terminal potential difference. As more lamps are connected in parallel to the power supply, the current increases and the lost volts, given by

lost volts = current × internal resistance

will increase. Thus the terminal potential difference decreases.

To return to the example of starting a car with its headlights switched on, a large current (perhaps 100 A) is supplied to the starter motor by the battery. There will then be a large potential difference across the internal resistance; that is, the lost voltage will be large. The terminal potential difference will drop and the lights will dim.

In the terminology of Figure 7.22, $V_R = IR$ and $V_r = Ir$, so $E = V_R + V_r$ becomes $E = IR + Ir$, or $E = I(R + r)$.

The maximum current that a power supply can deliver will be when its terminals are short-circuited by a wire of negligible resistance, so that $R = 0$. In this case, the potential difference across the internal resistance will equal the e.m.f. of the cell. The terminal p.d is then zero. **Warning: do not try out this experiment, as the wire gets very hot; there is also a danger of the battery exploding.**

Quite often, in problems, the internal resistance of a supply is assumed to be negligible, so that the potential difference V_R across the load is equal to the e.m.f. of the power supply.

Effect of internal resistance on power from a battery

The power delivered by a battery to a variable external load resistance can be investigated using the circuit of Figure 7.24. Readings of current I and potential difference V_R across the load are taken for different values of the variable load resistor. The product $V_R I$ gives the power dissipated in the load, and the quotient V_R/I gives the load resistance R.

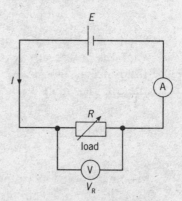

Figure 7.24 Circuit for investigating power transfer to an external load

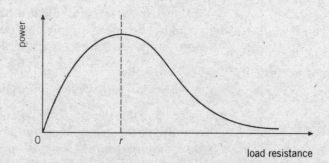

Figure 7.25 Graph of power delivered to external load against load resistance

Figure 7.25 shows the variation with load resistance R of the power $V_R I$ dissipated. The graph indicates that there is a maximum power delivered by the battery at one value of the external resistance. This value is equal to the internal resistance r of the battery.

A battery delivers maximum power to a circuit when the load resistance of the circuit is equal to the internal resistance of the battery.

Example

A high-resistance voltmeter reads 13.0 V when it is connected across the terminals of a battery. The voltmeter reading drops to 12.0 V when the battery delivers a current of 3.0 A to a lamp. State the e.m.f. of the battery. Calculate the potential difference across the internal resistance (the lost volts) when the battery is connected to the lamp. Calculate the internal resistance of the battery.

The e.m.f. is **13.0 V**, since this is the voltmeter reading when the battery is delivering negligible current.
   Using $V_r = E - V_R$, lost volts $= V_r = 13.0 - 12.0 = $ **1.0 V**.
   Using $V_r = Ir$, $r = 1.0/3.0 = $ **0.33 Ω**.

Now it's your turn

1 Three cells, each of e.m.f. 1.5 V, are connected in series to a 15 Ω light bulb. The current in the circuit is 0.27 A. Calculate the internal resistance of each cell.

2 A cell of e.m.f. 1.5 V has an internal resistance of 0.50 Ω. Calculate the maximum current it can deliver. Under what circumstances does it deliver this maximum current? Calculate also the maximum power it can deliver to an external load. Under what circumstances does it deliver this maximum power?

Section 7.3 Summary

■ At any junction in a circuit, the total current entering the junction is equal to the current leaving it. This is Kirchhoff's first law, and is a consequence of the law of conservation of charge.
■ The electromotive force (e.m.f.) of a supply measures the electrical energy gained per unit of charge passing through the supply.
■ The potential difference (p.d.) across a resistor measures the electrical energy converted per unit of charge passing through the resistor.
■ In any closed loop of a circuit, the sum of the electromotive forces is equal to the sum of the potential differences. This is Kirchhoff's second law, and is a consequence of the law of conservation of energy.
■ The voltage across the terminals of a supply (the terminal p.d.) is always less than the e.m.f. of the supply when the supply is delivering a current, because of the lost volts across the internal resistance.
■ For a supply of e.m.f. E which has internal resistance r, $E = I(R + r)$ where R is the external circuit resistance and I is the current in the supply.
■ A supply delivers maximum power to a load when the load resistance is equal to the internal resistance of the supply.

Section 7.3 Questions

1 The internal resistance of a dry cell increases gradually with age, even if the cell is not being used. However, the e.m.f. remains approximately constant. You can check the age of a cell by connecting a low-resistance ammeter across the cell and measuring the current. For a new 1.5 V cell of a certain type, the short-circuit current should be about 30 A.

(a) Calculate the internal resistance of a new cell.

(b) A student carries out this test on an older cell, and finds the short-circuit current to be only 5 A. Calculate the internal resistance of this cell.

2 A torch bulb has a power supply of two 1.5 V cells connected in series. The potential difference across the bulb is 2.2 V, and it dissipates energy at the rate of 550 mW. Calculate:

(a) the current through the bulb,

(b) the internal resistance of each cell,

(c) the heat energy dissipated in each cell in two minutes.

3 Two identical light bulbs are connected first in series, and then in parallel, across the same battery (assumed to have negligible internal resistance). Use Kirchhoff's laws to decide which of these connections will give the greater total light output.

7.4 More on series and parallel circuits

At the end of Section 7.4 you should be able to:

- derive, using Kirchhoff's laws, a formula for the combined resistance of two or more resistors in series
- solve problems using the formula for the combined resistance of two or more resistors in series
- derive, using Kirchhoff's laws, a formula for the combined resistance of two or more resistors in parallel
- solve problems using the formula for the combined resistance of two or more resistors in parallel
- apply Kirchhoff's laws to solve simple circuit problems
- show an understanding of the use of a potential divider circuit as a source of variable p.d.
- explain the use of thermistors and light-dependent resistors in potential dividers to provide a potential difference which is dependent on temperature and illumination respectively
- recall and solve problems using the principle of the potentiometer as a means of comparing potential differences.

Resistors in series

Figure 7.26 shows two resistors of resistances R_1 and R_2 connected in series, and a single resistor of resistance R equivalent to them. The current I in the resistors, and in their equivalent single resistor, is the same.

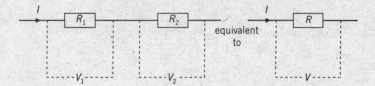

Figure 7.26 Resistors in series

The total potential difference V across the two resistors must be the same as that across the single resistor. If V_1 and V_2 are the potential differences across each resistor,

$$V = V_1 + V_2$$

But since potential difference is given by multiplying the current by the resistance,

$$IR = IR_1 + IR_2$$

Dividing by the current I,

$$R = R_1 + R_2$$

This equation can be extended so that the equivalent resistance R of several resistors connected in series is given by the expression

$$R = R_1 + R_2 + R_3 + \dots$$

Thus

> The combined resistance of resistors in series is the sum of all the individual resistances.

Resistors in parallel

Now consider two resistors of resistance R_1 and R_2 connected in parallel, as shown in Figure 7.27. The current through each will be different, but they will each have the same potential difference. The equivalent single resistor of resistance R will have the same potential difference across it, but the current will be the total current through the separate resistors.

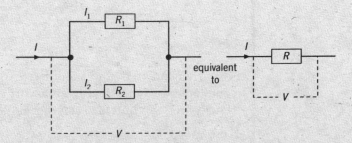

Figure 7.27 Resistors in parallel

By Kirchhoff's first law,

$$I = I_1 + I_2$$

and, using Ohm's law, $I = V/R$, so

$$V/R = V/R_1 + V/R_2$$

Dividing by the potential difference V,

$$1/R = 1/R_1 + 1/R_2$$

This equation can be extended so that the equivalent resistance R of several resistors connected in parallel is given by

$$\frac{1}{R} = \frac{1}{R_1} + \frac{1}{R_2} + \frac{1}{R_3} + \ldots$$

Thus

The reciprocal of the combined resistance of resistors in parallel is the sum of the reciprocals of all the individual resistances.

Note that:

1 For two identical resistors in parallel, the combined resistance is equal to half of the value of each one.
2 For resistors in parallel, the combined resistance is always less than the value of the smallest individual resistance.

Example

Calculate the equivalent resistance of the arrangement of resistors in Figure 7.28.

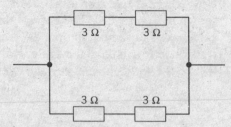

Figure 7.28

The arrangement is equivalent to two $6\,\Omega$ resistors in parallel, so the combined resistance R is given by $1/R = 1/6 + 1/6 = 2/6$.
(Don't forget to find the reciprocal of this value.)

Thus $R = 3\,\Omega$.

Now it's your turn

1 Calculate the equivalent resistance of the arrangement of resistors in Figure 7.29. *Hint*: First find the resistance of the parallel combination.

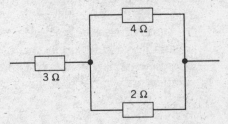

Figure 7.29

2 Calculate the effective resistance between the points A and B in the network in Figure 7.30.

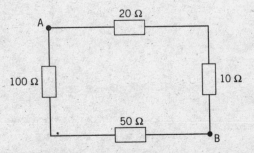

Figure 7.30

Potential dividers and potentiometers

Two resistors connected in series with a cell each have a potential difference. They may be used to divide the e.m.f. of the cell. This is illustrated in Figure 7.31.

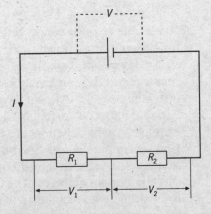

Figure 7.31 The potential divider

The current in each resistor is the same, because they are in series. Thus $V_1 = IR_1$ and $V_2 = IR_2$. Dividing the first equation by the second gives $V_1/V_2 = R_1/R_2$. The ratio of the voltages across the two resistors is the same as the ratio of their resistances. If the potential difference across the combination were 12 V and R_1 were equal to R_2, then each resistor would have 6 V across it. If R_1 were twice the magnitude of R_2, then V_1 would be 8 V and V_2 would be 4 V.

A **potentiometer** is a continuously variable potential divider. In Section 7.2, a variable voltage supply was used to vary the voltage across different circuit components. A variable resistor, or rheostat, may be used to produce a continuously variable voltage.

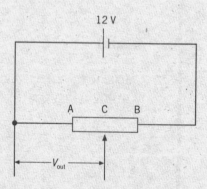

Figure 7.32 Potentiometer circuit

Such a variable resistor is shown in Figure 7.32. The fixed ends AB are connected across the battery so that there is the full battery voltage across the whole resistor. As with the potential divider, the ratio of the voltages across AC and CB will be the same as the ratio of the resistances of AC and CB. When the sliding contact C is at the end B, the output voltage V_{out} will be 12 V. When the sliding contact is at end A, then the output voltage will be zero. So, as the sliding contact is moved from A to B, the output voltage varies continuously from zero up to the battery voltage. In terms of the terminal p.d. V of the cell, the output V_{out} of the potential divider is given by

$$V_{out} = \frac{VR_1}{(R_1 + R_2)}$$

Figure 7.33 Internal and external views of a potentiometer

A variable resistor connected in this way is called a potentiometer. A type of potentiometer is shown in Figure 7.33. Note the three connections.

If a device with a variable resistance is connected in series with a fixed resistor, and the combination is connected to a cell or battery to make a potential divider, then we have the situation of a potential divider that is variable between certain limits. The device of variable resistance could be, for example, a light-dependent resistor or a thermistor, as shown on pages 172–173. Changes in the illumination or the temperature cause a change in the resistance of one component of the potential divider, so that the potential difference across this component changes. The change in the potential difference can be used to operate control circuitry if, for example, the illumination becomes too low or too high, or the temperature falls outside certain limits.

A potentiometer can also be used as a means of comparing potential differences. The circuit of Figure 7.34 illustrates the principle. In this circuit the variable potentiometer resistor consists of a length of uniform resistance wire, stretched along a metre rule. Contact can be made to any point on this wire using a sliding contact. Suppose that the cell A has a known e.m.f. E_A. This cell is switched into the circuit using the two-way switch. The sliding contact is then moved along the wire until the centre-zero galvanometer reads zero. The length l_A of the wire from the common zero end to the sliding contact is noted. Cell B has an unknown e.m.f. E_B. This cell is then switched into the circuit and the balancing process repeated. Suppose that the position at which the galvanometer reads zero is then a distance l_B from the common zero to the sliding contact. The ratio of the e.m.f.s is the ratio of the balance lengths; that is, $E_B/E_A = l_B/l_A$, and E_B can be determined in terms of the known e.m.f. E_A.

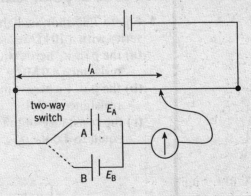

Figure 7.34 Potentiometer used to compare cell e.m.f.s

Example

1 A light-emitting diode (LED) is connected in series with a resistor to a 5.0 V supply.
  (a) Calculate the resistance of the series resistor required to give a current in the LED of 12 mA, with a voltage across it of 2.0 V.
  (b) Calculate the potential difference across the LED when the series resistor has resistance 500 Ω.

  (a) If the supply voltage is 5.0 V and the p.d. across the LED is 2.0 V, the p.d. across the resistor must be $5.0 - 2.0 = 3.0$ V. The current through the resistor is 12 mA as it is in series with the LED. Using $R = V/I$, the resistance of the resistor is $3.0/12 \times 10^{-3} = \mathbf{250\,\Omega}$.
  (b) The resistance of the LED is given by
   $R = V/I = 2.0/12 \times 10^{-3} = 167\,\Omega$. If this resistance is in series with a 500 Ω resistor and a 5.0 V supply, the p.d. across the LED is $5.0 \times 167/(167 + 500) = \mathbf{1.25\,V}$.

2 The e.m.f.s of two cells are compared using the slide-wire circuit of Figure 7.34. Cell A has a known e.m.f. of 1.02 V; using this cell, a balance point is obtained when the slider is 37.6 cm from the zero of the scale. Using cell B, the balance point is at 55.3 cm. ▶▶

(a) Calculate the e.m.f. of cell B.
(b) State the advantage of using this null method to compare the e.m.f.s.

(a) This is a straightforward application of the formula for the potentiometer, $E_B/E_A = l_B/l_A$. Substituting the values, $E_B = \mathbf{1.50\,V}$.

(b) When comparing the e.m.f.s of cells, it is necessary to arrange for the cells to be on open circuit so that there is no drop in terminal potential difference because of a current passing through the internal resistance. When the potentiometer is balanced, there is no current from the cell under test, which is exactly what is required.

Now it's your turn

1 Figure 7.35 shows a light-dependent resistor (LDR) connected in series with a $10\,k\Omega$ resistor and a 12 V supply. Calculate:
   (a) the p.d. V_L across the LDR when it is in the dark and has resistance $8.0\,M\Omega$,
   (b) the p.d. V_L across the LDR when it is in bright light and has resistance $500\,\Omega$,
   (c) the resistance of the LDR in lighting conditions which make V_L equal to 4.0 V.

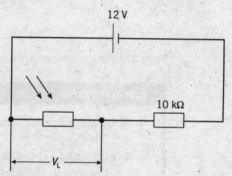

Figure 7.35

2 The thermistor in the potential divider circuit of Figure 7.36 has a resistance which varies between $100\,\Omega$ at $100\,°C$ and $6.0\,k\Omega$ at $0\,°C$. Calculate the potential difference across the thermistor at
(a) $100\,°C$, **(b)** $0\,°C$.

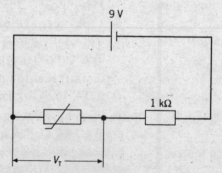

Figure 7.36

Section 7.4 Summary

- The equivalent resistance R of resistors connected in series is given by: $R = R_1 + R_2 + R_3 + \ldots$
- The equivalent resistance R of resistors connected in parallel is given by: $1/R = 1/R_1 + 1/R_2 + 1/R_3 + \ldots$
- Two resistors in series act as a potential divider, where $V_1/V_2 = R_1/R_2$. If V is the supply voltage: $V_{out} = VR_1/(R_1 + R_2)$
- A potentiometer is a variable resistor connected as a potential divider to give a continuously variable output voltage.

Section 7.4 Questions

1 You are given three resistors of resistance $22\,\Omega$, $47\,\Omega$ and $100\,\Omega$. Calculate:
(a) the maximum possible resistance,
(b) the minimum possible resistance,
that can be obtained by combining any or all of these resistors.

2 In the circuit of Figure 7.37, the currents I_1 and I_2 are equal. Calculate:
(a) the resistance R of the unknown resistor,
(b) the total current I_3.

3 Figure 7.38 shows a potential divider circuit, designed to provide p.d.s of $1.0\,V$ and $4.0\,V$ from a battery of e.m.f. $9.0\,V$ and negligible internal resistance.
(a) Calculate the value of resistance R.
(b) State and explain what happens to the voltage at terminal A when an additional $1.0\,\Omega$ resistor is connected between terminals B and C in parallel with the $5.0\,\Omega$ resistor. No calculations are required.

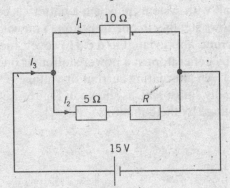

Figure 7.37

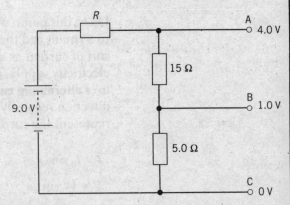

Figure 7.38

7.5 Alternating currents

At the end of Section 7.5 you should be able to:
- represent a sinusoidally alternating current or voltage by an equation of the form $x = x_0 \sin \omega t$
- show an understanding of and use the terms period, frequency, peak value and root-mean-square value as applied to an alternating current or voltage
- deduce that the mean power in a resistive load is half the maximum power for a sinusoidal alternating current
- distinguish between r.m.s. and peak values and recall and solve problems using the relationship $I_{rms} = I_0/\sqrt{2}$ for the sinusoidal case
- distinguish graphically between half-wave and full-wave rectification
- explain the use of a single diode for the half-wave rectification of an alternating current
- explain the use of four diodes (bridge rectifier) for the full-wave rectification of an alternating current
- analyse the effect of a single capacitor in smoothing, including the effect of the value of capacitance in relation to the load resistance.

Up to this point, we have dealt with systems in which a battery is connected to a circuit and the current flows steadily in one direction. You will know this sort of current as a direct current, abbreviated to d.c. However, the domestic electricity supply, produced by generators at a power station, is one which uses **alternating current (a.c.)**. An alternating current or voltage reverses its direction regularly and is usually sinusoidal, as shown in Figure 7.39. We can represent the current and voltage by the equations

$I = I_0 \sin \omega t$

$V = V_0 \sin \omega t$

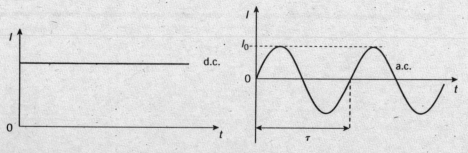

Figure 7.39 Direct and alternating currents

The time τ taken for one complete cycle of the a.c. is the period of the current $\tau = 2\pi/\omega$. The reciprocal of the period is the frequency f; that is, $f = 1/\tau$ ($f = \omega/2\pi$).

The frequency is the number of complete cycles of the supply in one second. The unit of frequency is the hertz, Hz. The **peak value** of the current or voltage is I_0 or V_0, the amplitude of the oscillating current or voltage. Sometimes the term peak-to-peak value is used; this means $2I_0$ or $2V_0$, or twice the amplitude.

It is clear from Figure 7.39 that the average value of an alternating current is zero. However, this does not mean that, when an a.c. source is connected to a resistor, no power is generated in the resistor. An alternating current in a wire can be thought of as electrons moving backwards and forwards, and passing on their energy by collision. The power generated in a resistance R is given by the usual formula

$$P = I^2R$$

but here the current I must be written as

$$I = I_0 \sin \omega t$$

Thus

$$P = I_0^2 R \sin^2 \omega t$$

Because I_0^2 and $\sin^2 \omega t$ are always positive, we see that the power P is also always positive.

The expression we have just derived for P gives the power at any instant. What is much more useful is the average or mean power. This is the quantity which must be used in assessing the power generated in the resistor. Because I_0 and R are constants, the average value of P will depend on the average value of $\sin^2 \omega t$, which is $\frac{1}{2}$. So the average power delivered to the resistor is

$$\langle P \rangle = \tfrac{1}{2} I_0^2 R = \tfrac{1}{2} V_0^2 / R$$

This is half the maximum power. We could use the average value of the square of the current or the voltage in these relations, since

$$\langle I^2 \rangle = \tfrac{1}{2} I_0^2 \qquad \text{and} \qquad \langle V^2 \rangle = \tfrac{1}{2} V_0^2$$

The square root of $\langle I^2 \rangle$ is called the **root-mean-square**, or **r.m.s.**, value of the current, and similarly for the voltage. The numerical relations are

$$I_{\text{rms}} = \sqrt{\langle I^2 \rangle} = I_0 / \sqrt{2} = 0.707 \, I_0$$
$$V_{\text{rms}} = \sqrt{\langle V^2 \rangle} = V_0 / \sqrt{2} = 0.707 \, V_0$$

The r.m.s. values are useful because they represent the *effective* values of current and voltage in an a.c. circuit. A direct current with a value of I equal to the r.m.s current I_{rms} of an a.c. circuit will produce exactly the same heating effect in a resistor. In specifying a domestic supply voltage, it is the r.m.s. value that is quoted, not the peak value.

> The r.m.s. value of the current or voltage is that value of the direct current or voltage that would produce heat at the same rate in a resistor.

Example

A 1.5 kW heater is connected to the domestic supply, which is quoted as 240 V. Calculate the peak current in the heater, and its resistance.
The r.m.s version of the power/current/voltage equation is
$I_{rms} = \langle P \rangle / V_{rms}$. This gives $I_{rms} = 1.5 \times 10^3/240 = 6.3$ A. The peak current $I_0 = \sqrt{2}\ I_{rms} =$ **8.8 A**. The resistance is $R = V_{rms}/I_{rms} = 240/6.3 =$ **38 Ω**.

Now it's your turn

A heater of resistance 40 Ω is connected to a domestic supply of 240 V.
(a) Calculate the average power in the resistor.
(b) What are the maximum and minimum values of the instantaneous power in the resistor?

Rectifiers

It is sometimes necessary to convert an alternating current into a direct current. This is because most electronic devices require direct current, whereas the domestic supply is alternating. The conversion can be done by a process known as **rectification**.

Suppose a single diode (see page 173) is connected in the a.c circuit of Figure 7.40. We know that the diode allows current to flow in one direction only. This means that the output voltage across the resistor will consist only of the positive half-cycles of the input voltage, as shown in Figure 7.41. The diode has rejected the negative part of the input, producing a unidirectional current which fluctuates considerably, rather than a constant direct current. Nevertheless, we have achieved what is called **half-wave rectification**.

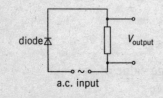

Figure 7.40 Single-diode circuit for half-wave rectification

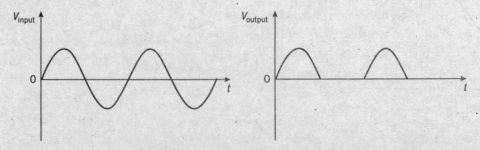

Figure 7.41 Half-wave rectification

It is more satisfactory also to make use of the negative half-cycles of the input and reverse their polarity, as shown in Figure 7.42. This process is called **full-wave rectification**. One circuit used for full-wave rectification is illustrated in Figure 7.43. It uses four diodes arranged in a diamond pattern; it is called a **bridge rectifier** circuit. The input terminals are P and Q. If P is positive during the first half-cycle, rectifiers 1 and 2 on opposite sides of the diamond will conduct. In the next half-cycle Q is positive, and rectifiers 3 and 4 conduct. Thus the resistor acting as a load will always have its upper terminal positive and its lower terminal negative.

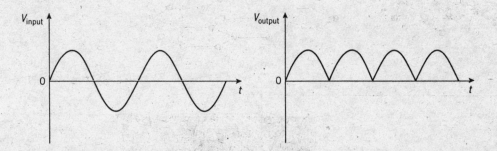

Figure 7.42 Full-wave rectification

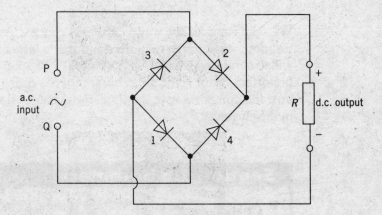

Figure 7.43 Four-diode (bridge) circuit for full-wave rectification

The circuit has produced unidirectional current, but the output is still not a good approximation to the steady direct current required for most electronic equipment. We can improve the situation by inserting a capacitor across the output terminals of the bridge circuit (Figure 7.44). The capacitor charges up on the rising part of the half-cycle, and then discharges as the output voltage falls. The effect is to reduce the fluctuations in the unidirectional output. This process is called **smoothing**.

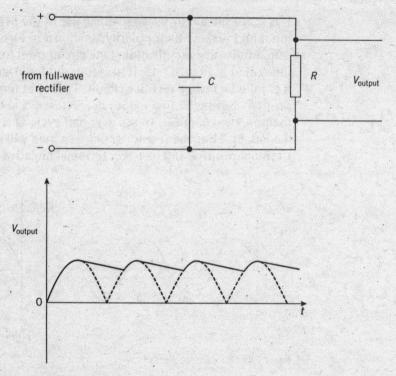

Figure 7.44 Smoothing by capacitor

We shall consider the way a capacitor charges and discharges in Chapter 11. For now, it is enough to understand that a larger value of the capacitance will generally give better smoothing. (In Chapter 11, we will see that the important factor is the *time constant* of the resistor–capacitor circuit, and how this compares with the time of the half-period covering the fluctuation in the current.)

Section 7.5 Summary

- Alternating current or voltage is usually sinusoidal: $I = I_0 \sin \omega t$ or $V = V_0 \sin \omega t$
- The peak value is I_0 or V_0.
- Root-mean-square (r.m.s.) values represent the effective values (values of d.c. that give the same heating effect):

$$I_{rms} = \sqrt{\langle I^2 \rangle} = I_0/\sqrt{2} \quad \text{and} \quad V_{rms} = \sqrt{\langle V^2 \rangle} = V_0/\sqrt{2}$$

- Mean power developed $= I_{rms} V_{rms} = I_0 V_0/2 = \frac{1}{2}$ peak power
- A single diode gives half-wave rectification: negative half-cycles are blocked.
- A bridge circuit of four diodes can give full-wave rectification.
- A capacitor across the output reduces the fluctuations of the rectified output voltage.

Section 7.5 Questions

1 A stereo system has two output channels, each connected to a loudspeaker of effective resistance $8.0\,\Omega$. Each channel can deliver a maximum average power output of 50 W to its speaker. Calculate the r.m.s. voltage and the r.m.s. current fed to the speaker at this maximum power. Assume that the loudspeaker can be treated as a simple resistance.

2 The half-wave rectifier circuit of Figure 7.40 is used to rectify an a.c. input voltage of 240 V r.m.s. The output resistor has resistance $25\,k\Omega$.
 (a) Calculate the peak value of the input voltage.
 (b) Estimate the average current in the output resistor.

Exam-style Questions

1 A student designs an electrical method to monitor the position of a steel sphere rolling on two parallel rails. Each rail is made from bare wire of length 30 cm and resistance $20\,\Omega$. The position-sensing circuit is shown in Figure 7.45. The resistance of the steel sphere and the internal resistance of the battery are negligible.

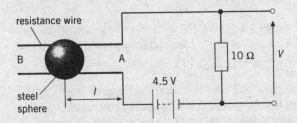

Figure 7.45

 (a) State the voltage across the $10\,\Omega$ resistor when the sphere is at A, where $I = 0$.
 (b) With the sphere at end B of the rails, calculate:
  (i) the total resistance of the circuit,
  (ii) the current in the $10\,\Omega$ resistor,
  (iii) the output voltage V.

2 Two equations for the power P dissipated in a resistor are $P = I^2R$ and $P = V^2/R$. The first suggests that the greater the resistance R of the resistor, the more power is dissipated. The second suggests the opposite: the greater the resistance, the less the power. Explain this inconsistency.

3 State the minimum number of resistors, each of the same resistance and power rating of 0.5 W, which must be used to produce an equivalent $1.2\,k\Omega$, 5 W resistor. Calculate the resistance of each, and state how they should be connected.

4 In the circuit shown in Figure 7.46 the current in the battery is 1.5 A. The battery has internal resistance $1.0\,\Omega$. Calculate:
 (a) the combined resistance of the resistors that are connected in parallel in the circuit of Figure 7.46,
 (b) the total resistance of the circuit,
 (c) the resistance of resistor Y,
 (d) the current through the $6\,\Omega$ resistor.

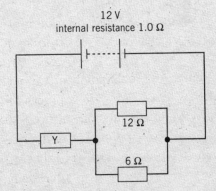

Figure 7.46

5 The current in the starter motor of a car is 160 A when starting the engine. The connecting cable has total length 1.3 m, and consists of fifteen strands of wire, each of diameter 1.2 mm. The resistivity of the metal of the strands is $1.4 \times 10^{-8}\,\Omega\,m$.
(a) Calculate:
  (i) the resistance of each strand,
  (ii) the total resistance of the cable,
  (iii) the power loss in the cable.
(b) When the starter motor is used to start the car, 700 C of charge pass through a given cross-section of the cable.
  (i) Assuming that the current is constant at 160 A, calculate for how long the charge flows.
  (ii) Calculate the number of electrons which pass a given cross-section of the cable in this time. The electron charge e is $-1.6 \times 10^{-19}\,C$.
(c) The e.m.f. of the battery is 13.6 V and its internal resistance is $0.012\,\Omega$. Calculate:
  (i) the potential difference across the battery terminals when the current in the battery is 160 A,
  (ii) the rate of production of heat energy in the battery.

6 A copper wire of length 16 m has a resistance of $0.85\,\Omega$. The wire is connected across the terminals of a battery of e.m.f. 1.5 V and internal resistance $0.40\,\Omega$.
(a) Calculate the potential difference across the wire and the power dissipated in it.
(b) In an experiment, the length of this wire connected across the terminals of the battery is gradually reduced.
  (i) Sketch a graph to show how the power dissipated in the wire varies with the connected length.
  (ii) Calculate the length of the wire when the power dissipated in the wire is a maximum.
  (iii) Calculate the maximum power dissipated in the wire.

7 From power stations, electrical energy is transmitted over long distances to homes and business, at 400 kV. An alternative might be to transmit at the domestic supply voltage of 230 V. Explain how the high-voltage option reduces power losses in the cables.

8 A small laboratory water heater is connected to the 240 V a.c. supply. The heater has resistance $40\,\Omega$ Calculate:
(a) the average power used,
(b) the maximum instantaneous power used,
(c) the minimum instantaneous power used.
Sketch a labelled graph to show at which points in the cycle of the alternating current the maximum and minimum instantaneous power is used.

8. Waves

The aim of this chapter is to introduce some general properties of waves. We shall meet two broad classifications of waves, transverse and longitudinal, based on the direction in which the particles vibrate relative to the direction in which the wave transmits energy. We shall define terms such as amplitude, wavelength and frequency for a wave, and derive the relationship between speed, frequency and wavelength. We shall look at demonstrations of some properties of waves, such as reflection, refraction, diffraction and interference. In connection with interference, we shall use the principle of superposition, which tells us how waves which come together at the same place interact.

8.1 Transverse and longitudinal waves

At the end of Section 8.1 you should be able to:
- show an understanding that energy is transferred due to a progressive wave
- compare transverse and longitudinal waves
- describe what is meant by wave motion as illustrated by vibration in ropes, springs and ripple tanks
- show an understanding and use the terms displacement, amplitude, phase difference, period, frequency, wavelength and speed
- analyse and interpret graphical representations of transverse and longitudinal waves
- recall and use the relationship intensity is proportional to (amplitude)^2
- deduce, from the definitions of speed, frequency and wavelength, the equation $v = f\lambda$
- recall and use the equation $v = f\lambda$.

Wave motion is a means of moving energy from place to place. For example, electromagnetic waves from the Sun carry the energy that plants need to survive and grow. The energy carried by sound waves causes our ear drums to vibrate. The energy carried by seismic waves (earthquakes) can devastate vast areas, causing land to move and buildings to collapse. Waves which move energy from place to place are called **progressive waves**.

Vibrating objects act as sources of waves. For example, a vibrating tuning fork sets the air close to it into oscillation, and a sound wave spreads out from the fork. For a radio wave, the vibrating objects are electrons.

There are two main groups of waves. These are **transverse waves** and **longitudinal waves**.

A transverse wave is one in which the vibrations of the particles in the wave are at right angles to the direction in which the energy of the wave is travelling.

Figure 8.1 shows a transverse wave moving along a rope. The particles of the rope vibrate up and down, whilst the energy travels at right angles to this, from A to B. There is no transfer of matter from A to B. Examples of transverse waves include light waves, surface water waves and secondary seismic waves (S-waves).

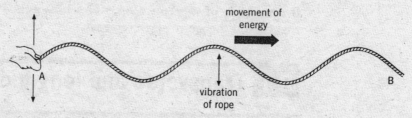

Figure 8.1 Transverse wave on a rope

A longitudinal wave is one in which the direction of the vibrations of the particles in the wave is along the direction in which the energy of the wave is travelling.

Figure 8.2 shows a longitudinal wave moving along a stretched spring (a 'slinky'). The coils of the spring vibrate along the length of the spring, whilst the energy travels along the same line, from A to B. Note that the spring itself does not move from A to B. Examples of longitudinal waves include sound waves and primary seismic waves (P-waves).

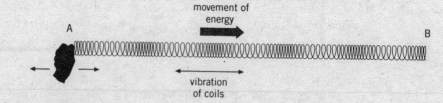

Figure 8.2 Longitudinal wave on a slinky spring

Graphical representation of waves

The **displacement** of a particle on a wave is its distance from its rest position.

Displacement is a vector quantity; it can be positive or negative. A transverse wave may be represented by plotting displacement y on the y-axis against distance x along the wave, in the direction of energy travel, on the x-axis. This is shown in Figure 8.3. It can be seen that the graph is a snapshot of what is actually observed to be a transverse wave.

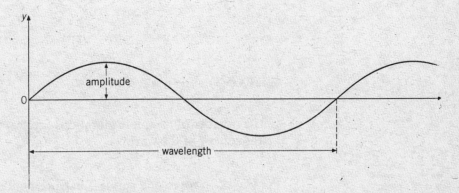

Figure 8.3 Displacement–distance graph for a transverse wave

For a longitudinal wave, the displacement of the particles is along the direction of energy travel. However, if these displacements are plotted on the y-axis of a graph of displacement against distance, the graph has exactly the same shape as for a transverse wave (see Figure 8.3). This is very useful, in that one graph can represent both types of wave. Using this graph, wave properties may be treated without reference to the type of wave.

The **amplitude** of the wave motion is defined as the maximum displacement of a particle in the wave.

Also, it can be seen that the wave repeats itself. That is, the wave can be constructed by repeating a section of the wave. The length of the smallest repetition unit is called the **wavelength**.

One wavelength is the distance between two neighbouring peaks or two neighbouring troughs, or two neighbouring points which are vibrating together in exactly the same way (in phase). It is the distance moved by the wave during one oscillation of the source of the waves.

The usual symbol for wavelength is λ, the Greek letter lambda.

Another way to represent both waves is to plot a graph of displacement y against time t. This is shown in Figure 8.4. Again, the wave repeats itself after a certain interval of time. The time for one complete vibration is called the **period** of the wave T.

The period of the wave is the time for a particle in the wave to complete one vibration, or one cycle.

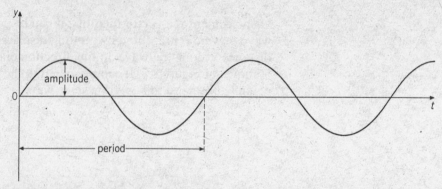

Figure 8.4 Displacement–time graph for a wave

The number of complete vibrations (cycles) per unit time is called the frequency *f* of the wave.

For waves on ropes and springs, displacement and amplitude are measured in mm, m or other units of length. Period is measured in seconds (s). Frequency has the unit per second (s^{-1}) or hertz (Hz).

A term used to describe the relative positions of the crests or troughs of two waves of the same frequency is **phase**. When the crests and troughs of the two waves are aligned, the waves are said to be **in phase**. When a crest is aligned with a trough, the waves are **out of phase**. When used as a quantitative measure, phase has the unit of angle (radians or degrees). Thus, when waves are out of phase, one wave is half a cycle behind the other. Since one cycle is equivalent to 2π radians or 360°, the **phase difference** between waves that are exactly out of phase is π radians or 180°.

Consider Figure 8.5, in which there are two waves of the same frequency, but with a phase difference between them. The period *T* corresponds to a phase angle of 2π rad or 360°. The two waves are out of step by a time *t*. Thus, phase difference is equal to $2\pi(t/T)$ rad $= 360(t/T)°$. A similar argument may be used for waves of wavelength λ which are out of step by a distance *x*. In this case the phase difference is $2\pi(x/\lambda)$ rad $= 360(x/\lambda)°$.

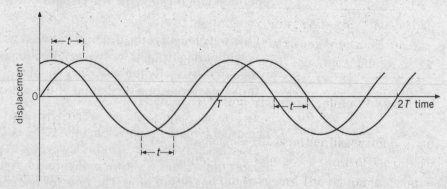

Figure 8.5 Phase difference

One of the characteristics of a progressive wave is that it carries energy. The amount of energy passing through unit area per unit time is called the **intensity** of the wave. The intensity is proportional to the square of the amplitude of a wave. Thus, doubling the amplitude of a wave increases the intensity of the wave by a factor of four. The intensity also depends on the frequency: intensity is proportional to the square of the frequency.

For a wave of amplitude A and frequency f, the intensity I is proportional to A^2f^2.

If the waves from a point source spread out equally in all directions, we have what is called a **spherical wave**. As the wave travels further from the source, the energy it carries passes through an increasingly large area. Since the surface area of a sphere is $4\pi r^2$, the intensity is $W/4\pi r^2$, where W is the power of the source. The intensity of the wave thus decreases with increasing distance from the source. The intensity I is proportional to $1/r^2$, where r is the distance from the source.

This relationship assumes that there is no absorption of wave energy.

Wave equation

From the definition of wavelength λ, in one cycle of the source the wave energy moves a distance λ. In f cycles, the wave moves a distance $f\lambda$. If f is the frequency of the wave, then f cycles are produced in unit time. Therefore $f\lambda$ is the distance moved in unit time. Referring to Chapter 3, speed v is the distance moved per unit time. Therefore

$$v = f\lambda$$

or

$$speed = frequency \times wavelength$$

This is an important relationship between the speed of a wave and its frequency and wavelength.

Examples

1 A tuning fork of frequency 170 Hz produces sound waves of wavelength 2.0 m. Calculate the speed of sound.

Using $v = f\lambda$, we have $v = 170 \times 2.0 = \mathbf{340\,m\,s^{-1}}$.

2 The amplitude of a wave in a rope is 15 mm. If the amplitude were changed to 20 mm, keeping the frequency the same, by what factor would the power carried by the rope change?

Intensity is proportional to the square of the amplitude. Here the amplitude has been increased by a factor of 20/15, so the power carried by the wave increases by a factor of $(20/15)^2 = \mathbf{1.8}$.

Properties of wave motions

Although there are many different types of waves (light waves, sound waves, electrical waves, mechanical waves, etc.) there are some basic properties which they all have in common. All waves can be reflected, refracted, diffracted, and can produce interference patterns.

These properties may be demonstrated using a ripple tank similar to that shown in Figure 8.6. As the motor turns, the wooden bar vibrates, creating ripples on the surface of the water. The ripples are lit from above. This creates shadows on the viewing screen. The shadows show the shape and movement of the waves. Each shadow corresponds to a particular point on the wave, and is referred to as a **wavefront**.

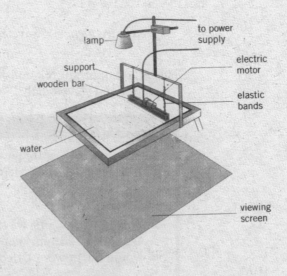

Figure 8.6 Ripple tank

Figure 8.7 illustrates the pattern of wavefronts produced by a low-frequency vibrator and one of higher frequency. Note that for the higher frequency the wavelength is less, since wave speed is constant and $v = f\lambda$.

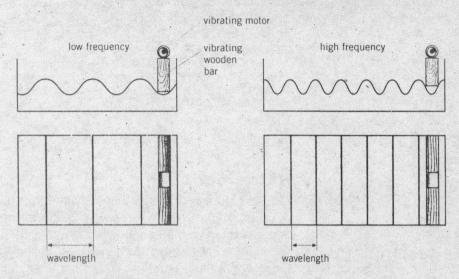

Figure 8.7 Ripple tank patterns for low- and high-frequency vibrations

Circular waves may be produced by replacing the vibrating bar with a small dipper, or by allowing drops of water to fall into the ripple tank. A circular wave is illustrated in Figure 8.8. This pattern is characteristic of waves spreading out from a point source.

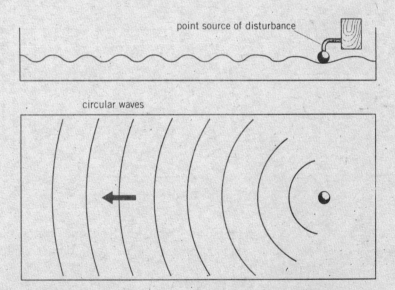

Figure 8.8 Ripple tank pattern for a point source

We shall now see how the ripple tank may be used to demonstrate the wave properties of reflection, refraction, diffraction and interference.

Reflection

As the waves strike a plane barrier placed in the water, they are reflected. The angle of reflection equals the angle of incidence, and there is no change in wavelength (see Figure 8.9a). If a curved barrier is used, the waves can be made to converge or diverge (Figure 8.9b).

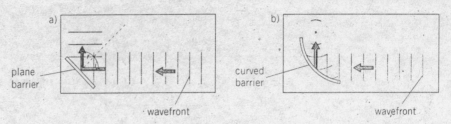

Figure 8.9 Ripple tank pattern showing reflection at a) a plane surface and b) a curved one

Refraction

If a glass block is submerged in the water, this produces a sudden change in the depth of the water. The speed of surface ripples on water depends on the depth of the water: the shallower the water, the slower the speed. Thus, the waves move more slowly as they pass over the glass block. The frequency of the waves remains constant, and so the wavelength decreases. If the waves are incident at an angle to the submerged block, they will change direction, as shown in Figure 8.10.

> The change in direction of a wave due to a change in speed is called **refraction**.

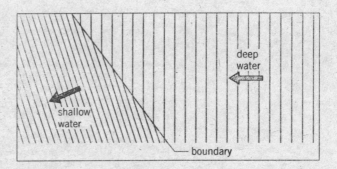

Figure 8.10 Ripple tank pattern showing refraction

As the waves re-enter the deeper water, their speed increases to the former value and they change direction once again.

Diffraction

When waves pass through a narrow gap, they spread out. This spreading out is called **diffraction**. The extent of diffraction depends on the width of the gap compared with the wavelength. It is most noticeable if the width of the gap is approximately equal to the wavelength. Diffraction is illustrated in Figure 8.11. Note that diffraction may also occur at an edge.

Diffraction will be considered further in Section 8.5.

> Diffraction is defined as the spreading of a wave into regions where it would not be seen if it moved only in straight lines.

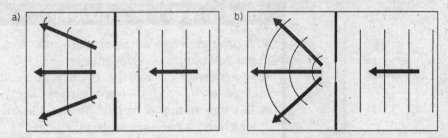

Figure 8.11 Ripple tank pattern showing diffraction at a) a wide gap, b) a narrow gap

Interference

If two or more waves overlap, the resultant displacement is the sum of the individual displacements. Remember that displacement is a vector quantity. The overlapping waves are said to **interfere**. This may lead to a resultant wave of either a larger or a smaller displacement than either of the two component waves.

Interference can be demonstrated in the ripple tank by using two point sources. Figure 8.13 shows such an interference pattern, and Figure 8.12 shows how it arises.

Interference will be discussed further in Section 8.3.

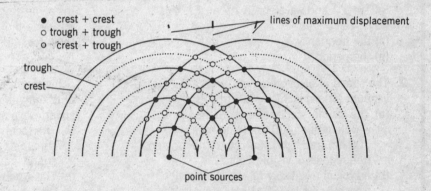

Figure 8.12 Two-source interference of circular waves in a ripple tank

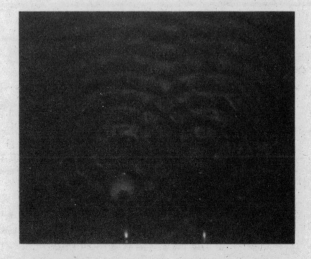

Figure 8.13 The resulting interference pattern

Section 8.1 Summary

- A progressive wave travels outwards from the source, carrying energy but without transferring matter.
- In a transverse wave, the oscillations are at right angles to the direction in which the wave carries energy.
- In a longitudinal wave, the oscillations are along the direction in which the wave carries energy.
- The intensity of a wave is the energy passing through unit area per unit time. Intensity is proportional to the square of the amplitude (and to the square of the frequency).
- The speed v, frequency f and wavelength λ of a wave are related by: $v = f\lambda$
- Properties of wave motion (reflection, refraction, diffraction and interference) can be observed in a ripple tank.

Section 8.1 Questions

1 A certain sound wave in air has a speed $340 \, \text{m s}^{-1}$ and wavelength 1.7 m. For this wave, calculate:
  (a) the frequency,
  (b) the period.

2 The speed of electromagnetic waves in vacuum (or air) is $3.0 \times 10^8 \, \text{m s}^{-1}$.
  (a) The visible spectrum extends from a wavelength of 400 nm (blue light) to 700 nm (red light). Calculate the range of frequencies of visible light.
  (b) A typical frequency for v.h.f. television transmission is 250 MHz. Calculate the corresponding wavelength.

3 Two waves travel with the same speed and have the same amplitude, but the first has twice the wavelength of the second. Calculate the ratio of the intensities transmitted by the waves.

4 A student stands at a distance of 5.0 m from a point source of sound, which is radiating uniformly in all directions. The intensity of the sound wave at her ear is $6.3 \times 10^{-6} \, \text{W m}^{-2}$.
  (a) The receiving area of the student's ear canal is $1.5 \, \text{cm}^2$. Calculate how much energy passes into her ear in one minute.
  (b) The student moves to a point 1.8 m from the source. Calculate the new intensity of the sound.

8.2 Polarisation of waves

At the end of Section 8.2 you should be able to:
- show an understanding that polarisation is a phenomenon associated with transverse waves.

Think about generating waves in a rope by moving your hand holding the stretched rope up and down, or from side to side. The transverse vibrations of the rope will be in just one plane – the vertical plane if your hand is moving up and down, or the horizontal plane if it moves from side to side. The vibrations are said to be **plane-polarised** in either a vertical plane or a horizontal plane. However, if the rope passes through a vertical slit, then only if you move your hand up and down will the wave pass through the slit. This is illustrated in Figure 8.14. If the rope passes through a horizontal slit, only waves generated by a side-to-side motion of the hand will be transmitted. If the rope passes through a slit of one orientation followed by one of the other orientation, the wave is totally blocked whether it was initially in the vertical or the horizontal plane.

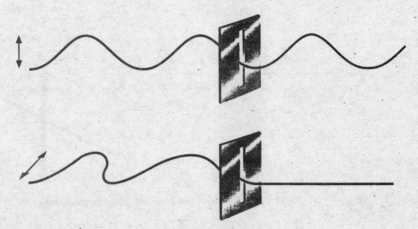

Figure 8.14 Transverse waves on a rope are plane-polarised.

The condition for a wave to be plane-polarised is for the vibrations to be in just one direction normal to the direction in which the wave is travelling.

Clearly, polarisation can apply only to transverse waves. Longitudinal waves vibrate along the direction of wave travel, and whatever the orientation of the slit, it would make no difference to the transmission of the waves.

The fact that light can be polarised was understood only in the early 1800s. This was a most important discovery. It showed that light is a transverse wave motion, and opened the way, 50 years later, to Maxwell's theory of light as electromagnetic radiation. Maxwell described light in terms of oscillating electric and magnetic fields, at right-angles to each other and at right-angles to the direction of travel of the wave energy (see Figure 9.15, page 250). When we talk about the direction of polarisation of a light wave, we refer to the direction of the electric field component of the electromagnetic wave.

The Sun and domestic light bulbs emit **unpolarised** light; that is, the vibrations take place in many directions at once, instead of in the single plane associated with plane-polarised radiation (see Figure 8.15).

Polarising light waves

Some transparent materials, such as a Polaroid sheet, allow vibrations to pass through in one direction only. A Polaroid sheet contains long chains of organic molecules aligned parallel to each other. When unpolarised light arrives at

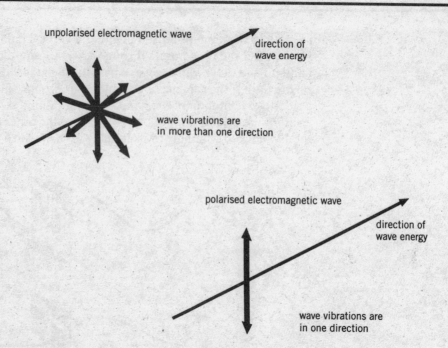

Figure 8.15 Unpolarised and polarised electromagnetic waves

the sheet, the component of the electric field of the incident radiation which is parallel to the molecules is strongly absorbed, whereas radiation with its electric field perpendicular to the molecules is transmitted through the sheet. The Polaroid sheet acts as a **polariser**, producing plane-polarised light from light that was originally unpolarised. Figure 8.16 illustrates unpolarised light entering the polariser, and polarised light leaving it.

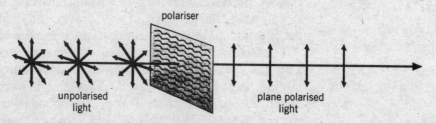

Figure 8.16 Action of a polariser

If you try to view plane-polarised light through a second sheet of Polaroid which is placed so that its polarising direction is at right-angles to the polarising direction of the first sheet, it will be found that no light is transmitted. In this arrangement, the Polaroids are said to be crossed. The second Polaroid sheet is acting as an **analyser**. If the two Polaroids have their polarising directions parallel, then plane-polarised light from the first Polaroid can pass through the second. These two situations are illustrated in Figure 8.17. Although the action of the Polaroid sheet is not that of a simple slit, the arrangement of the crossed Polaroids has the same effect as the crossed slits in the rope-and-slits experiment (Figure 8.14).

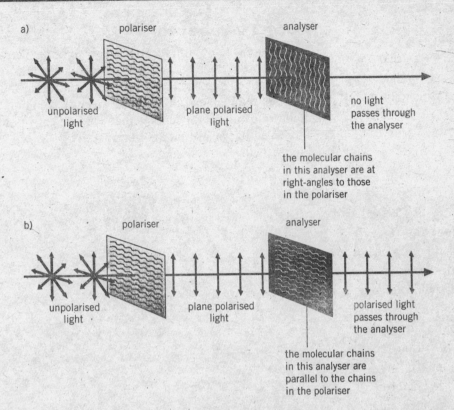

Figure 8.17 Polariser and analyser in a) crossed and b) parallel situations

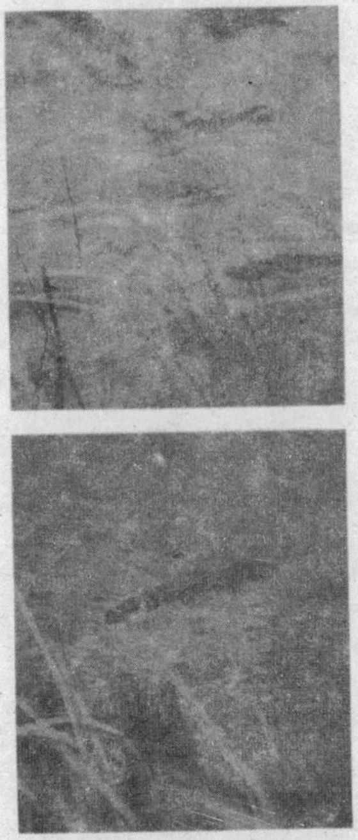

Applications of polarisation

See the light but not the glare!

A very familiar application of the polarisation of light arises in the use of Polaroid sunglasses. Light reflected from a non-metallic surface, such as the smooth surface of a swimming-pool, is partially polarised parallel to the surface. Wearing sunglasses with Polaroid lenses, arranged with the polarising direction vertical, will prevent the transmission of much of the light reflected from the water. Reduction of the glare of the reflected light allows you to see objects more clearly. Fishermen find that Polaroid sunglasses eliminate glare from the surface of a stream or pond, so that they can see beneath the water more easily (Figure 8.18).

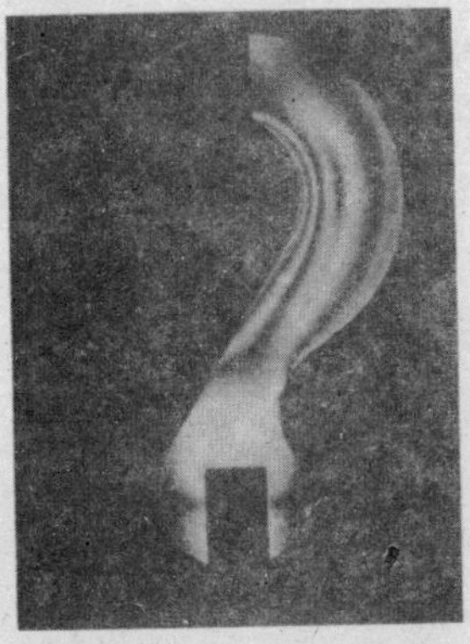

Stress analysis

When polarised light passes through some transparent materials, the plane of polarisation is rotated. If the material is put under stress, the amount of stress affects the degree of rotation. If the incident polarised light is white, each of its component colours is rotated by a different amount, creating a pattern of coloured fringes when viewed through the analyser Polaroid. Engineers make use of this phenomenon by making models of structural components in a suitable transparent material, such as Perspex, stressing the model, and examining the pattern of fringes to identify regions of high stress. Where possible, the structure is re-designed to remove these danger points. Figure 8.19 illustrates a Perspex model being examined between crossed Polaroids.

Figure 8.18 Wearing Polaroid sunglasses helps fishermen catch their prey!

Figure 8.19 Stress analysis by polarised light

Poor TV reception?

In hilly areas television reception can be poor. To correct this problem the signal is often boosted by a local relay station before being re-transmitted. Sometimes this introduces further difficulties because of interference between the incident and the re-transmitted signal. This can be avoided if the main transmitter emits waves that are vertically polarised, and the relay station emits waves that are horizontally polarised. This is illustrated in Figure 8.20.

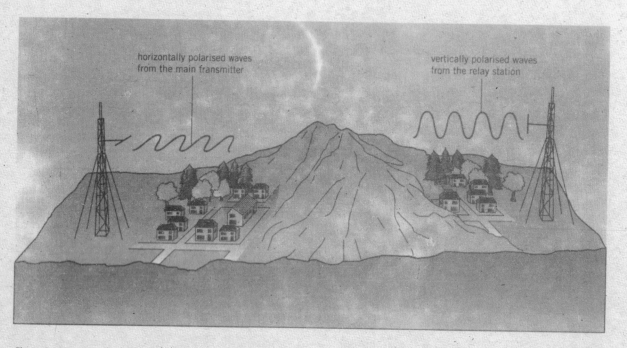

Figure 8.20 Avoiding interference between television signals

Section 8.2 Summary

- In a plane-polarised wave, the vibrations of the wave are in one direction only, which is normal to the direction of travel of the wave.
- Transverse waves can be polarised; longitudinal waves cannot.
- Plane-polarised light can be produced from unpolarised light by using a polariser, such as a sheet of Polaroid.
- Rotating an analysing Polaroid in a beam of plane-polarised light prevents transmission of the polarised light. This occurs when the polarising directions of polariser and analyser are at right-angles.

Section 8.2 Question

You have been sold a pair of sunglasses, and are not sure whether they have Polaroid lenses. How could you find out?

8.3 Interference

At the end of Section 8.3 you should be able to:

- explain and use the principle of superposition in simple applications
- show an understanding of the terms interference and coherence
- show an understanding of experiments which demonstrate two-source interference using water, light and microwaves
- show an understanding of the conditions required if two-source interference fringes are to be observed
- recall and solve problems using the equation $\lambda = ax/D$ for double-slit interference using light.

Any moment now the unsuspecting fisherman in Figure 8.21 is going to experience the effects of interference. The amplitude of oscillation of his boat will be significantly affected by the two approaching waves and their interaction when they reach his position.

Figure 8.21

We have already met the idea of interference through demonstrations of wave properties with the ripple tank. In this section we shall look at the phenomenon more closely and obtain some important results.

Figure 8.22 shows two waves arriving at a point at the same time. If they arrive **in phase** – that is, if their crests arrive at exactly the same time – they will interfere **constructively**. A resultant wave will be produced which has crests much higher than either of the two individual waves, and troughs which are much deeper. If the two incoming waves have the same frequency and equal amplitude A, the resultant wave produced by constructive interference has an amplitude of $2A$. The frequency of the resultant is the same as that of the incoming waves.

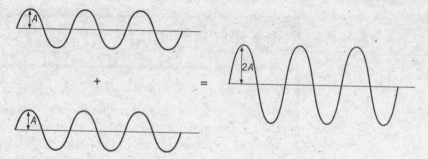

Figure 8.22 Constructive interference

If the two waves arrive **out of phase** (with a phase difference of π radians or 180°) – that is, if the peaks of one wave arrive at the same time as the troughs from the other – they will interfere **destructively**. The resultant wave will have a smaller amplitude. In the case shown in Figure 8.23, where the incoming waves have equal amplitude, the resultant wave has zero amplitude.

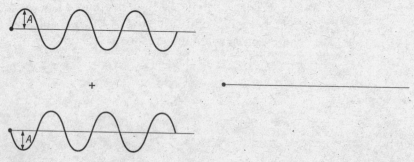

Figure 8.23 Destructive interference

This situation is an example of the **principle of superposition of waves**. The principle describes how waves, which meet at the same point in space, interact.

> The principle of superposition states that, when two or more waves meet at a point, the resultant displacement at that point is equal to the sum of the displacements of the individual waves at that point.

Because displacement is a vector, we must remember to add the individual displacements taking account of their directions. The principle applies to all types of wave.

If we consider the effect of superposition at a number of points in space, we can build up a pattern showing some areas where there is constructive interference, and hence a large wave disturbance, and other areas where the interference is destructive, and there is little or no wave disturbance.

Figure 8.24 illustrates the interference of waves from two point sources A and B. The point C is equidistant from A and B: a wave travelling to C from A moves through the same distance as a wave travelling to C from B. If the waves started out in phase at A and B, they will arrive in phase at C. They combine constructively, producing a maximum disturbance at C.

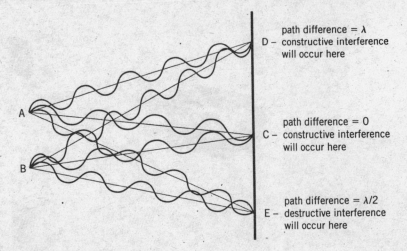

path difference = λ
D – constructive interference will occur here

path difference = 0
C – constructive interference will occur here

path difference = λ/2
E – destructive interference will occur here

Figure 8.24 Producing an interference pattern

At other places, such as D, the waves will have travelled different distances from the two sources. There is a **path difference** between the waves arriving at D. If this path difference is a whole number of wavelengths (λ, 2λ, 3λ, etc.) the waves arrive in phase and interfere constructively, producing maximum disturbance again. However, at places such as E the path difference is an odd number of half-wavelengths ($\lambda/2$, $3\lambda/2$, $5\lambda/2$, etc.). The waves arrive at E out of phase, and interference is destructive, producing a minimum resultant disturbance. This collection of maxima and minima produced by the superposition of overlapping waves is called an **interference pattern**. One was shown in Figure 8.13.

Producing an interference pattern with sound waves

Figure 8.25 shows an experimental arrangement to demonstrate interference with sound waves from two loudspeakers connected to the same signal generator and amplifier, so that each speaker produces a note of the same frequency. The demonstration is best carried out in the open air (on playing-fields, for example) to avoid reflections from walls, but it should be a windless day. Moving about in the space around the speakers, you pass through places where the waves interfere constructively and you can hear a loud sound. In places where the waves interfere destructively, the note is much quieter than elsewhere in the pattern.

Producing an interference pattern with light waves

If you try to set up a demonstration with two separate light sources, such as car headlights, you will find that it is not possible to produce an observable interference pattern (Figure 8.26). A similar demonstration works with sound waves from two separate loudspeakers. What has gone wrong?

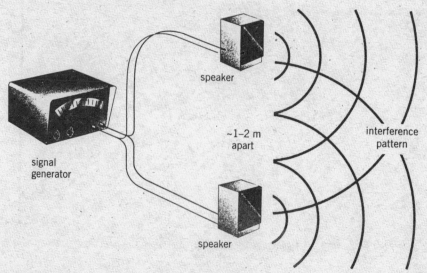

Figure 8.25 Demonstration of interference with sound waves

Figure 8.26 Failure of an interference demonstration with light

To produce an observable interference pattern, the two wave sources must have the same **single frequency**, not a mixture of frequencies as is the case for light from car headlights. They must also have a **constant phase relationship**. In the sound experiment, the waves from the two loudspeakers have the same frequency and a constant phase relationship because the loudspeakers are connected to the same oscillator and amplifier. If the waves emitted from the speakers are in phase when the experiment begins, they stay in phase for the whole experiment.

> Wave sources which maintain a constant phase relationship are described as **coherent** sources.

This is illustrated in Figure 8.27.

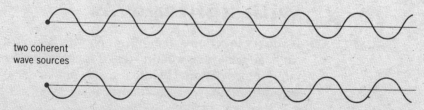

Figure 8.27 Coherent wave trains

Light is emitted from sources as a series of pulses or packets of energy. These pulses last for a very short time, about a nanosecond (10^{-9}s). Between each pulse there is an abrupt change in the phase of the waves. Waves from two separate sources may be in phase at one instant, but out of phase in the next nanosecond. The human eye cannot cope with such rapid changes, so the pattern is not observable. So separate light sources, even of the same frequency, produce incoherent waves (Figure 8.28).

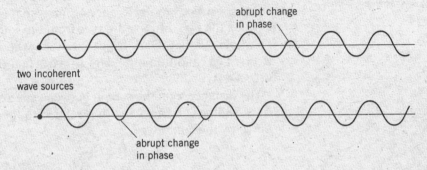

Figure 8.28 Incoherent light trains

To obtain observable interference patterns, it is not essential for the amplitudes of the waves from the two sources to be the same. However, if the amplitudes are not equal, a completely dark fringe will never be obtained, and the contrast of the pattern is reduced.

Young's double-slit experiment

In 1801 Thomas Young demonstrated how light waves could produce an interference pattern. The experimental arrangement is shown in Figure 8.29 (not to scale).

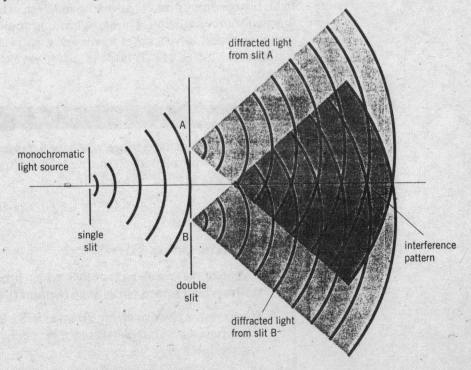

Figure 8.29 Young's double-slit experiment

Figure 8.30 Fringe pattern in Young's experiment

A monochromatic light source (a source of one colour, and hence one wavelength λ) is placed behind a single slit to create a small, well defined source of light. Light from this source is diffracted at the slit, producing two light sources at the double slit A and B. Because these two light sources originate from the same primary source, they are coherent and create a sustained and observable interference pattern, as seen in the photograph of the dark and bright interference fringes in Figure 8.30. Bright fringes are seen where constructive interference occurs – that is, where the path difference between the two diffracted waves from the sources A and B is $n\lambda$, where n is a whole number. Dark fringes are seen where destructive interference occurs. The condition for a dark fringe is that the path difference should be $(n + \frac{1}{2})\lambda$.

The distance x on the screen between successive bright fringes is called the fringe width. The fringe width is related to the wavelength λ of the light source by the equation

$$x = \frac{\lambda D}{a} \quad \text{or} \quad \lambda = \frac{ax}{D}$$

where D is the distance from the double slit to the screen and a is the distance between the centres of the slits. Note that, because the wavelength of light is so small (of the order of 10^{-7} m), to produce observable fringes D needs to be large and a as small as possible. (This is another reason why you could never see an interference pattern from two sources such as car headlamps.) See the Example below.

Although Young's original double-slit experiment was carried out with light, the conditions for constructive and destructive interference apply for any two-source situation. The same formula applies for all types of wave, including sound waves, water waves and microwaves, provided that the fringes are detected at a distance of many wavelengths from the two sources.

Example

Calculate the observed fringe width for a Young's double-slit experiment using light of wavelength 600 nm and slits 0.50 mm apart. The distance from the slits to the screen is 0.80 m.

Using $x = \lambda D/a$, $x = 600 \times 10^{-9} \times 0.80/0.50 \times 10^{-3} = 9.6 \times 10^{-4}$ m
$= 0.96$ mm

Now it's your turn

1 Calculate the wavelength of light which produces fringes of width 0.50 mm on a screen 60 cm from two slits 0.75 mm apart.

2 Radar waves of wavelength 50 mm create a fringe pattern 1.0 km from the aerials. Calculate the distance between the aerials if the fringe spacing is 80 cm.

White-light fringes

If the two slits in Young's experiment are illuminated with white light, each of the different wavelengths making up the white light produces its own fringe pattern. At the centre of the pattern, where the path difference for all waves is zero, there will be a white maximum with a black fringe on each side. Thereafter, the maxima and minima of the different colours overlap in such a way as to produce a pattern of coloured fringes. Only a few will be visible; a short distance from the centre so many wavelengths overlap that they combine to produce what is effectively white light again.

Figure 8.31 White-light interference fringes

Section 8.3 Summary

■ The principle of superposition of waves states that when waves meet at the same point in space, the resultant displacement is given by the sum of the displacements of the individual waves.

■ Constructive interference is obtained when the waves that meet are completely in phase, so that the resultant wave is of greater amplitude than any of its constituents.

■ Destructive interference is obtained when the waves that meet are completely out of phase.

■ To produce a sustained and observable interference pattern the sources must be monochromatic and coherent (have a constant phase relationship).

■ Coherent sources have a constant phase difference between them.

■ Young's double-slit experiment:
  – condition for constructive interference: path difference = $n\lambda$
  – condition for destructive interference: path difference = $(n + \frac{1}{2})\lambda$
  – fringe width $x = \lambda D/a$, where a is the separation of the source slits and D is the distance of the screen from the slits.

Section 8.3 Questions

1 Compare a two-source experiment to demonstrate the interference of sound waves with a Young's double-slit experiment using light. What are the similarities and differences between the two experiments?

2 (a) Explain the term *coherence* as applied to waves from two sources.
  (b) Describe how you would produce two coherent sources of light.
  (c) A double-slit interference pattern is produced using slits separated by 0.45 mm, illuminated with light of wavelength 633 nm from a laser. The pattern is projected on to a wall 2.50 m from the slits. Calculate the fringe separation.

3 Figure 8.32 (page 220) shows the arrangement for obtaining interference fringes in a Young's double-slit experiment. Describe and explain what will be seen on the screen if the arrangement is altered in each of the following ways:
  (a) the slit separation a is halved,
  (b) the distance D from slits to screen is doubled,
  (c) the monochromatic light source is replaced with a white-light source,
  (d) a piece of Polaroid is placed in front of each slit, the polarising directions of the Polaroids being vertical for one slit and horizontal for the other.

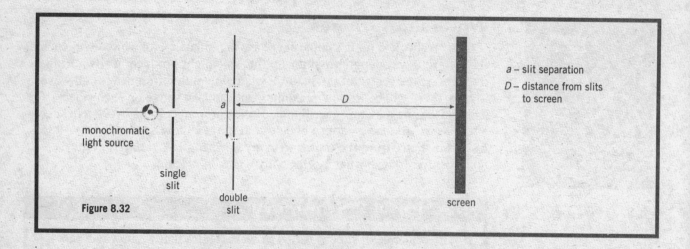

a – slit separation

D – distance from slits to screen

Figure 8.32

8.4 Stationary waves

At the end of Section 8.4 you should be able to:
- show an understanding of experiments which demonstrate the formation of stationary waves using microwaves, stretched strings and air columns
- explain the formation of a stationary wave using a graphical method, and identify nodes and antinodes
- determine the wavelength of sound using stationary waves.

The notes we hear from a cello are created by the vibrations of its strings (Figure 8.33). The wave patterns on the vibrating strings are called **stationary waves** (or **standing waves**). The waves in the air which carry the sound to our ears transfer energy, and so are called **progressive waves**.

Figure 8.34 shows a single transverse pulse travelling along a 'slinky' spring. The pulse is reflected when it reaches the fixed end. If a second pulse is sent along the slinky (Figure 8.35), the reflected pulse will pass through the outward-going pulse, creating a new pulse shape. Interference will take place between the outward and reflected pulses.

If the interval between outward pulses is reduced, a progressive wave is generated. When the wave reaches the fixed end, it is reflected. We now have two progressive waves of equal frequency and amplitude travelling in opposite directions on the same spring. The waves interfere, producing a wave pattern (Figure 8.36) in which the crests and troughs do not move. This pattern is called a **stationary** or **standing wave**, because it does not move. A stationary wave is the result of interference between two waves of equal frequency and amplitude, travelling along the same line with the same speed but in opposite directions.

Figure 8.33 Cello being bowed

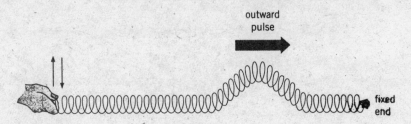

outward
pulse

fixed
end

Figure 8.34 Single transverse pulse travelling along a slinky

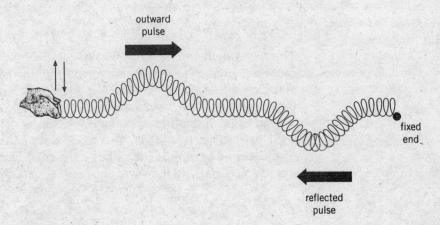

outward
pulse

fixed
end

reflected
pulse

Figure 8.35 Reflected pulse about to meet an outward-going pulse

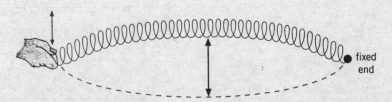

fixed
end

Figure 8.36 A stationary wave is created when two waves travelling in opposite directions interfere.

Stationary waves on strings

If a string is plucked and allowed to vibrate freely, there are certain frequencies at which it will vibrate. The amplitude of vibration at these frequencies is large. This is known as a **resonance** effect.

It is possible to investigate stationary waves in a more controlled manner using a length of string under tension and a vibrator driven by a signal generator. As the frequency of the vibrator is changed, different standing wave patterns are formed. Some of these are shown in Figure 8.37.

Figure 8.38 shows the simplest way in which a stretched string can vibrate. The wave pattern has a single loop. This is called the **fundamental mode** of vibration, or the **first harmonic**. At the ends of the string there is no vibration. These points are called **nodes**. At the centre of the string, the amplitude is a maximum. A point of maximum amplitude is called an **antinode**. Nodes and antinodes do not move along the string.

Figure 8.37 First four modes of vibration of a string

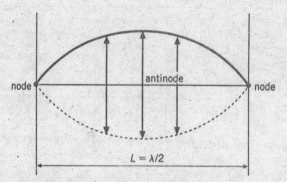

Figure 8.38 Fundamental mode of vibration of a stretched string

The wavelength λ of the standing wave in the fundamental mode is $2L$. From the wave equation $c = f\lambda$, the frequency f_1 of the note produced by the string vibrating in its fundamental mode is given by $f_1 = c/2L$, where c is the speed of the progressive waves which have interfered to produce the stationary wave.

Figure 8.39 shows the second mode of vibration of the string. The stationary wave pattern has two loops. This mode is sometimes called the **first overtone**, or the **second harmonic** (don't be confused!). The wavelength of this second mode is L. Applying the wave equation, the frequency f_2 is found to be c/L.

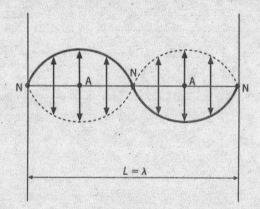

Figure 8.39 Second mode of vibration of a stretched string

Figure 8.40 shows the third mode (the second overtone, or third harmonic). This is a pattern with three loops. The wavelength is $2L/3$, and the frequency f_3 is $3c/2L$.

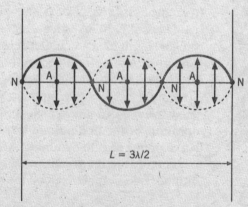

Figure 8.40 Third mode of vibration of a stretched string

The general expression for the frequency f_n of the nth mode (or the nth harmonic, or $(n-1)$th overtone) is

$$f_n = \frac{cn}{2L} \quad n = 1, 2, 3, \dots$$

The key features of a stationary wave pattern on a string, which distinguish it from a progressive wave, are as follows.

■ The nodes and antinodes do not move along the string, whereas in a progressive wave, the crests and troughs do move along it.
■ The amplitude of vibration varies with position along the string: it is zero at a node, and maximum at an antinode. In a progressive wave, all points have the same amplitude.
■ Between adjacent nodes, all points of the stationary wave vibrate in phase. That is, all particles of the string are at their maximum displacement at the same instant. In a progressive wave, phase varies continuously along the wave.

Stationary waves explained by interference

Let us explain the formation of a stationary wave using the principle of superposition. The set of graphs in Figure 8.41 represents two progressive waves of equal amplitude and frequency travelling in opposite directions. The red wave is travelling from left to right, and the blue wave is going from right to left.

The top graph catches the waves at an instant at which they are in phase. Superposition gives the purple curve, which has twice the amplitude of either of the progressive graphs. The second graph is the situation a quarter of a period (cycle) later, when the two progressive waves have each moved a quarter of a wavelength in opposite directions. This has brought them to a situation where the movement of one relative to the other is half a wavelength, so that the waves are exactly out of phase. The resultant, obtained by superposition, is zero everywhere. In the fourth graph, half a period from the start, the waves are again in phase, with maximum displacement for the resultant. The process continues through the fourth graph, showing the next out-of-phase situation, with zero displacement of the resultant everywhere. Finally, the fifth graph, one period on from the first, brings the waves into phase again.

We can see how there are some positions, the nodes N, where the displacement of the resultant is zero *throughout* the cycle. The displacement of the resultant at the antinodes A fluctuates from a maximum value when the two progressive waves are in phase to zero when they are out of phase.

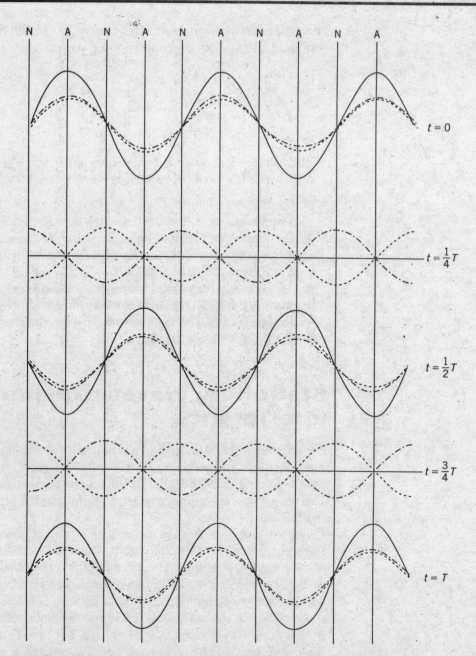

Figure 8.41 Formation of a stationary wave by superposition of two progressive waves travelling in opposite directions

Stationary waves in air

Figure 8.42 shows an experiment to demonstrate the formation of stationary waves in air. A fine, dry powder (such as cork dust or lycopodium powder) is sprinkled evenly along the transparent tube. A loudspeaker powered by a signal generator is placed at the open end. The frequency of the sound from the loudspeaker is gradually increased. At certain frequencies, the powder forms itself into evenly spaced heaps along the tube. A stationary wave has been set up in the air, caused by the interference of the sound wave from the loudspeaker and the wave reflected from the closed end of the tube. At nodes

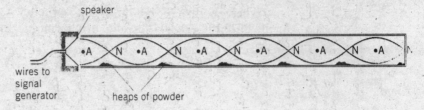

Figure 8.42 Demonstration of standing waves in air

(positions of zero amplitude) there is no disturbance, and the powder can settle into a heap. At antinodes the disturbance is at a maximum, and the powder is dispersed.

For stationary waves in a closed pipe, the air cannot move at the closed end, and so this must always be a node N. However, the open end is a position of maximum disturbance, and this is an antinode A. (In fact, the antinode is slightly outside the open end. The distance of the antinode from the end of the tube is called the end-correction. The value of the end-correction depends on the diameter of the tube.)

Figure 8.43 shows the simplest way in which the air in a pipe, closed at one end, can vibrate. Figure 8.43a illustrates the motion of some of the air particles in the tube. Their amplitude of vibration is zero at the closed end, and increases with distance up the tube to a maximum at the open end. This representation is tedious to draw, and Figure 8.43b is the conventional way of showing the amplitude of vibration: the amplitudes along the axis of the tube are plotted as a continuous curve. One danger of using diagrams like Figure 8.43b is that they give the impression that the sound wave is transverse rather than longitudinal. So be warned! The mode illustrated in Figure 8.43 is the fundamental mode (the first harmonic). The wavelength of this stationary wave (ignoring the end-correction) is $4L$, where L is the length of the pipe. Using the wave equation, the frequency f_1 of the fundamental mode is given by $f_1 = c/4L$, where c is the speed of the sound in air.

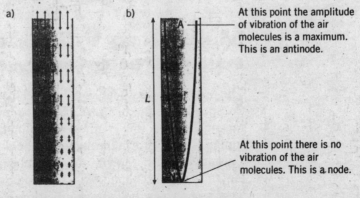

a)

b)

At this point the amplitude of vibration of the air molecules is a maximum. This is an antinode.

L

At this point there is no vibration of the air molecules. This is a node.

Figure 8.43 Fundamental mode of vibration of air in a closed pipe

$L = \dfrac{3\lambda}{4}$

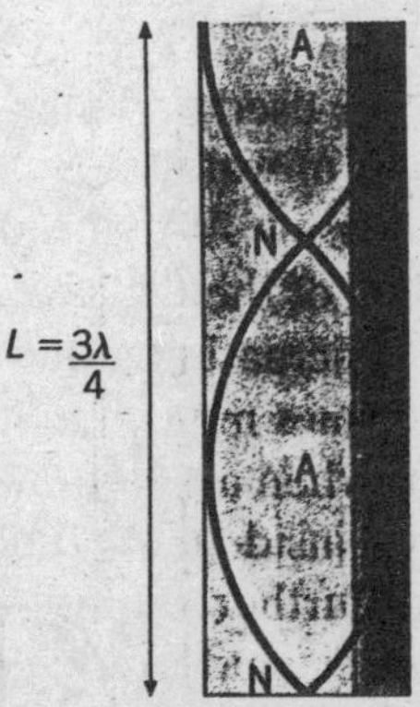

Figure 8.44 Second mode of vibration of air in a closed pipe

Other modes of vibration are possible. Figures 8.44 and 8.45 show the second mode (the first overtone, or second harmonic) and the third mode (the second overtone, or third harmonic). The corresponding wavelengths are $4L/3$ and $4L/5$, and the frequencies are $f_2 = 3c/4L$ and $f_3 = 5c/4L$.

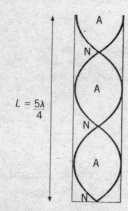

Figure 8.45 Third mode of vibration of air in a closed pipe

$$L = \frac{5\lambda}{4}$$

The general expression for the frequency f_n of the nth mode of vibration of the air in the closed tube (the nth harmonic, or the $(n-1)$th overtone) is

$$f_n = \frac{(2n-1)c}{4L}$$

This is another example of resonance. The particular frequencies at which stationary waves are obtained in the pipe are the resonant frequencies of the pipe.

Stationary waves using microwaves

Stationary waves can be demonstrated using microwaves. A source of microwaves faces a metal reflecting plate, as shown in Figure 8.46. A small detector is placed between source and reflector. The reflector is moved towards or away from the source until the signal picked up by the detector fluctuates regularly as it is moved slowly back and forth. The minima are nodes of the stationary wave pattern, and the maxima are antinodes. The distance moved by the detector between successive nodes is half the wavelength of the microwaves.

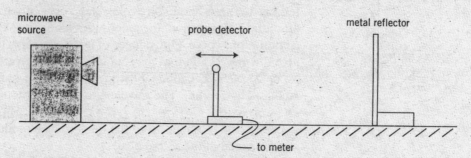

Figure 8.46 Using microwaves to demonstrate stationary waves

Measuring the speed of sound using stationary waves

The principle of resonance in a tube closed at one end can be used to measure the speed of sound in air. A glass tube is placed in a cylinder of water. By raising the tube, the length of the column of air can be increased. A vibrating tuning fork of known frequency f is held above the open end of the glass tube, causing the air in it to vibrate. The tube is gradually raised, increasing the length of the air column. At a certain position the note becomes much louder. This is known as the first position of resonance, and occurs when a stationary wave corresponding to the fundamental mode is established inside the tube. The length L_1 of the air column is noted. The tube is raised further until a second resonance position is found. This corresponds to the second mode of vibration. The length L_2 at this position is also noted. The two resonance positions are illustrated in Figure 8.47.

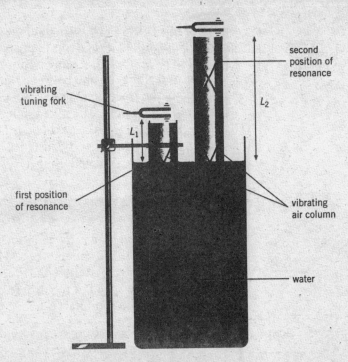

Figure 8.47 Speed of sound by the resonance tube method

At the first position of resonance, $\lambda/4 = L_1 + e$, where e is the end-correction of the tube (to allow for the fact that the antinode is slightly above the open end of the tube). At the second position of resonance, $3\lambda/4 = L_2 + e$. By subtracting these equations, we can eliminate e to give

$$\lambda/2 = L_2 - L_1$$

From the wave equation, the speed of sound c is given by $c = f\lambda$. Thus

$$c = 2f(L_2 - L_1)$$

Figure 8.48 illustrates a method of measuring the speed of sound using stationary waves in free air, rather than in a resonance tube. The signal generator and loudspeaker produce a note of known frequency f. The reflector

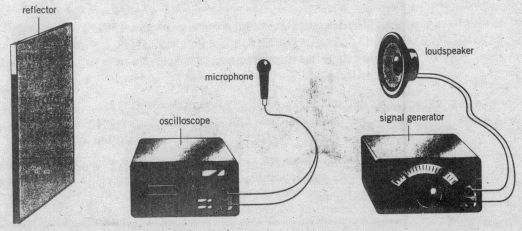

Figure 8.48 Speed of sound using stationary waves in free air

is moved slowly back and forth until the trace on the oscilloscope has a minimum amplitude. When this happens, a stationary wave has been set up with one of its nodes in the same position as the microphone. The microphone is now moved along the line between the loudspeaker and the reflector. The amplitude of the trace on the oscilloscope will increase to a maximum, and then decrease to a minimum. The microphone has been moved from one node, through an antinode, to the next node. The distance d between these positions is measured. We know that the distance between nodes is $\lambda/2$. The speed of sound can then be calculated using $c = f\lambda$, giving $c = 2fd$.

Examples

1 A string 75 cm long is fixed at one end. The other end is moved up and down with a frequency of 15 Hz. This frequency gives a stationary wave pattern with three complete loops on the string. Calculate the speed of the progressive waves which have interfered to produce the stationary wave.

The three-loop pattern corresponds to the situation where the length L of the string is $3\lambda/2$ (see Figure 8.40). The wavelength λ is thus $2 \times 0.75/3 = 0.50$ m. The frequency of the wave is 15 Hz, so by the wave equation $c = f\lambda = 15 \times 0.50 = \textbf{7.5 m s}^{-1}$.

2 Find the fundamental frequency and first two overtones for an organ pipe which is 0.17 m long and closed at one end. The speed of sound in air is 340 m s^{-1}.

The frequencies of the fundamental and first two overtones of a tube of length L, closed at one end, are $c/4L$, $3c/4L$ and $5c/4L$ (see Figures 8.43–8.45). The frequencies are thus $340/4 \times 0.17 = \textbf{500 Hz}$, $3 \times 340/4 \times 0.17 = \textbf{1500 Hz}$ and $5 \times 340/4 \times 0.17 = \textbf{2500 Hz}$.

Now it's your turn

1 A violin string vibrates with a fundamental frequency of 440 Hz. What are the frequencies of its first two overtones?

2 The speed of waves on a certain stretched string is 48 m s^{-1}. When the string is vibrated at frequency of 64 Hz, stationary waves are set up. Find the separation of successive nodes in the stationary wave pattern.

3 You can make an empty lemonade bottle resonate by blowing across the top. What fundamental frequency of vibration would you expect for a bottle 25 cm deep? The speed of sound in air is 340 m s^{-1}.

4 A certain organ pipe, closed at one end, can resonate at consecutive frequencies of 640 Hz, 896 Hz and 1152 Hz. Deduce its fundamental frequency.

Section 8.4 Summary

- A stationary wave is the result of interference between two progressive waves of equal frequency and amplitude travelling along the same line with the same speed, but in opposite directions.
- Points of zero amplitude on a stationary wave are called nodes; points of maximum amplitude are called antinodes.
- For stationary waves on a stretched string, frequency f_n of the nth mode is given by $f_n = cn/2L$, where c is the speed of progressive waves on the string and L is the length of the string.
- For stationary waves in air in a tube closed at one end, frequency f_n of the nth mode is given by $f_n = (2_n - 1)c/4L$, where c is the speed of sound in air and L is the length of the tube.

Section 8.4 Questions

1 State and explain four ways in which stationary waves differ from progressive waves.

2 A source of sound of frequency 2000 Hz is placed in front of a flat wall. When a microphone is moved away from the source towards the wall, a series of maxima and minima are detected.

(a) Explain what has happened to create these maxima and minima.

(b) The speed of sound in air is 340 m s^{-1}. Calculate the distance between successive minima.

3 A string is stretched between two fixed supports separated by 1.20 m. Stationary waves are generated on the string. It is observed that two stationary wave frequencies are 180 Hz and 135 Hz; there is no resonant frequency between these two. Calculate:

(a) the speed of progressive waves on the stretched string,

(b) the lowest resonant frequency of the string.

8.5 Diffraction

At the end of Section 8.5 you should be able to:
- explain the meaning of the term diffraction
- show an understanding of experiments which demonstrate diffraction including the diffraction of water waves in a ripple tank with both a wide gap and a narrow gap
- recall and solve problems using the formula $d \sin \theta = n\lambda$ and describe the use of a diffraction grating to determine the wavelength of light.

Although we often hear the statement 'light travels in straight lines', there are occasions when this appears not to be the case. Newton tried to explain the fact that when light travels through an aperture, or passes the edge of an obstacle, it deviates from the straight-on direction and appears to spread out.

We have seen from the ripple tank demonstration (Figure 8.11) that water waves spread out when they pass through an aperture. This effect is called **diffraction.** The fact that light undergoes diffraction is powerful evidence that light has wave properties. Newton's attempt to explain diffraction was not, in fact, based on a wave theory of light. The Dutch scientist Christian Huygens, a contemporary of Newton, favoured the wave theory, and used it to account for reflection, refraction and diffraction. (It was not until 1815 that the French scientist Augustin Fresnel developed the wave theory of light so as to explain diffraction in detail.)

The experiment illustrated in Figure 8.11 showed that the degree to which waves are diffracted depends upon the size of the obstacle or aperture and the wavelength of the wave. The greatest effects occur when the wavelength is about the same size as the aperture. The wavelength of light is very small (green light has wavelength 5×10^{-7} m), and therefore diffraction effects can be difficult to detect.

Huygens' explanation of diffraction

If we let a single drop of water fall into a ripple tank, it will create a circular wavefront which will spread outwards from the disturbance (Figure 8.8). Huygens put forward a wave theory of light which was based on the way in which circular wavefronts advance. He suggested that at any instant all points on a wavefront could be regarded as secondary disturbances, giving rise to their own outward-spreading circular wavelets. The envelope, or tangent curve, of the wavefronts produced by the secondary sources gives the new position of the original wavefront. This construction is illustrated in Figure 8.49 for a circular wavefront.

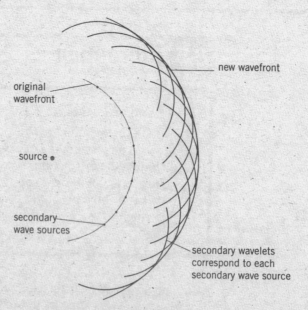

Figure 8.49 Huygens' construction for a circular wavefront

Figure 8.50 Diffraction of light at a single slit

Now think about a plane (straight) wavefront. If the wavefront is restricted in any way, for example by passing through an aperture in the form of a slit, some of the wavelets making up the wavefront are removed, causing the edges of the wavefront to be curved. If the wavelength is small compared with the size of the aperture, the wavefronts which pass through the aperture show curvature only at their ends, and the diffraction effect is relatively small. If the aperture is comparable with the wavelength, the diffracted wavefronts become circular, centred on the slit. Note that there is no change of wavelength of diffraction. This effect is illustrated in Figure 8.11.

Figure 8.50 shows the diffraction pattern created by a single slit illuminated by monochromatic light. The central region of the pattern is a broad, bright area with narrow, dark fringes on either side. Beyond these is a further succession of bright and dark areas. The bright areas become less and less intense as we move away from the centre.

This single-slit diffraction pattern has many features that we associate with an interference pattern. But how can a single slit produce an interference-type pattern? The explanation that follows is based on Huygens' wavelet idea.

Figure 8.51 shows plane wavefronts arriving at a single slit of width a. Each point on the wavefront passing through the slit can be considered to be a source of secondary wavelets. One such source is at A, at the top edge of the slit, and a second is at B, at the centre of the slit, a distance $a/2$ along the wavefront from A. These two sources behave like the sources in a two-source interference experiment. The wavelets spreading out from these points overlap and create an interference pattern. In the straight-on direction, there is no path difference between the waves from A and B. Constructive interference occurs in this direction, giving a bright fringe in the centre of the pattern. To either side of the central fringe there are directions where the path difference between the waves from A and B is an odd number of half-wavelengths. This is the condition for destructive interference, resulting in dark fringes. The condition for constructive interference is that the path difference should be a whole number of wavelengths. Thus, the dark fringes alternate with bright fringes.

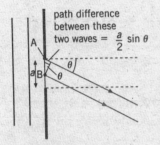

Figure 8.51 Light leaving a single slit

This argument can be applied to the whole of the slit. Every wavelet spreading out from a point in the top half of the slit can be paired with one coming from a point $a/2$ below it in the lower half of the slit. When wavelets from points right across the aperture are added up, we find that there are certain directions in which constructive interference occurs, and other directions in which the interference is destructive. Figure 8.52 is a graph of the intensity of the diffraction pattern as a function of the angle θ at which the light is viewed. It shows that most of the intensity is in the central area, and that this is flanked by dark and bright fringes.

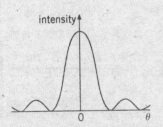

Figure 8.52 Light intensity graph for single slit diffraction

We can use Figure 8.51 to derive an expression for the angle at which the first dark fringe is obtained. Remember that the condition for destructive interference is that the path difference between the two rays should be half a wavelength. The path difference between the two rays shown in Figure 8.51 is $\frac{1}{2}a \sin \theta$. If this is to be $\frac{1}{2}\lambda$, we have

$$\sin \theta = \frac{\lambda}{a}$$

This is the condition to observe the first dark fringe at angle θ. More generally,

$$\sin \theta = \frac{n\lambda}{a}$$

where n is a whole number called the *order* of the dark fringe being considered, counting outwards from the centre.

Although we have been concentrating on a diffraction pattern obtained with light, the derivation above applies to any type of wave passing through a rectangular aperture.

The wavelength of light is generally small compared with the width of slits or other apertures, so the diffraction angle θ is also small. Provided that θ is only a few degrees (less than about 5°), the approximation $\sin \theta = \theta$ may be used (remember that θ must be in radians!). Very often, the single-slit diffraction equation for light is expressed in the form

$$\theta = \frac{n\lambda}{a}$$

making use of the $\sin \theta$ approximation. But take care! This approximate form may not apply for the diffraction of other types of wave, such as sound or water waves, where the wavelength may be closer in magnitude to the width of the aperture, and diffraction angles are larger.

Example

Calculate the angle between the centre of the diffraction pattern and the first minimum when light of wavelength 600 nm passes through a slit 0.10 mm wide.

Using $a \sin \theta = n\lambda$, we have $\sin \theta = n\lambda/a$.
Substituting, $\sin \theta = 1 \times 6.0 \times 10^{-7}/1.0 \times 10^{-4}$ (don't forget to convert the nm and mm to m) $= 0.0060$, and $\theta = $ **0.34°**.

Using the $\sin \theta = \theta$ approximation, we would have obtained $\theta = n\lambda/a$ $= 1 \times 6.0 \times 10^{-7}/1.0 \times 10^{-4} = $ **0.0060 rad** (which is equal to 0.34°).

Now it's your turn

1 Calculate the angle between the centre of the diffraction pattern and the first minimum when a sound wave of wavelength 1.0 m passes through a door 1.2 m wide.

2 Calculate the wavelength of water waves which, on passing through a gap 50 cm wide, create a diffraction pattern such that the angle between the centre of the pattern and the second-order minimum is 60°.

The diffraction grating

A **diffraction grating** is a plate on which there is a very large number of parallel, identical, very closely spaced slits. If monochromatic light is incident on this plate, a pattern of narrow bright fringes is produced.

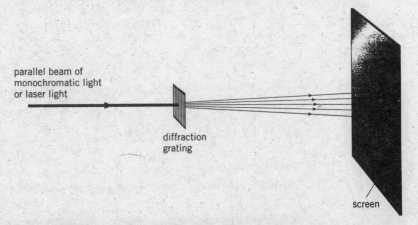

Figure 8.53 Arrangement for obtaining a fringe pattern with a diffraction grating

Although the device is called a *diffraction* grating, we shall use straight-forward superposition and interference ideas in obtaining an expression for the angles at which the maxima of intensity are obtained.

Figure 8.54 shows a parallel beam of light incident normally on a diffraction grating in which the spacing between adjacent slits is d. Consider first rays 1 and 2 which are incident on adjacent slits. The path difference between these rays when they emerge at an angle θ is $d \sin \theta$. To obtain constructive interference in this direction from these two rays, the condition is that the path difference should be an integral number of wavelengths. The path difference between rays 2 and 3, 3 and 4, and so on, will also be $d \sin \theta$. The condition for constructive interference is the same. Thus, the condition for a maximum of intensity at angle θ is

$$d \sin \theta = n\lambda$$

where λ is the wavelength of the monochromatic light used, and n is a whole number.

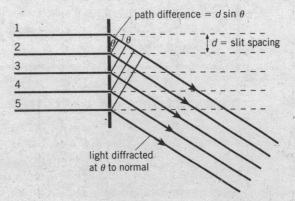

Figure 8.54

When $n = 0$, $\sin\theta = 0$ and θ is also zero; this gives the straight-on direction, or what is called the zero-order maximum. When $n = 1$, we have the first-order diffraction maximum, and so on (Figure 8.55).

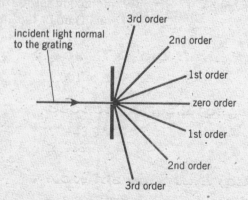

Figure 8.55 Maxima in the diffraction pattern of a diffraction grating

Example

Monochromatic light is incident normally on a grating with 7.00×10^5 lines per metre. A second-order maximum is observed at an angle of diffraction of 40.0°. Calculate the wavelength of the incident light.

The slits on a diffraction grating are created by drawing parallel lines on the surface of the plate. The relationship between the slit spacing d and the number N of lines per metre is $d = 1/N$. For this grating, $d = 1/7.00 \times 10^5 = 1.43 \times 10^{-6}$ m. Using $n\lambda = d\sin\theta$, $\lambda = (d/n)\sin\theta = (1.43 \times 10^{-6}/2)\sin 40.0° = \textbf{460 nm}$.

Now it's your turn

1 Monochromatic light is incident normally on a grating with 5.00×10^5 lines per metre. A third-order maximum is observed at an angle of diffraction of 78.0°. Calculate the wavelength of the incident light.

2 Light of wavelength 5.90×10^{-7} m is incident normally on a diffraction grating with 8.00×10^5 lines per metre. Calculate the diffraction angles of the first- and second-order diffraction images.

3 Light of wavelength 590 nm is incident normally on a grating with spacing 1.67×10^{-6} m. How many orders of diffraction maxima can be obtained?

The diffraction grating with white light

If white light is incident on a diffraction grating, each wavelength λ making up the white light is diffracted by a different amount, as described by the equation $n\lambda = d\sin\theta$. Red light, because it has the longest wavelength in the visible spectrum, is diffracted through the largest angle. Blue light has the shortest wavelength, and is diffracted the least. Thus, the white light is split into its component colours, producing a continuous spectrum (Figure 8.56).

The spectrum is repeated in the different orders of the diffraction pattern. Depending on the grating spacing, there may be some overlapping of different orders. For example, the red component of the first-order image may overlap with the blue end of the second-order spectrum.

An important use of the diffraction grating is in a **spectrometer**, a piece of apparatus used to investigate spectra. By measuring the angle at which a particular diffracted image appears, the wavelength of the light producing that image may be determined.

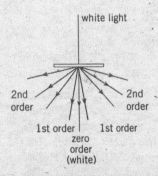

Figure 8.56 Production of the spectrum of white light with a diffraction grating

Section 8.5 Summary

- Diffraction is the spreading out of waves after passing through an aperture or meeting an obstacle. It is most obvious when the size of the aperture and the wavelength of the wave are approximately the same.
- Single slit diffraction: the condition for minima is $a \sin \theta = n\lambda$, where a is the width of the slit.
- The condition for a diffraction maximum in a diffraction grating pattern is $d \sin \theta = n\lambda$, where d is the grating spacing, θ is the angle at which the diffraction maximum is observed, n is an integer (the order of the image), and λ is the wavelength of the light.

Section 8.5 Questions

1 Blue and red light, with wavelengths 450 nm and 650 nm respectively, is incident normally on a diffraction grating which has 4.0×10^5 lines per metre.
   (a) Calculate the grating spacing.
   (b) Calculate the angle between the second-order maxima for these wavelengths.
   (c) For each wavelength, find the maximum order that can be observed.

2 Discuss any difference between the interference patterns formed by (a) two parallel slits 1 μm apart, (b) a diffraction grating with grating spacing 1 μm, when illuminated with monochromatic light.

3 Light of wavelength 633 nm passes through a slit 50 μm wide. Calculate the angular separation between the central maximum and the first minimum of the diffraction pattern.

Exam-style Questions

1 Assume that waves spread out uniformly in all directions from the epicentre of an earthquake. The intensity of a particular earthquake wave is measured as 5.0×10^6 W m^{-2} at a distance of 40 km from the epicentre. What is the intensity at a distance of only 2 km from the epicentre?

2 A string is stretched between two fixed supports 3.5 m apart. Stationary waves are generated by disturbing the string. One possible mode of vibration of the stationary waves is shown in Figure 8.57. The nodes and antinodes are labelled N and A respectively.

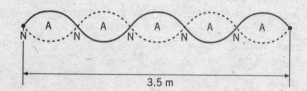

3.5 m

Figure 8.57

(a) Distinguish between a node and an antinode in a stationary wave.
(b) State the phase difference between the vibrations of particles of the string at any two neighbouring antinodes.
(c) Calculate the ratio of the frequency of the mode of vibration shown in Figure 8.57 to the frequency of the fundamental mode of vibration of the string.
(d) The frequency of the mode of vibration shown in Figure 8.57 is 160 Hz. Calculate the speed of the progressive waves which produced this stationary wave.

3 Explain why light will not pass through two sheets of Polaroid when they are arranged in contact with their polarising directions at right-angles. What would happen if a third sheet of Polaroid, with its polarising direction at 45° to the polarising directions of the other two, were placed between them? (If you can get hold of Polaroid filters, it is worth trying this experiment, starting with thinking about how you would determine the polarising direction of each!)

4 A vibrating tuning fork of frequency 320 Hz is held over the open end of a resonance tube. The other end of the tube is immersed in water. The length of the air column is gradually increased until resonance first occurs. Taking the speed of sound in air as 340 m s^{-1}, calculate the length of the air column. (Neglect any end-correction.)

5 We can hear sounds round corners. We cannot see round corners. Both sound and light are waves. Explain why sound and light seem to behave differently.

6 Figure 8.58 shows a narrow beam of monochromatic laser light incident normally on a diffraction grating. The central bright spot is formed at O.

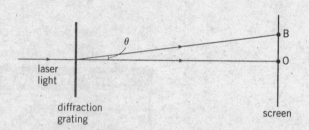

Figure 8.58

(a) Write down the relationship between the wavelength λ of the light and the angle θ for the first diffraction image formed at B. Identify any other symbol used.
(b) The screen is 1.1 m from the diffraction grating and the grating has 300 lines per mm. The laser light has wavelength 6.3×10^{-7} m. Find the distance OB from the central spot to the first bright image at B.
(c) The diffraction grating is now replaced by one which has 600 lines per mm. For this second grating, calculate the distance from the central spot to the first bright image.

9. Photons, electrons and atoms

In the last chapter we spent a long time in describing properties of waves, and in emphasising experiments which showed that light is a wave motion. We now look at phenomena which, during the last hundred years, have completely revolutionised physics. A number of experiments around the 1900s showed that light behaves more like a stream of particles than as a wave. Later, it was shown experimentally that electrons, which had always been thought of as particles, could also behave as waves. This idea of *wave–particle duality* is one of the most important concepts of modern physics. We start by describing the photoelectric effect, the phenomenon that demonstrated that light had a particle nature. This is the work which Einstein first explained in 1905, and for which he received the 1921 Nobel Prize. We shall then describe the diffraction of electrons. Remember that diffraction is a characteristic of waves, yet electrons have always been regarded as particles!

9.1 The photoelectric effect

At the end of Section 9.1 you should be able to:
- show an appreciation of the particulate nature of electromagnetic radiation
- show an understanding that the photoelectric effect provides evidence for a particulate nature of electromagnetic radiation, while phenomena such as interference and diffraction provide evidence for a wave nature
- recall and use $E = hf$
- recall the significance of threshold frequency
- explain photoelectric phenomena in terms of photon energy and work function energy
- explain why the maximum kinetic energy of photoelectrons is independent of intensity, whereas the photoelectric current is proportional to intensity
- recall, use and explain the significance of $hf = \phi + \frac{1}{2}mv_{max}^2$

Some of the electrons in a metal are free to move around in it. (It is these free electrons that form the electric current when a potential difference is applied across the ends of a metal wire.) However, to remove free electrons from a metal requires energy, because they are held in the metal by the electrostatic attraction of the positively charged nuclei. If an electron is to escape from the surface of a metal, work must be done on it. The electron must be given energy. When this energy is in the form of light energy, the phenomenon is called **photoelectric emission**.

> Photoelectric emission is the release of electrons from the surface of a metal when electromagnetic radiation is incident on its surface.

Demonstration of photoelectric emission

A clean zinc plate is placed on the cap of a gold-leaf electroscope. The electroscope is then charged negatively, and the gold leaf deflects, proving that the zinc plate is charged. This is illustrated in Figure 9.1. If visible light of any colour is shone on to the plate, the leaf does not move. Even when the intensity (the brightness) of the light is increased, the leaf remains in its deflected position. However, when ultra-violet radiation is shone on the plate the leaf falls immediately, showing that it is losing negative charge. This means that electrons are being emitted from the zinc plate. These electrons are called **photoelectrons**. If the intensity of the ultra-violet radiation is increased, the leaf falls more quickly, showing that the rate of emission of electrons has increased.

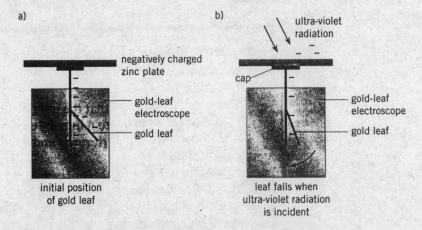

Figure 9.1 Demonstration of photoelectric emission

The difference between ultra-violet radiation and visible light is that ultra-violet radiation has a shorter wavelength and a higher frequency than visible light.

> Further investigations with apparatus like this lead to the following conclusions:
>
> - If photoemission takes place, it does so instantaneously. There is no delay between illumination and emission.
> - Photoemission takes place only if the frequency of the incident radiation is above a certain minimum value called the **threshold frequency** f_0.

- Different metals have different threshold frequencies.
- Whether or not emission takes place depends only on whether the frequency of the radiation used is above the threshold for that surface. It does not depend on the intensify of the radiation.
- For a given frequency, the rate of emission of photoelectrons is proportional to the intensity of the radiation.

Another experiment, using the apparatus shown in Figure 9.2, can be carried out to investigate the energies of the photoelectrons. If ultra-violet radiation of a fixed frequency (above the threshold) is shone on to the metal surface A, it emits photoelectrons. Some of these electrons travel from A to B. Current is detected using the microammeter. If a potential difference is applied between A and B, with B negative with respect to A, any electron going from A to B will gain potential energy as it moves against the electric field. The gain in potential energy is at the expense of the kinetic energy of the electron. That is,

loss in kinetic energy = gain in potential energy
= charge of electron × potential difference

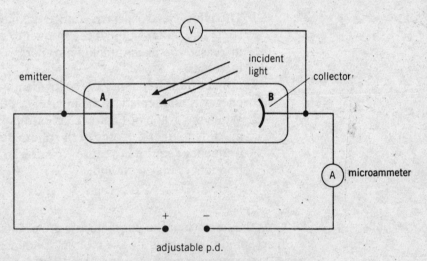

Figure 9.2 Experiment to measure the maximum kinetic energy of photoelectrons

If the voltage between A and B is gradually increased, the current registered on the microammeter decreases and eventually falls to zero. The minimum value of the potential difference necessary to stop the electron flow is known as the stopping potential. It measures the maximum kinetic energy with which the photoelectrons are emitted. The fact that there is a current in the microammeter at voltages less than the stopping potential indicates that there is range of kinetic energies for these electrons.

If the experiment is repeated with radiation of greater intensity but the same frequency, the maximum current in the microammeter increases, but the value of the stopping potential is unchanged.

The experiment can be repeated using ultra-violet radiation of different frequencies, measuring the stopping potential for each frequency. When the maximum kinetic energy of the photoelectrons is plotted against the frequency of the radiation, the graph of Figure 9.3 is obtained.

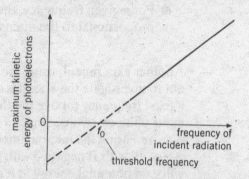

Figure 9.3 Graph of maximum kinetic energy of photoelectrons against frequency of radiation

The following conclusions are drawn from this experiment:

1 The photoelectrons have a range of kinetic energies, from zero up to some maximum value. If the frequency of the incident radiation is increased the maximum kinetic energy of the photoelectrons also increases.

2 For constant frequency of the incident radiation, the maximum kinetic energy is unaffected by the intensity of the radiation.

3 When the graph of Figure 9.3 is extrapolated to the point where the maximum kinetic energy of the photoelectrons is zero, the minimum frequency required to cause emission from the surface (the threshold frequency) may be found.

At the time when the photoelectric effect was first being studied, it was fully accepted that light is a wave motion. Evidence for this came from observed interference and diffraction effects. The conclusions of experiments on photoemission produced doubt as to whether light is a continuous wave. One of the main problems concerns the existence of a threshold frequency.

Classical wave theory predicts that when an electromagnetic wave (that is, light) interacts with an electron, the electron will absorb energy from it. So, if an electron absorbs enough energy it should be able to escape from the metal. Remember from Chapter 8 that the energy carried by a wave depends on its amplitude and its frequency. Thus, even if we have a low-frequency wave, its energy can be boosted by increasing the amplitude (that is, by increasing the brightness of the light). So, according to wave theory, we ought to be able to cause photoemission by using any frequency of light, provided we make it bright enough. Alternatively, we could use less bright light and shine it on the metal for a longer time, until enough energy to cause emission has been delivered. But this does not happen. The experiments we have described above showed conclusively that radiation of frequency below the threshold, no matter how intense or for how long it is used, does not produce photoelectrons. The classical wave theory of electromagnetic radiation leads to the following predictions:

1 Whether an electron is emitted or not should depend on the power of the incident wave; that is, on its intensity. A very intense wave, of any frequency, should cause photoemission.
2 The maximum kinetic energy of the photoelectrons should be greater if the radiation intensity is greater.
3 There is no reason why photoemission should be instantaneous.

These predictions, based on wave theory, do not match the observations. A new approach, based on an entirely new concept, the **quantum theory**, was used to explain these findings.

Einstein's theory of photoelectric emission

In 1901, the German physicist Max Planck had suggested that the energy carried by electromagnetic radiation might exist as discrete packets called **quanta**. The energy E carried in each quantum is given by

$$E = hf$$

where f is the frequency of the radiation and h is a constant called the Planck constant. The value of the Planck constant is 6.63×10^{-34} J s.

In 1905, Albert Einstein developed the theory of quantised energy to explain all the observations associated with photoelectric emission. He proposed that light radiation consists of a stream of energy packets called **photons**.

A photon is the special name given to a quantum of energy when the energy is in the form of electromagnetic radiation.

When a photon interacts with an electron, it transfers all its energy to the electron. It is only possible for a single photon to interact with a single electron; the photon cannot share its energy between several electrons. This transfer of energy is instantaneous.

The photon theory of photoelectric emission is as follows. If the frequency of the incident radiation is less than the threshold frequency for the metal, the energy carried by each photon is insufficient for an electron to escape the surface of the metal. If the photon energy is insufficient for an electron to escape, it is converted to thermal energy in the metal.

The minimum amount of energy necessary for an electron to escape from the surface is called the **work function energy** ϕ.

Figure 9.4 Albert Einstein

Some values for the work function energy ϕ and threshold frequency f_0 of different metals are given in Table 9.1.

Table 9.1 Work function energies and threshold frequencies

metal	ϕ/J	ϕ/eV	f_0/Hz
sodium	3.8×10^{-19}	2.4	5.8×10^{14}
calcium	4.6×10^{-19}	2.9	7.0×10^{14}
zinc	5.8×10^{-19}	3.6	8.8×10^{14}
silver	6.8×10^{-19}	4.3	1.0×10^{15}
platinum	9.0×10^{-19}	5.6	1.4×10^{15}

Remember: $1\,\text{eV} = 1.6 \times 10^{-19}\text{J}$.

If the frequency of the incident radiation is equal to the threshold frequency, the energy carried by each photon is just sufficient for electrons at the surface to escape. If the frequency of the incident radiation is greater than the threshold frequency, surface electrons will escape and have surplus energy in the form of kinetic energy. These electrons will have the maximum kinetic energy. If a photon interacts with an electron below the surface, some energy is used to take the electron to the surface, so that it is emitted with less than the maximum kinetic energy. This gives rise to a range of values of kinetic energy.

Einstein used the principle of conservation of energy to derive the photoelectric equation

photon energy = work function energy + maximum kinetic energy of photoelectron

or

$$hf = \phi + \tfrac{1}{2} m_e v_{max}^2$$

For radiation incident at the threshold frequency, $\tfrac{1}{2} m_e v_{max}^2 = 0$, so that $hf_0 = \phi$. The photoelectric equation can then be written

$$hf = hf_0 + \tfrac{1}{2} m_e v_{max}^2$$

Example

The work function energy of platinum is 9.0×10^{-19} J. Calculate:
(a) the threshold frequency for the emission of photoelectrons from platinum,
(b) the maximum kinetic energy of a photoelectron when radiation of frequency 2.0×10^{15} Hz is incident on a platinum surface.

(a) Using $hf_0 = \phi$, $f_0 = \phi/h$, so

$$f_0 = 9.0 \times 10^{-19}/6.6 \times 10^{-34} = \mathbf{1.4 \times 10^{15}\ Hz}$$

(b) Using $hf = hf_0 + \frac{1}{2}m_e v_{max}^2$, $hf - hf_0 = \frac{1}{2}m_e v_{max}^2$ and

$$\tfrac{1}{2}m_e v_{max}^2 = 6.6 \times 10^{-34}\,(2.0 \times 10^{15} - 1.4 \times 10^{15}) = \mathbf{4.0 \times 10^{-19}\,J}$$

Now it's your turn

1 The work function energy of silver is 4.3 eV. Show that the threshold frequency is about 1.0×10^{15} Hz.

2 Electromagnetic radiation of frequency 3.0×10^{15} Hz is incident on the surface of sodium metal. The emitted photoelectrons have a maximum kinetic energy of 1.6×10^{-18} J. Calculate the threshold frequency for photoemission from sodium.

Section 9.1 Summary

- Electrons may be emitted from metal surfaces if the metal is illuminated by electromagnetic radiation. This phenomenon is called photoelectric emission.
- Photoelectric emission cannot be explained by the wave theory of light. It is necessary to use the quantum theory, in which electromagnetic radiation is thought of as consisting of packets of energy called photons.
- The energy E of a photon is given by: $E = hf$, where h is the Planck constant and f is the frequency of the radiation.
- The work function energy ϕ of a metal is the minimum energy needed to free an electron from the surface of the metal.
- The Einstein photoelectric equation is: $hf = \phi + \frac{1}{2}m_e v_{max}^2$
- The threshold frequency f_0 is given by: $hf_0 = \phi$

Section 9.1 Questions

1 Calculate the energy range, in eV, of photons in the visible region of the electro-magnetic spectrum (wavelengths 400 nm–700 nm).
(Planck constant $h = 6.6 \times 10^{-34}$ J s; speed of light $c = 3.0 \times 10^8$ m s^{-1}; electron charge $e = -1.6 \times 10^{-19}$ C)

2 The threshold frequency for photoemission from a certain metal is 8.8×10^{14} Hz.
(a) To what wavelength does this frequency correspond?
(b) Calculate the maximum kinetic energy of the emitted photoelectrons when the metal is illuminated with ultra-violet radiation of wavelength 240 nm.
(Planck constant $h = 6.6 \times 10^{-34}$ J s; speed of light $c = 3.0 \times 10^8$ m s^{-1})

3 In a stopping-potential experiment using ultra-violet radiation of wavelength 265 nm, the photocurrent from a certain metal is reduced to zero at an applied potential difference of 1.18 V. Calculate the work function energy of the metal.
(Planck constant $h = 6.63 \times 10^{-34}$ J s; speed of light $c = 3.00 \times 10^8$ m s^{-1}; electron charge $e = -1.60 \times 10^{-19}$ C)

9.2 Wave–particle duality

At the end of Section 9.2 you should be able to:
- describe and explain qualitatively the evidence provided by electron diffraction for the wave nature of particles
- recall and use the relation for the de Broglie wavelength $\lambda = h/p$

Figure 9.5 X-ray diffraction pattern of a metal foil

Figure 9.6 Electron diffraction pattern of graphite

If light waves can behave like particles (photons), perhaps moving particles can behave like waves?

If a beam of X-rays of a single wavelength is directed at a thin metal foil, a diffraction pattern is produced (Figure 9.5). This is a similar effect to the diffraction pattern produced when light passes through a diffraction grating (see Chapter 8). The foil contains many tiny crystals. The gaps between neighbouring planes of atoms in the crystals act as slits, creating a diffraction pattern.

If a beam of electrons is directed at a thin metal foil, a similar diffraction pattern is produced, as shown in Figure 9.6. The electrons, which we normally consider to be particles, are exhibiting a property we would normally associate with waves. Remember that, to observe diffraction, the wavelength of the radiation should be comparable with the size of the aperture. The separation of planes of atoms in crystals is of the order of 10^{-10} m. The fact that diffraction is observed with electrons suggest that they have a wavelength of about the same magnitude.

In 1924 the French physicist Louis de Broglie suggested that all moving particles have a wave-like nature. Using ideas based upon the quantum theory and Einstein's theory of relativity, he suggested that the momentum p of a particle and its associated wavelength λ are related by the equation

$$\lambda = \frac{h}{p}$$

where h is the Planck constant. λ is known as the **de Broglie wavelength**.

Example

Calculate the de Broglie wavelength of an electron travelling with a speed of $1.0 \times 10^7 \text{ m s}^{-1}$.
(Planck constant $h = 6.6 \times 10^{-34}$ J s; electron mass $m_e = 9.1 \times 10^{-31}$ kg)

Using $\lambda = h/p$ and $p = mv$,
$\lambda = 6.6 \times 10^{-34}/9.1 \times 10^{-31} \times 1.0 \times 10^7 = \mathbf{7.3 \times 10^{-11}}$ **m**

Now it's your turn

1 Calculate the de Broglie wavelength of an electron travelling with a speed of $5.5 \times 10^7 \text{ m s}^{-1}$.
(Planck constant $h = 6.6 \times 10^{-34}$ J s; electron mass $m_e = 9.1 \times 10^{-31}$ kg)

2 Calculate the de Broglie wavelength of an electron which has been accelerated from rest through a potential difference of 100 V.
(Planck constant $h = 6.6 \times 10^{-34}$ J s; electron mass $m_e = 9.1 \times 10^{-31}$ kg; electron charge $e = -1.6 \times 10^{-19}$ C)

Section 9.2 Summary

- Moving particles show wave-like properties.
- The de Broglie wavelength $\lambda = h/p$, where p is the momentum of the particle and h is the Planck constant.

Section 9.2 Questions

1 Calculate the de Broglie wavelength of an electron with energy 1.0 keV.
(Planck constant $h = 6.6 \times 10^{-34}$ J s; electron mass $m_e = 9.1 \times 10^{-31}$ kg; 1 eV = 1.6×10^{-19} J)

2 Calculate the speed of a neutron with de Broglie wavelength 1.5×10^{-10} m.
(Planck constant $h = 6.6 \times 10^{-34}$ J s; neutron mass $m_n = 1.7 \times 10^{-27}$ kg)

9.3 Emission spectra

At the end of Section 9.3 you should be able to:
- show an understanding of the existence of discrete electron energy levels in isolated atoms (e.g atomic hydrogen) and deduce how this leads to spectral lines
- distinguish between emission and absorption line spectra
- recall and solve problems using the relation $hf = E_2 - E_1$
- state that all electromagnetic waves travel with the same speed in free space and recall the orders of magnitude of the principal radiations from radio waves to γ-rays.

Continuous and line spectra

If white light from a tungsten filament lamp is passed through a prism, the light is dispersed into its component colours, as illustrated in Figure 9.7. The band of different colours is called a **continuous spectrum**. A continuous spectrum has all colours (and wavelengths) between two limits. In the case of white light, the colour and wavelength limits are violet (about 400 nm) and red (about 700 nm). Since this spectrum has been produced by the emission of light from the tungsten filament lamp, it is referred to as an **emission spectrum**. Finer detail of emission spectra than is obtained using a prism may be achieved using a diffraction grating.

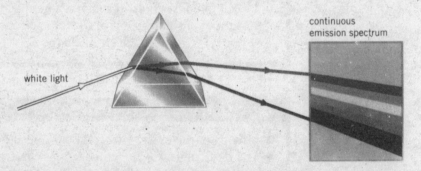

continuous
emission spectrum

white light

Figure 9.7 Continuous spectrum of white light from a tungsten filament lamp

A discharge tube is a transparent tube containing a gas at low pressure. When a high potential difference is placed across two electrodes in the tube, light is emitted. Examination of the light with a diffraction grating shows that the emitted spectrum is no longer continuous, but consists of a number of bright lines (Figure 9.8).

Such a spectrum is known as a **line spectrum**. It consists of a number of separate colours, each colour being seen as the image of the slit in front of the source. The wavelengths corresponding to the lines of the spectrum are characteristic of the gas which is in the discharge tube.

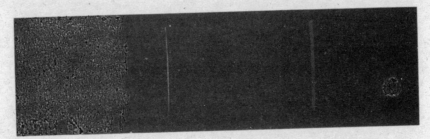

Figure 9.8 Line spectrum of hydrogen from a discharge tube

Electron energy levels in atoms

To explain how line spectra are produced we need to understand how electrons in atoms behave. Electrons in an atom can have only certain specific energies. These energies are called the **electron energy levels** of the atom. The energy levels may be represented as a series of lines against a vertical scale of energy, as illustrated in Figure 9.9. The electron in the hydrogen atom can have any of these energy values, but cannot have energies between them.

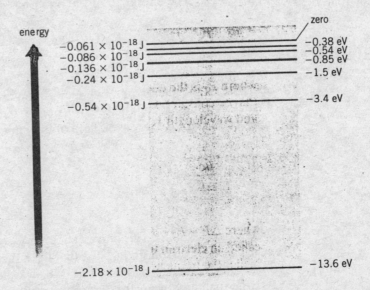

Figure 9.9 Electron energy levels for the hydrogen atom

Normally electrons occupy the lowest energy levels available. Under these conditions the atom and its electrons are said to be in the **ground state**. Figure 9.10a represents a hydrogen atom with its single electron in the lowest energy state.

If, however, the electron absorbs energy, perhaps by being heated, or by collision with another electron, it may be promoted to a higher energy level. The energy absorbed is exactly equal to the difference in energy of the two levels. Under these conditions the atom is described as being in an **excited state**. This is illustrated in Figure 9.10b.

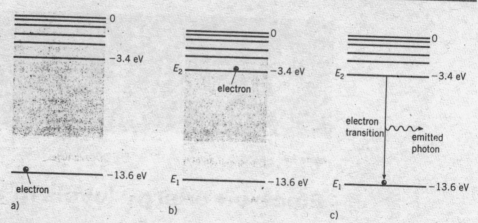

Figure 9.10 Hydrogen atom a) in its ground state, b) in an excited state, c) returning to its ground state with photon emission

An excited atom is unstable. After a short time, the excited electron will return to a lower level. To achieve this, the electron must lose energy. It does so by emitting a photon of electromagnetic radiation, as illustrated in Figure 9.10c. The energy hf of the photon is given by

$$hf = E_2 - E_1$$

where E_2 is the energy of the higher level and E_1 that of the lower, and h is the Planck constant. Using the relation between the speed c of light, frequency f and wavelength λ, the wavelength of the emitted radiation is given by

$$\lambda = \frac{hc}{\Delta E}$$

where $\Delta E = E_2 - E_1$. This movement of an electron between energy levels is called an **electron transition**. Note that the larger the energy of the transition, the higher the frequency (and the shorter the wavelength) of the emitted radiation.

Note that this downward transition results in the **emission** of a photon. The atom can be raised to an excited state by the **absorption** of a photon, but the photon must have just the right energy, corresponding to the difference in energy of the excited state and the initial state. So, a downward transition corresponds to photon emission, and an upward transition to photon absorption.

Spectroscopy

Figure 9.11 shows some of the possible transitions that might take place when electrons in an excited atom return to lower energy levels. Each of the transitions results in the emission of a photon with a particular wavelength. For example, the transition from E_4 to E_1 results in light with the highest frequency and shortest wavelength. On the other hand, the transition from E_4 to E_3 gives the lowest frequency and longest wavelength.

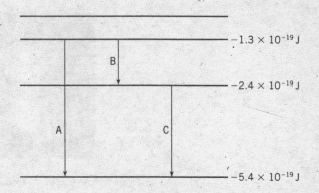

Figure 9.11 Some possible electron transitions

Because all elements have different energy levels, the energy differences are unique to each element. Consequently, each element produces a different and characteristic line spectrum. Spectra can be used to identify the presence of a particular element. The line spectrum of mercury is shown in Figure 9.12. The study of spectra is called **spectroscopy**, and instruments used to measure the wavelengths of spectra are **spectrometers**. Spectrometers for accurate measurement of wavelength make use of diffraction gratings (see Chapter 8) to disperse the light.

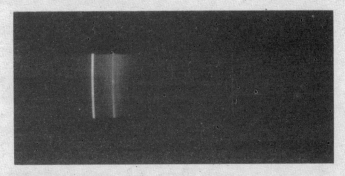

Figure 9.12 Mercury line spectrum

Continuous spectra

Whilst the light emitted by isolated atoms such as those in low-pressure gases produces line spectra, the light emitted by atoms in a solid, a liquid, or a gas at high pressure produces a continuous spectrum. This happens because of the proximity of the atoms to each other. Interaction between the atoms results in a broadening of the electron energy levels. Consequently, transitions of a wide range of magnitudes of energy are possible, and light of a broad spread of wavelengths may be emitted. This is seen as a **continuous spectrum**.

Absorption spectra

If white light passes through a low-pressure gas and the spectrum of the white light is then analysed, it is found that light of certain wavelengths is missing. In their place are dark lines. This type of spectrum is called an **absorption spectrum**; one is shown in Figure 9.13. As the white light passes through the

Figure 9.13 An absorption spectrum

gas, some electrons absorb energy and make transitions to higher energy levels. The wavelengths of the light they absorb correspond exactly to the energies needed to make the particular upward transitions. When these excited electrons return to lower levels, the photons are emitted in all directions, rather than in the original direction of the white light. Thus, some wavelengths appear to be missing. It follows that the wavelengths missing from an absorption spectrum are those present in the emission spectrum of the same element. This is illustrated in Figure 9.14.

Figure 9.14 Relation between an absorption spectrum and the emission spectrum of the same element: a) spectrum of white light, b) absorption spectrum of element, c) emission spectrum of same element

The electromagnetic spectrum

Visible light is just a small region of the **electromagnetic spectrum**. All electromagnetic waves are transverse waves, consisting of electric and magnetic fields which oscillate at right angles to each other and to the direction in which the wave is travelling. This is illustrated in Figure 9.15.

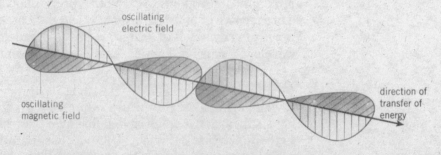

oscillating
electric field

oscillating
magnetic field

direction of
transfer of
energy

Figure 9.15 Oscillating electric and magnetic fields in an electromagnetic wave

Electromagnetic (e.m.) waves show all the properties common to wave motions (see Chapter 8): they can be reflected, refracted, and diffracted; they obey the principle of superposition and produce interference patterns. Because they are transverse waves, they can, in addition, be polarised (see Section 8.2). In a vacuum all electromagnetic waves travel at the same speed, 3.00×10^8 m s^{-1}.

The complete electromagnetic spectrum is divided into a series of regions based on the properties of electromagnetic waves in these regions, as illustrated in Figure 9.16. It should be noted that there is no clear boundary between regions.

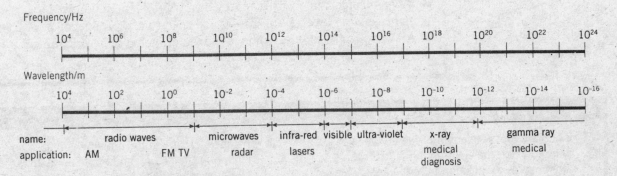

Figure 9.16 The electromagnetic spectrum

Example

Calculate the wavelength of the radiation emitted when the electron in a hydrogen atom makes a transition from the energy level at -0.54×10^{-18} J to the level at -2.18×10^{-18} J.
(Planck constant $h = 6.62 \times 10^{-34}$ J s; speed of light $c = 3.00 \times 10^8$ m s^{-1})

Here $\Delta E = E_2 - E_1 = -0.54 \times 10^{-18} - (-2.18 \times 10^{-18}) = 1.64 \times 10^{-18}$ J
Using $\lambda = hc/\Delta E$,
$\lambda = 6.62 \times 10^{-34} \times 3.00 \times 10^8/1.64 \times 10^{-18} = 1.21 \times 10^{-7}$ m = **121 nm**

Now it's your turn

1 Calculate the wavelength of the radiation emitted when the electron in a hydrogen atom makes a transition from the energy level at -3.4×10^{-18} J to the level at -8.5×10^{-18} J.
(Planck constant $h = 6.6 \times 10^{-34}$ J s; speed of light $c = 3.0 \times 10^8$ m s^{-1})

2 The electron in a hydrogen atom makes a transition from the energy level at -13.58 eV to the level at -0.38 eV when a photon is absorbed. Calculate the frequency of the radiation absorbed.
(Planck constant $h = 6.62 \times 10^{-34}$ J s; 1 eV = 1.60×10^{-19} J)

Section 9.3 Summary

- Electrons in atoms can have only certain energies. These energies may be represented in an energy level diagram.
- Electrons in a given energy level may absorb energy and make a transition to a higher energy level.
- Excited electrons may return to a lower level with the emission of a photon. The frequency f of the emitted radiation is given by $E_2 - E_1 = hf$, where E_2 and E_1 are the energies of the upper and lower levels and h is the Planck constant; the wavelength λ is given by $\lambda = c/f$, where c is the speed of light.
- When an electron absorbs energy from white light and moves to a higher energy level, a line absorption spectrum is produced.
- All wavelengths of electromagnetic radiation have the same speed $c = 3.00 \times 10^8 \text{ m s}^{-1}$ in a vacuum.

Section 9.3 Questions

1 Calculate the wavelength of the radiation emitted when the electron in the hydrogen atom makes a transition from the energy level at -0.54 eV to the level at -3.39 eV. (Planck constant $h = 6.62 \times 10^{-34}$ J s; speed of light $c = 3.00 \times 10^8 \text{ m s}^{-1}$; 1 eV $= 1.60 \times 10^{-19}$ J)

2 The energy required to completely remove an electron in the ground state from an atom is called the *ionisation energy*. This energy may be supplied by the absorption of a photon, in which case the process is called *photo-ionisation*. Use information from Figure 9.9 to deduce the wavelength of radiation required to achieve photo-ionisation of hydrogen. (Planck constant $h = 6.62 \times 10^{-34}$ J s; speed of light $c = 3.00 \times 10^8 \text{ m s}^{-1}$)

Exam-style Questions

1 A zinc plate is placed on the cap of a gold leaf electroscope and charged negatively. The gold leaf is seen to deflect. Explain fully the following observations.
  (a) When the zinc plate is illuminated with red light, the gold leaf remains deflected.
  (b) When the zinc plate is irradiated with ultra-violet radiation, the leaf collapses.
  (c) When the intensity of the ultra-violet radiation is increased, the leaf collapses more quickly.
  (d) If the zinc plate is initially charged positively, the gold leaf remains deflected regardless of the nature of the incident radiation.

2 A beam of monochromatic light of wavelength 630 nm transports energy at the rate of 0.25 mW. Calculate the number of photons passing a given cross-section of the beam each second. (Planck constant $h = 6.6 \times 10^{-34}$ J s; speed of light $c = 3.0 \times 10^8 \text{ m s}^{-1}$)

3 The work function energy of a certain metal is 4.0×10^{-19} J.
  (a) Calculate the longest wavelength for which photoemission is obtained.
  (b) This metal is irradiated with ultra-violet radiation of wavelength 250 nm. Calculate, for the emitted electrons:
     (i) the maximum kinetic energy,
     (ii) the maximum speed.
  (Planck constant $h = 6.6 \times 10^{-34}$ J s; speed of light $c = 3.0 \times 10^8 \text{ m s}^{-1}$; electron mass $m_e = 9.1 \times 10^{-31}$ kg)

4 Calculate the de Broglie wavelengths of:
   (a) a ball of mass 0.30 kg moving at $50 \, m \, s^{-1}$,
   (b) a bullet of mass 50 g moving at $500 \, m \, s^{-1}$,
   (c) an electron of mass 9.1×10^{-31} kg moving at $3.0 \times 10^7 \, m \, s^{-1}$,
   (d) a proton of mass 1.7×10^{-27} kg moving at $3.0 \times 10^6 \, m \, s^{-1}$.
   (Planck constant $h = 6.6 \times 10^{-34}$ J s)

5 Atoms in the gaseous state (for example, low-pressure gas in a discharge tube) produce an emission spectrum consisting of a series of separate lines. The wavelengths of these lines are characteristic of the particular atoms involved. Hot atoms in the solid state (for example, a hot metal filament in an electric light bulb) produce a continuous emission spectrum which is characteristic of the temperature of the filament rather than of the atoms involved. Suggest reasons for this difference.

6 Figure 9.17 shows some transitions which may take place as excited electrons return to lower energy levels. Calculate the wavelengths of the light emitted in each transition.
   (Planck constant $h = 6.6 \times 10^{-34}$ J s; speed of light $c = 3.0 \times 10^8 \, m \, s^{-1}$)

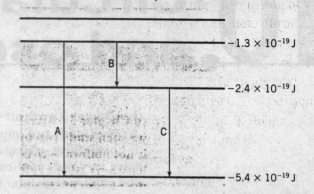

Figure 9.17

7 When the visible spectrum emitted by the Sun is observed closely it is noted that light of certain frequencies is missing and in their place are dark lines.
   (a) Explain how the cool outer gaseous atmosphere of the Sun could be responsible for the absence of these frequencies.
   (b) Suggest how an analysis of this spectrum could be used to determine which gases are present in the Sun's atmosphere.

10. Circular motion and oscillations

In Chapter 3 we dealt with uniformly accelerated motion. In this chapter we shall study two important examples of motion in which the acceleration is not uniform – circular motion and simple harmonic oscillations. Both have very wide applications in physics and in everyday life. Understanding the physics of circular motion, for example, has enabled scientists to launch satellites into orbit around the Earth and to design theme park rides.

10.1 Going round in circles

At the end of Section 10.1 you should be able to:
- describe qualitatively motion in a curved path due to a perpendicular force, and understand the centripetal acceleration in the case of uniform motion in a circle
- express angular displacement in radians
- understand and use the concept of angular velocity to solve problems
- recall and use $v = r\omega$ to solve problems
- recall and use centripetal acceleration $a = r\omega^2$, $a = v^2/r$
- recall and use centripetal force $F = mr\omega^2$, $F = mv^2/r$

Centripetal force

Newton's first law of motion (see Chapter 5) tells us that an object with a resultant force of zero acting on it will either not be moving at all, or will be moving in a straight line at constant speed (that is, its velocity does not change). The object is said to be in equilibrium. (The full conditions of equilibrium require there to be no resultant force or resultant moment acting on the body.)

An object travelling in a circle may have a constant speed, but it is not travelling in a straight line. Therefore its velocity is changing, since velocity is speed in a certain direction. A change in velocity means the object is accelerating.

In Chapter 3 we dealt with objects moving along a straight line, where acceleration related to changes of speed. In circular motion, objects accelerate because their *direction* changes. The acceleration a of an object moving in a circle of radius r with speed v is given by the expression

$$a = \frac{v^2}{r}$$

We shall derive this later. This acceleration is towards the centre of the circle. It is called the **centripetal acceleration**. In order to make an object accelerate, there must be a resultant force acting on it. This force is called the **centripetal force**. From Newton's second law of motion (see Chapter 5), the force F to give to mass m an acceleration a is

$$F = ma$$

By substituting for a, the centripetal force is

$$F = \frac{mv^2}{r}$$

The centripetal force acts towards the centre of the circle.

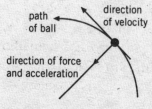

path of ball

direction of velocity

direction of force and acceleration

Figure 10.1 A ball swung in a circle on the end of a string

This means that the centripetal force acts at right angles to the instantaneous velocity of the object.

Consider a ball on a string which is being swung in a horizontal circle. The tension in the string provides the centripetal force. At any instant the direction of the ball's velocity is along the tangent to the circle, as shown in Figure 10.1. If the string breaks or is released, there is no longer any centripetal force. The ball will travel in the direction of the tangent to the circle at the moment of release.

Example

An aircraft in a display team makes a turn in a horizontal circle of radius $500\,\text{m}$. It is travelling at a speed of $100\,\text{m s}^{-1}$.
(a) Calculate the centripetal acceleration of the aircraft.
(b) The mass of the pilot in the aircraft is $80\,\text{kg}$. Calculate the centripetal force acting on the pilot.

(a) From $a = v^2/r$, the centripetal acceleration $a = 100^2/500 = \textbf{20 m s}^{-2}$.
(b) From $F = ma$, the centripetal force on the pilot $F = 80 \times 20 = \textbf{1600 N}$.

Now it's your turn

1 In a model of the hydrogen atom, the electron moves round the nucleus in a circular orbit of radius $5.3 \times 10^{-11}\,\text{m}$. The speed of the electron is $2.2 \times 10^6\,\text{m s}^{-1}$. Calculate the centripetal acceleration of the electron.

2 A satellite orbits the Earth $200\,\text{km}$ above its surface. The acceleration towards the centre of the Earth is $9.2\,\text{m s}^{-2}$. The radius of the Earth is $6400\,\text{km}$. Calculate:
(a) the speed of the satellite,
(b) the time to complete one orbit.

Radians and degrees

In circular motion, it is convenient to measure angles in **radians** rather than degrees. One degree is by tradition equal to the angle of a circle divided by 360.

> One radian (rad) is defined as the angle subtended at the centre of a circle by an arc equal in length to the radius.

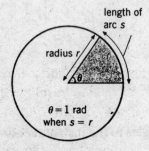

length of arc s

radius r

$\theta = 1$ rad when $s = r$

Figure 10.2 θ in radians = arc/radius

Thus, to obtain an angle in radians, we divide the length of the arc by the radius of the circle (see Figure 10.2).

$$\theta = \frac{length\ of\ arc}{radius\ of\ circle}$$

The angle in radians in a complete circle would be

$$\theta = \frac{circumference\ of\ the\ circle}{radius} = \frac{2\pi r}{r} = 2\pi$$

Since the angle of a complete circle is 360°, then

$$2\pi\ \text{rad} = 360°$$

or

$$1\ \text{rad} = 57.3°$$

Angular velocity

For an object moving in a circle:

> The **angular speed** is defined as the angle swept out by the radius per second.

The **angular velocity** is the angular speed in a given direction (for example, clockwise). The unit of angular speed and angular velocity is the radian per second (rad s⁻¹).

$$angular\ speed\ \omega = \frac{\Delta\theta}{\Delta t}$$

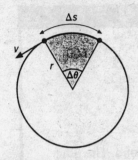

Figure 10.3 Angular velocity $\omega = v/r$

Figure 10.3 shows an object travelling at constant speed v in a circle of radius r. In a time Δt the object moves along an arc of length Δs and sweeps out an angle $\Delta\theta$. From the definition of the radian,

$$\Delta\theta = \Delta s/r \quad \text{or} \quad \Delta s = r\Delta\theta$$

Dividing both sides of this equation by Δt,

$$\Delta s/\Delta t = r\Delta\theta/\Delta t$$

By definition, $\Delta s/\Delta t$ is the linear speed v of the object, and $\Delta\theta/\Delta t$ is the angular speed ω. Hence,

$$v = r\omega$$

Derivation of expression for centripetal acceleration

Figure 10.4 shows an object which has travelled at constant speed v in a circular path from A to B in time Δt. At A, its velocity is v_A, and at B the velocity is v_B. Both v_A and v_B are vectors.

The change in velocity Δv may be seen in the vector diagram of Figure 10.5. A vector Δv must be added to v_A in order to give the new velocity v_B.

The angle between the two radii OA and OB is $\Delta\theta$. This angle is also equal to the angle between the vectors v_A and v_B, because triangles OAB and CDE are similar. Consider angle $\Delta\theta$ to be so small that the arc AB may be approximated to a straight line. Then, using similar triangles, DE/CD = AB/OA, and $\Delta v/v_A = \Delta s/r$ or

$$\Delta v = \Delta s(v_A/r)$$

The time to travel either the distance Δs or the angle $\Delta\theta$ is Δt. Dividing both sides of the equation by Δt,

$$\Delta v/\Delta t = (\Delta s/\Delta t)(v/r)$$

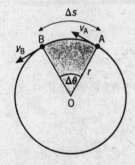

Figure 10.4 Diagram for proof of $a = v^2/r$

and from the definitions of acceleration ($a = \Delta v/\Delta t$) and speed ($v = \Delta s/\Delta t = v_A = v_B$) we have

$$a = v(v/r) \quad \text{or} \quad a = v^2/r$$

This expression can be written in terms of angular speed ω. Since $v = r\omega$,

$$\text{centripetal acceleration} = \frac{v^2}{r} = r\omega^2$$

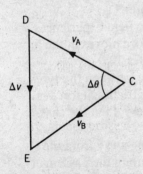

Figure 10.5 Vector diagram for proof of $a = v^2/r$

and, since $F = ma$,

$$\text{centripetal force} = \frac{mv^2}{r} = mr\omega^2$$

Example

The drum of a spin dryer has a radius of 20 cm and rotates at 600 revolutions per minute.
(a) Show that the angular speed of the drum is about 63 rad s^{-1}.
(b) Calculate, for a point on the edge of the drum:
   (i) its linear speed,
   (ii) its acceleration towards the centre of the drum.

(a) 600 revolutions per minute is 10 revolutions per second. The time for one revolution is thus 0.10 s. Each revolution is 2π rad, so the angular speed $\omega = \theta/t = 2\pi/0.10 = \textbf{63 rad s}^{-1}$.
(b) (i) Using $v = r\omega$, $v = 0.20 \times 63 = \textbf{13 m s}^{-1}$ **(12.6 m s^{-1})**.
   (ii) Using $a = v^2/r$, $a = (13)^2/0.20 = \textbf{800 m s}^{-2}$.

Now it's your turn

A toy train moves round a circular track of diameter 0.70 m, completing one revolution in 10 seconds. Calculate, for this train:
(a) the linear speed,
(b) the angular speed,
(c) the centripetal acceleration.

Examples of circular motion

When a ball is whirled round on the end of a string, you can see clearly that the string is making the ball accelerate towards the centre of the circle. However, in other examples it is not always so easy to see what force is providing the centripetal acceleration.

A satellite in Earth orbit experiences gravitational attraction towards the centre of the Earth. This attractive force provides the centripetal force and causes the satellite to accelerate towards the centre of the Earth, and so it moves in a circle. We shall return to this in detail in Chapter 11.

A charged particle moving at right-angles to a magnetic field experiences a force at right-angles to its direction of motion, and therefore moves in the arc of a circle. This will be considered in more detail in Chapter 12.

For a car travelling in a curved path, the frictional force between the tyres and the road surface provides the centripetal force. If this frictional force is not large enough, for example if the road is oily and slippery, then the car carries on moving in a straight line – it skids.

A passenger in a car that is cornering appears to be flung away from the centre of the circle. The centripetal force required to maintain the passenger in circular motion is provided through the seat of the car. This force is below the centre of mass of the passenger, causing rotation about the centre of mass (Figure 10.6). The effect is that the upper part of the passenger moves outwards unless another force acts on the upper part of the body, preventing rotation.

Figure 10.6 Passenger in a car rounding a corner

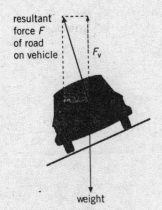

Figure 10.7 Cornering on a banked track

For cornering which does not rely only on friction, the road can be banked (see Figure 10.7). The road provides a resultant force normal to its surface through contact between the tyres and the road. This resultant force F is at an angle to the vertical, and can be resolved into a vertical component F_v and a horizontal component F_h. F_v is equal to the weight of the vehicle, thus maintaining equilibrium in the vertical direction. The horizontal component F_h provides the centripetal force towards the centre of the circle. Many roads are banked for greater road safety, so as to reduce the chance of loss of control of vehicles due to skidding outwards on the corner, and for greater passenger comfort.

An aircraft has a lift force caused by the different rates of flow of air above and below the wings. The lift force balances the weight of the aircraft when it flies on a straight, level path (Figure 10.8a). In order to change direction, the aircraft is banked so that the wings are at an angle to the horizontal (Figure 10.8b). The lift force now has a horizontal component which provides a centripetal force to change the aircraft's direction.

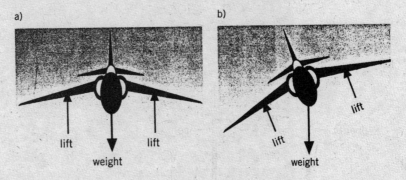

Figure 10.8 An aircraft a) in straight, level flight and b) banking

A centrifuge (Figure 10.9) is a device used to spin objects at high speed about an axis. It is used to separate particles in mixtures. More massive particles require larger centripetal forces in order to maintain circular motion than do less massive ones. As a result, the more massive particles tend to separate from less massive particles, collecting further away from the axis of rotation.

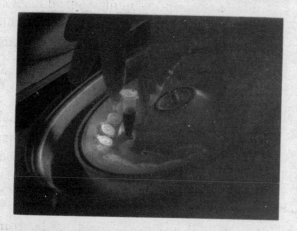

Figure 10.9 Separation of a solid from a liquid in a laboratory centrifuge

Figure 10.10 Centrifuge testing the effect of acceleration on the human body

Space research centres, such as NASA, use centrifuges which are large enough to rotate a person (Figure 10.10). Their purpose is to investigate the effects of large accelerations on the human body.

Motion in a vertical circle

Some theme park rides involve rotation in a vertical circle. A person on such a ride must have a resultant force acting towards the centre of the circle.

The forces acting on the person are the person's weight, which always acts vertically downwards, and the reaction from the seat, which acts at right angles to the seat.

Figure 10.11 Ferris wheel at a theme park

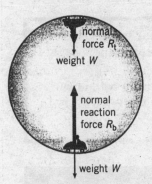

Figure 10.12 Forces on a person on a circular ride

Consider a person moving round a vertical circle at speed v. At the bottom of the ride, the reaction R_b from the seat must provide the centripetal force as well as overcoming the weight W of the person. Figure 10.12 illustrates the situation. Hence the centripetal force is given by

$$mv^2/r = R_b - W$$

At the top of the ride, the weight W and the reaction force R_t both act downwards towards the centre of the circle. The centripetal force is now given by

$$mv^2/r = R_t + W$$

This means that the force R_t from the seat at the top of the ride is less than the force R_b at the bottom. If the speed v is not large, then at the top of the circle the weight may be greater than the centripetal force. The person would lose contact with the seat and fall inwards.

Example

A rope is tied to a bucket of water, and the bucket is swung in a vertical circle of radius 1.2 m. What must be the minimum speed of the bucket at the highest point of the circle if the water is to stay in the bucket throughout the motion?

This example is similar to the problem of the theme park ride. Water will fall out of the bucket if its weight is greater than the centripetal force. The critical speed v is given by $mv^2/r = mg$ or $v^2 = gr$. Here $v = \sqrt{(9.8 \times 1.2)} = \mathbf{3.4\,m\,s^{-1}}$.

Now it's your turn

At an air show, an aircraft diving at a speed of $170\,m\,s^{-1}$ pulls out of the dive by moving in the arc of a circle at the bottom of the dive.
(a) Calculate the minimum radius of this circle if the centripetal acceleration of the aircraft is not to exceed five times the acceleration of free fall.
(b) The pilot has mass 85 kg. What is the resultant force upwards on him at the instant when the aircraft is at its lowest point?

Section 10.1 Summary

- A resultant force acting towards the centre of the circle, called the centripetal force, is required to make an object move in a circle.
- Angles may be measured in radians (rad). One radian is the angle subtended at the centre of a circle by an arc of the circle equal in length to its radius.
- Angular speed ω is the angle swept out per unit time by a line rotating about a point.
- A particle moving along a circle of radius r with linear speed v has angular velocity ω given by: $v = r\omega$

> ■ An object moving along a circle of radius r with linear speed v and angular speed ω has an acceleration a towards the centre (the centripetal acceleration) given by: $a = v^2/r = r\omega^2$
> ■ For an object of mass m moving along a circle of radius r with linear speed v and angular speed ω, the centripetal force F is given by: $F = mv^2/r = mr\omega^2$

Section 10.1 Questions

1 State how the centripetal force is provided in the following examples:
(a) a planet orbiting the Sun,
(b) a child on a playground roundabout,
(c) a train on a curved track,
(d) a passenger in a car going round a corner.

2 NASA's 20-G centrifuge is used for testing space equipment and the effect of acceleration on humans. The centrifuge consists of an arm of length 17.8 m, rotating at constant speed and producing an acceleration equal to 20 times the acceleration of free fall. Calculate:
(a) the angular speed required to produce a centripetal acceleration of 20g,
(b) the rate of rotation of the arm.
($g = 9.8$ m s^{-2})

10.2 Oscillations

At the end of Sections 10.2–10.4 you should be able to:
■ investigate the motion of an oscillator using experimental and graphical methods
■ understand and use the terms amplitude, period, frequency, angular frequency and phase difference, and express the period in terms of both frequency and angular frequency
■ recognise and use the equation $a = -\omega^2 x$ as the defining equation of simple harmonic motion
■ recall and use $x = x_0 \sin \omega t$ as a solution to the equation $a = -\omega^2 x$
■ recognise and use $v = v_0 \cos \omega t$ and $v = \pm \omega \sqrt{(x_0^2 - x^2)}$
■ describe, with graphical examples, the changes in displacement, velocity and acceleration during simple harmonic motion
■ describe simple examples of free oscillations

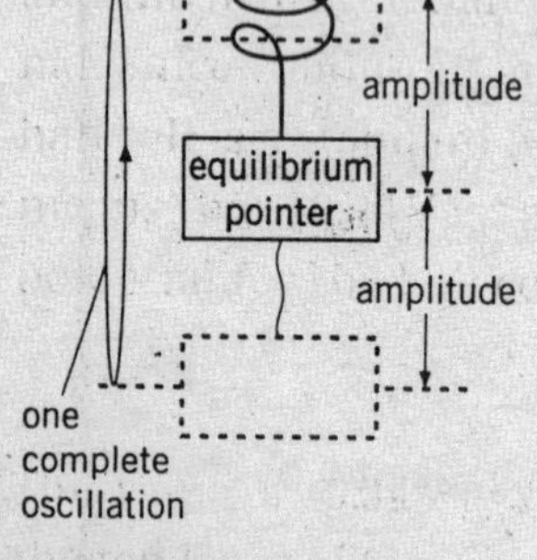

Figure 10.13 Oscillation of a mass on a spring

Some movements involve repetitive to-and-fro motion, such as a pendulum, the beating of a heart, the motion of a child on a swing, and the vibrations of a guitar string. Another example would be a mass bouncing up and down on a spring, as illustrated in Figure 10.13. One complete movement from the starting or rest position, up, then down and finally back up to the rest position, is known as an **oscillation**.

The time taken for one complete oscillation is referred to as the **period T** of the oscillation.

The oscillations repeat themselves.

The number of oscillations per unit time is the **frequency f**.

Frequency may be measured in hertz (Hz), where one hertz is one oscillation per second ($1\,\text{Hz} = 1\,\text{s}^{-1}$). However, frequency may also be measured in min^{-1}, hour^{-1}, etc. For example, it would be appropriate to measure the frequency of the tides in h^{-1}.

Since period T is the time for one oscillation then

frequency $f = 1/T$

As the mass oscillates, it moves from its rest position.

The distance from the equilibrium position is known as the **displacement**.

This is a vector quantity, since the displacement may be on either side of the equilibrium position.

The **amplitude** (a scalar quantity) is the maximum displacement.

Some oscillations maintain a constant period even when the amplitude of the oscillation changes. Galileo discovered this fact for a pendulum. He timed the swings of an oil lamp in Pisa Cathedral, using his pulse as a measure of time. Oscillators that have a constant time period are called isochronous, and may be made use of in timing devices. For example, in quartz watches the oscillations of a small quartz crystal provide constant time intervals. Galileo's experiment was not precise, and we now know that a pendulum swinging with a large amplitude is not isochronous.

The quantities period, frequency, displacement and amplitude should be familiar from our study of waves in Chapter 8. It should not be a surprise to meet them again, as the idea of oscillations is vital to the understanding of waves.

Displacement–time graphs

It is possible to plot displacement–time graphs (see Chapter 3) for oscillators. One experimental method is illustrated in Figure 10.14. A mass on a spring oscillates above a position sensor that is connected to a computer through a datalogging interface causing a trace to appear on the monitor.

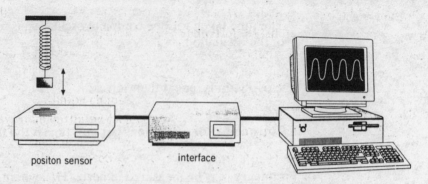

positon sensor   interface

Figure 10.14 Apparatus for plotting displacement–time graphs for a mass on a spring

The graph describing the variation of displacement with time may have different shapes, depending on the oscillating system. For many oscillators the graph is approximately a sine (or cosine) curve. A sinusoidal displacement–time graph is a characteristic of an important type of oscillation called **simple harmonic motion** (s.h.m.). Oscillators which move in s.h.m. are called **harmonic** oscillators. We shall analyse simple harmonic motion in some detail, because it successfully describes many oscillating systems, both in real life and in theory. Fortunately, the mathematics of s.h.m. can be approached through a simple defining equation. The properties of the motion can be deduced from the relations between graphs of displacement against time, and velocity against time, which we have already met in Chapter 3.

10.3 Simple harmonic motion

Simple harmonic motion is defined as the motion of a particle about a fixed point such that its acceleration a is proportional to its displacement x from the fixed point, and is directed towards the point.

Mathematically, we write this definition as

$$a = -\omega^2 x$$

where ω^2 is a constant. We take the constant as a squared quantity, because this will ensure that the constant is always positive (the square of a positive number, or of a negative number, will always be positive). Why worry about keeping the constant positive? This is because the minus sign in the equation must be preserved. It has a special significance, because it tells us that the acceleration a is always in the opposite direction to the displacement x. Remember that both acceleration and displacement are vector quantities, so the minus sign is shorthand for the idea that the acceleration is always directed towards the fixed point from which the displacement is measured. This is illustrated in Figure 10.15.

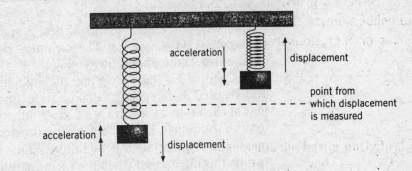

Figure 10.15 Directions of displacement and acceleration are always opposite.

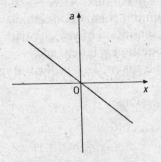

Figure 10.16 Graph of the defining equation for simple harmonic motion

The defining equation is represented in a graph of a against x as a straight line, of negative gradient, through the origin, as shown in Figure 10.16. The gradient is negative because of the minus sign in the equation. Note that both positive and negative values for the displacement should be considered.

The square root of the constant ω^2 (that is, ω) is known as the **angular frequency** of the oscillation. This angular frequency ω is related to the frequency f of the oscillation by the expression

$$\omega = 2\pi f$$

By Newton's second law, the force acting on a body is proportional to the acceleration of the body. The defining equation for simple harmonic motion can thus be related to the force acting on the particle. If the acceleration of the particle is proportional to its displacement from a fixed point, the resultant force acting on the particle is also proportional to the displacement. We can bring in the idea of the direction of the acceleration by specifying that the force is always acting towards the fixed point, or by calling it a **restoring** force.

Solution of equation for simple harmonic motion

In order to find the displacement–time relation for a particle moving in a simple harmonic motion, we need to solve the equation $a = -\omega^2 x$. To derive the solution requires mathematics which is beyond the requirements of A/AS Physics. However, you need to know the form of the solution. This is

$$x = x_0 \sin \omega t$$

or

$$x = x_0 \cos \omega t$$

where x_0 is the amplitude of the oscillation. The solution $x = x_0 \sin \omega t$ is used when, at time $t = 0$, the particle is at its equilibrium position where $x = 0$.

a)

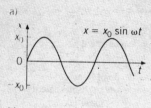

b)

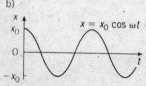

Figure 10.17 Displacement–time curves for the two solutions to the s.h.m. equation

a)

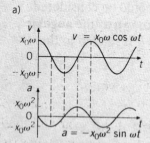

b)

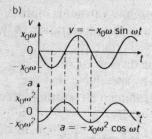

Figure 10.18 Velocity-time and acceleration-time graphs for the two solutions to the s.h.m. equation

Conversely, if at time $t = 0$ the particle is at its maximum displacement, $x = x_0$, the solution is $x = x_0 \cos \omega t$. The variation with time t of the displacement x for the two solutions is shown in Figure 10.17.

In Chapter 3 it was shown that the gradient of a displacement–time graph may be used to determine velocity. Referring to Figure 10.17, it can be seen that at each time at which $x = x_0$, the gradient of the graph is zero (this applies to both solutions). Thus, the velocity is zero whenever the particle has its maximum displacement. If we think about a mass vibrating up and down on a spring, this means that when the spring is fully stretched and the mass has its maximum displacement, the mass stops moving downwards and has zero velocity. Also from Figure 10.17, we can see that the gradient of the graph is at a maximum whenever $x = 0$. This means that when the spring is neither under or over-stretched the speed of the mass is at a maximum. After passing this point the spring forces the mass to slow down until it changes direction.

If a full analysis is carried out, it is found that the variation of velocity with time is cosinusoidal if the sinusoidal displacement solution is taken, and sinusoidal if the cosinusoidal displacement solution is taken. This is illustrated in Figure 10.18.

The velocity v of the particle is given by the expressions

$$v = x_0\omega \cos \omega t \quad \text{when} \quad x = x_0 \sin \omega t$$

and

$$v = -x_0\omega \sin \omega t \quad \text{when} \quad x = x_0 \cos \omega t$$

In each case, there is a phase difference between velocity and displacement. The velocity curve is $\pi/2$ rad ahead of the displacement curve. (If the phase angle is not considered, the variation with time of the velocity is the same in each case.) The maximum speed v_0 is given by

$$v_0 = x_0\omega$$

There is an alternative expression for the velocity.

$$v = \pm \omega\sqrt{(x_0^2 - x^2)}$$

which is derived on pages 271–272.

For completeness, Figure 10.18 also shows the variation with time of the acceleration a of the particle. This could be derived from the velocity–time graph by taking the gradient. The equations for the acceleration are

$$a = -x_0\omega^2 \sin \omega t \quad \text{when} \quad x = x_0 \sin \omega t$$

and

$$a = -x_0\omega^2 \cos \omega t \quad \text{when} \quad x = x_0 \cos \omega t$$

Note that, for both solutions, these equations are consistent with the defining equation for simple harmonic motion, $a = -\omega^2 x$. You can easily prove this by eliminating $\sin \omega t$ from the first set of equations, and $\cos \omega t$ from the second.

Example

The displacement x at time t of a particle moving in simple harmonic motion is given by $x = 0.25 \cos 7.5t$, where x is in metres and t is in seconds.
(a) Use the equation to find the amplitude, frequency and period for the motion.
(b) Find the displacement when $t = 0.50$ s.

(a) Compare the equation with $x = x_0 \cos \omega t$. The amplitude $x_0 = \textbf{0.25 m}$. The angular frequency $\omega = 7.5$ rad s^{-1}. Remember that $\omega = 2\pi f$, so the frequency $f = \omega/2\pi = 7.5/2\pi = \textbf{1.2 Hz}$. The period $T = 1/f = 1/1.2 = \textbf{0.84 s}$.
(b) Substitute $t = 0.50$ s in the equation, remembering that the angle ωt is in radians and not degrees. $\omega t = 7.5 \times 0.50 = 3.75$ rad $= 215°$. So $x = 0.25 \cos 215° = \textbf{−0.20 m}$.

Now it's your turn

A mass oscillating on a spring has an amplitude of 0.10 m and a period of 2.0 s.
(a) Deduce the equation for the displacement x if timing starts at the instant when the mass has its maximum displacement.
(b) Calculate the time interval from $t = 0$ before the displacement is 0.08 m.

10.4 Examples of simple harmonic motion

By definition, an object whose acceleration is proportional to the displacement from the equilibrium position, and which is always directed towards the equilibrium position, is undergoing simple harmonic motion. We now look at two simple examples of oscillatory motion which approximate to this definition. They may easily be set up as demonstrations of s.h.m.

Mass on a helical spring

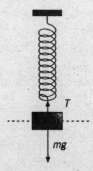

Figure 10.19 Mass on a helical spring

Figure 10.19 illustrates a mass m suspended from a spring. (This sort of spring is called a *helical* spring because it has the shape of a helix.)

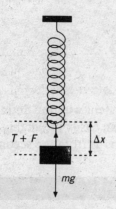

Figure 10.20 Mass on a helical spring: additional extension Δx

The weight mg of the mass is balanced by the tension T in the spring. When the spring is extended by an amount Δx, as in Figure 10.20, there is an additional upward force in the spring given by

$$F = -k\Delta x$$

where k is a constant for a particular spring, known as the spring constant. The spring constant is a measure of the stiffness of the spring. A stiff spring has a large value of k; a more flexible spring has a smaller value of k and, for the same force, would have a larger extension than one with a large spring constant. The spring constant k is given by the expression

$$k = \frac{F}{\Delta x}$$

and is measured in newtons per metre (N m^{-1}).

When the mass is released, the restoring force F pulls the mass towards the equilibrium position. (The minus sign in the expression for F shows the direction of this force.) The restoring force is proportional to the displacement. This means that the acceleration of the mass is proportional to the displacement from the equilibrium position and is directed towards the equilibrium position. This is the condition for simple harmonic motion.

The full theory shows that, for a mass m suspended from a light spring having spring constant k, the period T of the oscillations is given by

$$T = 2\pi\sqrt{\frac{m}{k}}$$

For oscillations to be simple harmonic, the spring must obey Hooke's law throughout – that is, the extensions must not exceed the limit of proportionality. Furthermore, for large amplitude oscillations, the spring may become slack. Ideally, the spring would have no mass, but if the suspended mass is more than about 20 times the mass of the spring, the error involved in assuming that the spring has no mass is less than 1%.

This example of simple harmonic motion is particularly useful in modelling the vibrations of molecules. A molecule containing two atoms oscillates as if the atoms were connected by a tiny spring. The spring constant of this spring depends upon the type of bonding between the atoms. The frequency of oscillation of the molecule can be measured experimentally using spectroscopy, and this gives direct information about the bonding. This model can be extended to solids, where atoms are often thought of as being connected to their neighbours by springs. Again, this leads to an experimental way of obtaining information about interatomic forces in the solid.

The simple pendulum

Figure 10.21 illustrates a simple pendulum. Ideally, a simple pendulum is a point mass m on a light, inelastic string. In real experiments we use a pendulum bob of finite size. When the bob is pulled aside through an angle θ

and then released, there will be a restoring force acting in the direction of the equilibrium position.

Because the simple pendulum moves in the arc of a circle, the displacement will be an angular displacement θ rather than the linear displacement x we have been using so far.

The two forces on the bob are its weight mg and the tension T in the string. The component of the weight along the direction of the string, $mg \cos \theta$, is equal to the tension in the string. The component of the weight at right angles to the direction of the string, $mg \sin \theta$, is the restoring force F. This makes the bob accelerate towards the equilibrium position.

The restoring force depends on $\sin \theta$. As θ increases, the restoring force is not proportional to the displacement (in this case θ), and so the motion is oscillatory but not simple harmonic. However, the situation is different if the angle θ is kept small (less than about 5°). For these small angles, θ is proportional to $\sin \theta$. In fact, if θ is measured in radians, then

$$\theta \text{ in radians} \approx \sin \theta$$

You can check this using your calculator. Some values of $\theta/°$, θ/rad and $\sin \theta$ are given in Table 10.1.

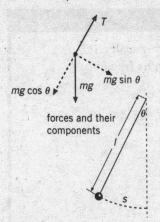

Figure 10.21 The simple pendulum

Table 10.1

$\theta/°$	θ/rad	$\sin \theta$
1.00	0.0175	0.0175
2.00	0.0349	0.0349
3.00	0.0524	0.0523
5.00	0.0873	0.0871
10.00	0.1745	0.1736

This means that, for small-amplitude oscillations (the angle of the string to the vertical should be less than about 5°), the pendulum bob oscillates with simple harmonic motion.

A full treatment of the theory shows that the period T is related to the length l of the pendulum (l is the distance between the centre of mass of the pendulum bob and the point of suspension – see Figure 10.21) by the expression

$$T = 2\pi \sqrt{\frac{l}{g}}$$

where g is the acceleration of free fall. An experiment in which the period of a simple pendulum is measured can be used to determine the acceleration of free fall. The experiment is repeated for different lengths of pendulum, and the gradient of a graph of T^2 against l is $4\pi^2/g$. This provides an alternative to dynamics experiments in which the time for a body to fall through measured distances is determined, and g is calculated from the equation of motion (see Chapter 3).

Example

A helical spring is clamped at one end and hangs vertically. It extends by 10 cm when a mass of 50 g is hung from its free end. Calculate:

(a) the spring constant of the spring,

(b) the period of small amplitude oscillations of the mass.

(a) Using $k = F/\Delta x$, the spring constant
$$k = 50 \times 10^{-3} \times 9.8/10 \times 10^{-2} = \textbf{4.9 N m}^{-1}$$

(b) Using $T = 2\pi\sqrt{(m/k)}$, the period
$$T = 2\pi\sqrt{(50 \times 10^{-3}/4.9)} = \textbf{0.63 s}$$

Now it's your turn

(a) The acceleration of free fall at the Earth's surface is $9.8 \, \text{m s}^{-2}$. Calculate the length of a simple pendulum which would have a period of 1.0 s.

(b) The acceleration of free fall on the Moon's surface is $1.6 \, \text{m s}^{-2}$. Calculate, for the pendulum in (a), its period of oscillation on the Moon.

Sections 10.2–10.4 Summary

- The period of an oscillation is the time taken to complete one oscillation.
- Frequency is the number of oscillations per unit time.
- Frequency f is related to period T by the expression: $f = 1/T$
- The displacement of a particle is its distance from the equilibrium position.
- Amplitude is the maximum displacement.
- Simple harmonic motion (s.h.m.) is defined as the motion of a particle about a fixed point such that its acceleration a is proportional to its displacement x from the fixed point, and is directed towards the fixed point: $a \propto -x$ or $a = -\omega^2 x$
- The constant ω in the defining equation for simple harmonic motion is known as the angular frequency.
- For a particle oscillating in s.h.m. with frequency f, then: $\omega = 2\pi f$
- Simple harmonic motion is described in terms of displacement x, amplitude x_0, frequency f, angular frequency ω by the following relations.

displacement:	$x = x_0 \sin \omega t$	or	$x = x_0 \cos \omega t$
velocity:	$v = x_0\omega \cos \omega t$	or	$v = -x_0\omega \sin \omega t$
acceleration:	$a = -x_0\omega^2 \sin \omega t$	or	$a = -x_0\omega^2 \cos \omega t$

Remember that $\omega = 2\pi f$, and equations may appear in either form.

Sections 10.2–10.4 Questions

1 A particle is oscillating in simple harmonic motion with period 4.5 ms and amplitude 3.0 cm. At time $t = 0$, the particle is at the equilibrium position. Calculate, for this particle:
   (a) the frequency,
   (b) the angular frequency,
   (c) the maximum speed,
   (d) the magnitude of the maximum acceleration,
   (e) the speed at time $t = 1.0$ ms.

2 A particle is oscillating in simple harmonic motion with frequency 50 Hz and amplitude 15 mm. Calculate the speed when the displacement from the equilibrium position is 12 mm.

3 Geologists use the fact that the period of oscillation of a simple pendulum depends on the acceleration of free fall to map variations of g. A geologist determines the frequency of oscillation of a test pendulum of effective length 515.6 mm to be 0.6948 Hz. Calculate the acceleration of free fall at this locality.

4 A spring stretches by 85 mm when a mass of 50 g is hung from it. The spring is then stretched a further distance of 15 mm from the equilibrium position, and the mass is released at time $t = 0$. Calculate:
   (a) the spring constant,
   (b) the amplitude of the oscillations,
   (c) the period,
   (d) the displacement at time $t = 0.20$ s.
   ($g = 9.8$ m s^{-2})

10.5 Energy changes in simple harmonic motion

At the end of Section 10.5 you should be able to:
- describe the interchange between kinetic and potential energy during simple harmonic motion.

Kinetic energy

In Section 10.3 we saw that the velocity of a particle vibrating with simple harmonic motion varies with time and, consequently, with the displacement of the particle. For the case where displacement x is zero at time $t = 0$, displacement and velocity are given by

$$x = x_0 \sin \omega t$$

and

$$v = x_0 \omega \cos \omega t$$

There is a trigonometrical relation between the sine and the cosine of an angle θ, which is $\sin^2 \theta + \cos^2 \theta = 1$. Applying this relation, we have

$$x^2/x_0^2 + v^2/x_0^2\omega^2 = 1$$

which leads to

$$v^2 = x_0^2\omega^2 - x^2\omega^2$$

and so

$$v = \pm\,\omega\sqrt{(x_0^2 - x^2)}$$

(If we had taken the displacement and velocity equations for the case when the displacement is a maximum at $t = 0$, exactly the same relation would have been obtained. Try it!)

The kinetic energy of the particle (of mass m) oscillating with simple harmonic motion is $\frac{1}{2}mv^2$. Thus, the kinetic energy E_k at displacement x is given by

$$E_k = \tfrac{1}{2}m\omega^2(x_0^2 - x^2)$$

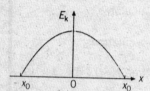

Figure 10.22 Variation of kinetic energy in s.h.m.

The variation with displacement of the kinetic energy is shown in Figure 10.22.

Potential energy

In Section 10.3 we stated that the defining equation for simple harmonic motion could be expressed in terms of the restoring force acting on the particle. Since $F = ma$ and $a = -\omega^2x$ then at displacement x, this force is

$$F_{res} = -m\omega^2x$$

where m is the mass of the particle. To find the change in potential energy of the particle when the displacement increases by Δx, we need to find the work done against the restoring force.

The work done in moving the point of application of a force F by a distance Δx is $F\Delta x$. In the case of the particle undergoing simple harmonic motion, we know that the restoring force is directly proportional to displacement. To calculate the work done against the restoring force in giving the particle a displacement x, we take account of the fact that F_{res} depends on x by taking the average value of F_{res} during this displacement. The average value is just $\frac{1}{2}m\omega^2x$, since the value of F_{res} is zero at $x = 0$ and increases linearly to $m\omega^2x$ at displacement x. Thus, the potential energy E_p at displacement x is given by *average restoring force × displacement*, or

$$E_p = \tfrac{1}{2}m\omega^2x^2$$

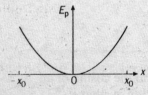

Figure 10.23 Variation of potential energy in s.h.m.

The variation with displacement of the potential energy is shown in Figure 10.23.

Total energy

The total energy E_{tot} of the oscillating particle is given by

$$E_{tot} = E_k + E_p$$
$$= \tfrac{1}{2}m\omega^2(x_0^2 - x^2) + \tfrac{1}{2}m\omega^2 x^2$$
$$E_{tot} = \tfrac{1}{2}m\omega^2 x_0^2$$

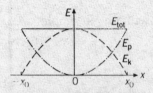

Figure 10.24 Energy variations in s.h.m.

This total energy is constant since m, ω and x_0 are all constant. We might have expected this result, as it merely expresses the law of conservation of energy.

The variations with displacement x of the total energy E_{tot}, the kinetic energy E_k and the potential energy E_p are shown in Figure 10.24.

Example

A particle of mass 60 g oscillates in simple harmonic motion with angular frequency 6.3 rad s^{-1} and amplitude 15 mm. Calculate:
(a) the total energy,
(b) the kinetic and potential energies at half-amplitude
    (at displacement $x = 7.5$ mm).

(a) Using $E_{tot} = \tfrac{1}{2}m\omega^2 x_0^2$,
$$E_{tot} = \tfrac{1}{2} \times 60 \times 10^{-3} \times 6.3^2 \times (15 \times 10^{-3})^2$$
$$= 2.7 \times 10^{-4}\,\text{J}$$

(don't forget to convert g to kg and mm to m)

(b) Using $E_k = \tfrac{1}{2}m\omega^2(x_0^2 - x^2)$,
$$E_k = \tfrac{1}{2} \times 60 \times 10^{-3} \times 6.3^2 \times [(15 \times 10^{-3})^2 - (7.5 - 10^{-3})^2]$$
$$= 2.0 \times 10^{-4}\,\text{J}$$

Using $E_p = \tfrac{1}{2}m\omega^2 x^2$,
$$E_p = \tfrac{1}{2} \times 60 \times 10^{-3} \times (7.5 \times 10^{-3})^2$$
$$= 0.7 \times 10^{-4}\,\text{J}$$

Note that $E_{tot} = E_k + E_p$, as expected.

Now it's your turn

A particle of mass 0.40 kg oscillates in simple harmonic motion with frequency 5.0 Hz and amplitude 12 cm. Calculate, for the particle at displacement 10 cm:
(a) the kinetic energy,
(b) the potential energy,
(c) the total energy.

- The kinetic energy E_k of a particle of mass m oscillating in simple harmonic motion with angular frequency ω and amplitude x_0 is:

$$E_k = \tfrac{1}{2}m\omega^2(x_0^2 - x^2)$$

where x is the displacement.

- The potential energy E_p of a particle of mass m oscillating in simple harmonic motion with angular frequency ω is:

$$E_p = \tfrac{1}{2}m\omega^2 x^2$$

where x is the displacement.

- The total energy E_{tot} of a particle of mass m oscillating in simple harmonic motion with angular frequency ω and amplitude x_0 is:

$$E_{tot} = \tfrac{1}{2}m\omega^2 x_0^2$$

- For a particle oscillating in simple harmonic motion:

$$E_{tot} = E_k + E_p$$

This expresses the law of conservation of energy.

Section 10.5 Questions

1 One particle oscillating in simple harmonic motion has ten times the total energy of another particle, but the frequencies and masses are the same. Calculate the ratio of the amplitudes of the two motions.

2 (a) Calculate the displacement, expressed as a fraction of the amplitude x_0, of a particle moving in simple harmonic motion with a speed equal to half the maximum value.

(b) Calculate the displacement at which the energy of the particle has equal amounts of kinetic and of potential energy.

10.6 Free and damped oscillations

At the end of Section 10.6 you should be able to:

- describe practical examples of damped oscillations with particular reference to the effects of the degree of damping and the importance of critical damping in cases such as a car suspension system
- describe practical examples of forced oscillations and resonance
- describe graphically how the amplitude of a forced oscillation changes with frequency near the natural frequency of the system, and understand qualitatively the factors which determine the frequency response and the sharpness of resonance
- show an appreciation that there are some circumstances in which resonance is useful and other circumstances in which resonance should be avoided.

A particle is said to be undergoing **free oscillations** when the only external force acting on it is the restoring force.

There are no forces to dissipate energy and so the oscillations have constant amplitude. Total energy remains constant. This is the situation we have been considering so far. Simple harmonic oscillations are free oscillations.

In real situations, however, frictional and other resistive forces cause the oscillator's energy to be dissipated, and this energy is converted eventually into heat energy. The oscillations are said to be **damped**.

The total energy of the oscillator decreases with time. The damping is said to be light when the amplitude of the oscillations decreases gradually with time. This is illustrated in Figure 10.25. The decrease in amplitude is, in fact, exponential with time. The period of the oscillation is slightly greater than that of the corresponding free oscillation.

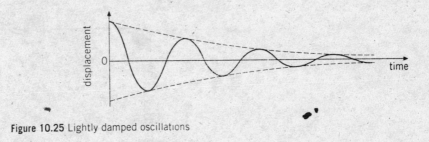

Figure 10.25 Lightly damped oscillations

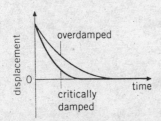

Figure 10.26 Critical damping and overdamping

Heavy damping causes the oscillations to die away more quickly. If the damping is increased further, then the system reaches **critical damping** point. Here the displacement decreases to zero in the shortest time, without any oscillation (Figure 10.26).

Any further increase in damping produces **overdamping**. The displacement decreases to zero in a longer time than for critical damping (Figure 10.26).

Damping is often useful in an oscillating system. For example, vehicles have springs between the wheels and the frame to give a smoother and more comfortable ride (Figure 10.27). If there was no damping, a vehicle would move up and down for some time after hitting a bump in the road. Dampers (shock absorbers) are connected in parallel with the springs so that the suspension has critical damping and comes to rest in the shortest time possible. Dampers often work through hydraulic action. When the spring is compressed, a piston connected to the vehicle frame forces oil through a small hole in the piston, so that the energy of the oscillation is dissipated as heat energy in the oil.

Many swing doors have a damping mechanism fitted to them. The purpose of the damper is so that the open door, when released, does not overshoot the closed position with the possibility of injuring someone approaching the door. Most door dampers operate in the overdamped mode.

Figure 10.27 Vehicle suspension system showing springs and dampers

Forced oscillations and resonance

When a vibrating body undergoes free (undamped) oscillations, it vibrates at its **natural frequency**. We have already met the idea of a natural frequency in Chapter 8, when talking about stationary waves on strings. The natural frequency of such a system is the frequency of the first mode of vibration; that is, the fundamental frequency. A practical example is a guitar string, plucked at its centre, which oscillates at a particular frequency which depends on the speed of progressive waves on the string and the length of the string. The speed of progressive waves on the string depends on the mass per unit length of the string and the tension in the string.

Vibrating objects may have periodic forces acting on them. These periodic forces will make the object vibrate at the frequency of the applied force, rather than at the natural frequency of the system. The object is then said to be undergoing **forced vibrations**. Figure 10.28 illustrates apparatus which may be used to demonstrate the forced vibrations of a mass on a helical spring. The vibrator provides the forcing (driving) frequency.

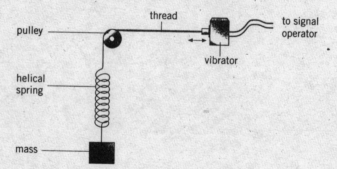

Figure 10.28 Demonstration of forced oscillations

As the frequency of the vibrator is gradually increased from zero, the mass begins to oscillate. At first the amplitude of the oscillations is small, but it increases with increasing frequency. When the driving frequency equals the natural frequency of oscillation of the mass–spring system, the amplitude of the oscillations reaches a maximum. The frequency at which this occurs is called the **resonant frequency**, and **resonance** is said to occur. If the driving frequency is increased further, the amplitude of oscillation of the mass decreases. The variation with driving frequency f of the amplitude A of vibration of the mass is illustrated in Figure 10.29. This graph is often called a **resonance curve**.

> Resonance occurs when the natural frequency of vibration of an object is equal to the driving frequency, giving a maximum amplitude of vibration.

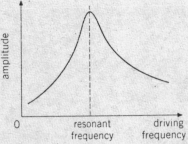

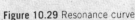

Figure 10.29 Resonance curve

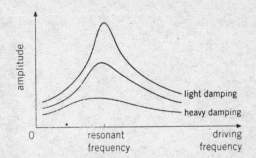

Figure 10.30 Effect of damping on the resonance curve

The effect of damping on the amplitude of forced oscillations can be investigated by attaching a light but stiff card to the mass in Figure 10.28. Movement of the card gives rise to air resistance and thus damping of the oscillations. The degree of damping may be varied by changing the area of the card. The effects of damping are illustrated in Figure 10.30. It can be seen that, as the degree of damping increases:

- the amplitude of oscillation at all frequencies is reduced
- the frequency at maximum amplitude shifts gradually towards lower frequencies
- the peak becomes flatter.

Barton's pendulums may also be used to demonstrate resonance and the effects of damping. The apparatus consists of a set of light pendulums, made (for example) from paper cones, and a more massive pendulum (the driver), all supported on a taut string. The arrangement is illustrated in Figure 10.31. The lighter pendulums have different lengths, but one has the same length as the driver. This has the same natural frequency as the driver and will, therefore, vibrate with the largest frequency of all the pendulums (Figure 10.32).

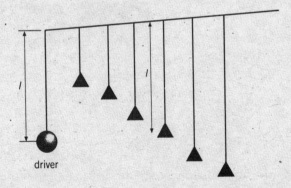

driver

Figure 10.31 Barton's pendulums

Figure 10.32 Time-exposure photographs of Barton's pendulums with light damping, taken end-on. The longest arc, in the middle, is the driver.

Adding weights to the paper cones reduces the effect of damping. With less damping, the amplitude of the resonant pendulum is much larger.

There are many examples of resonance in everyday life. One of the simplest is that of pushing a child on a swing (Figure 10.33). We push at the same frequency as the natural frequency of oscillation of the swing and child, so that the amplitude of the motion increases.

Figure 10.33 Pushing a child on a swing makes the swing go higher.

The operation of the engine of a vehicle causes a periodic force on the parts of the vehicle, which can cause them to resonate. For example, at particular frequencies of rotation of the engine, the mirrors may resonate. To prevent excessive vibration, the mountings of the mirrors provide damping.

Musical instruments rely on resonance to amplify the sound produced. The sound from a tuning fork is louder when it is held over a tube of just the right length, so that the column of air resonates. We have met this phenomenon in Chapter 8, in connection with the resonance tube method of measuring the speed of sound in air. Stringed instruments have a hollow wooden box with a hole under the strings which acts in a similar way. To amplify all notes from all of the strings, the sounding-box has to be a complex shape so that it resonates at many different frequencies.

A spectacular example of resonance that is often quoted is the failure of the first suspension bridge over the Tacoma Narrows in Washington State, USA.

Figure 10.34 The Tacoma Narrows bridge disaster

Wind caused the bridge to oscillate. It was used for months even though the roadway was oscillating with transverse vibrations. Approaching vehicles would appear, and then disappear, as the bridge deck vibrated up and down. One day strong winds set up twisting vibrations (Figure 10.34) and the amplitude of vibration increased due to resonance, until eventually the bridge collapsed. The driver of a car that was on the bridge managed to walk to safety before the collapse, although his dog could not be persuaded to leave the car!

Section 10.6 Summary

- Free oscillations are oscillations where there are no resistive forces acting on the oscillating system.
- Damping is produced by resistive forces which dissipate the energy of the vibrating system.
- Light damping causes the amplitude of vibration of the oscillation to decrease gradually. Critical damping causes the displacement to be reduced to zero in the shortest time possible, without any oscillation of the object. Overdamping also causes an exponential reduction in displacement, but over a greater time than for critical damping.
- The natural frequency of vibration of an object is the frequency at which the object will vibrate when allowed to do so freely.
- Forced oscillations occur when a periodic driving force is applied to a system which is capable of vibration.
- Resonance occurs when the driving frequency on the system is equal to its natural frequency of vibration. The amplitude of vibration is a maximum at the resonant frequency.

Section 10.6 Question

The apparatus of Figure 10.28 is used to demonstrate forced vibrations and resonance. A 50 g mass is suspended from the spring, which has a spring constant of $7.9 \, \text{N m}^{-1}$.

(a) Calculate the resonant frequency f_0 of the system.

(b) A student suggests that resonance should also be observed at a frequency of $2f_0$. Discuss this suggestion.

Exam-style Questions

1 All bodies moving in a circle experience centripetal acceleration.

(a) What does *centripetal* mean? Explain how the centripetal acceleration arises.

(b) The Moon's orbit around the Earth may be assumed to be circular, with radius 3.84×10^5 km. The Moon moves with constant speed in the orbit, and takes 27.3 days to complete one orbit. Calculate the magnitude of the centripetal acceleration of the Moon.

2 A bicycle is set up on a test rig with its front wheel clamped and its rear wheel resting on a roller. The pedals are turned so that, if the bicycle had been free, it would have been travelling at a speed of 8.3 m s^{-1}. The outside diameter of each tyre is 0.66 m.

(a) Calculate:
  (i) the angular velocity of the rear wheel of the bicycle,
  (ii) how many times per second the rear wheel rotates.

(b) A small pebble of mass 4.0 g is embedded in the tread of the rear tyre.
  (i) Calculate the magnitude, and state the direction, of the force needed to keep the pebble in circular motion.
  (ii) This force is equal to the maximum frictional force that the tyre can exert on the pebble. Describe and explain what happens when the rear wheel is rotated at a faster rate.

3 (a) Define simple harmonic motion.

(b) Figure 10.35a illustrates a U-tube of uniform cross-sectional area A containing liquid of density ρ. The total length of the liquid column is L. When the liquid is displaced by an amount Δx from its equilibrium position (see Figure 10.35b), it oscillates with simple harmonic motion.
  (i) The weight of liquid above AB in Figure 10.35b provides the restoring force. Write down an expression for the restoring force.
  (ii) Write down an expression for the acceleration of the liquid column caused by this force.
  (iii) Explain how this fulfils the condition for simple harmonic motion.
  (iv) Write down an expression for the frequency of the oscillations.

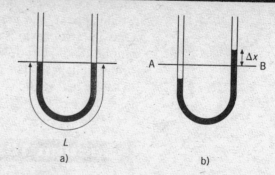

Figure 10.35

4 A 'baby bouncer' consists of a harness attached to a rubber cord. A baby of mass 6.5 kg is placed gently in the harness and the cord extends by 0.40 m. The baby is then pulled down another 0.10 m and, when released, begins to move with simple harmonic motion. Calculate:
(a) the period of the motion,
(b) the maximum force on the baby.
(acceleration of free fall $g = 9.8$ m s^{-2})

5 A simple model of a hydrogen molecule assumes that it consists of two oscillating hydrogen atoms connected by a spring of spring constant 1.1×10^3 N m^{-1}.
(a) The mass of a hydrogen atom is 1.7×10^{-27} kg. Calculate the frequency of oscillation of the hydrogen molecule.
(b) Explain why light of wavelength 2.3 μm would be strongly absorbed by this model of the hydrogen molecule.
(speed of light $c = 3.0 \times 10^8$ m s^{-1})

6 If you walk holding a full cup of coffee, you may find that, for a particular cup size and rate of taking steps, the coffee sloshes above the side of the cup, whereas if you walk more slowly, there is no risk of spillage. Explain this phenomenon, and estimate the critical combination of cup diameter and rate of taking steps.

7 As a party trick, the operatic tenor Enrico Caruso used to shatter a wine glass by singing a note of just the right frequency at full volume. If Caruso had been a physicist, what would he have understood by 'just the right frequency'? Why might he not have been successful if the glass had still contained some wine?

11. Fields

The aim of this chapter is to investigate the nature of electric and gravitational fields. Electric forces hold electrons in atoms, and bind atoms together in molecules and in solids. Gravitational forces are responsible for the birth of stars in giant molecular clouds in space. Both involve fields in which forces act over a distance. We shall study the similarities and differences between these fields, and develop an appreciation of the fields in our physical world.

11.1 Electric charges

At the end of Section 11.1 you should be able to:

- recall and use Coulomb's law in the form $F = Q_1Q_2/4\pi\epsilon_0 r^2$ for the force between two point charges in free space or air
- show an understanding of the concept of an electric field as an example of a field of force and define electric field strength as force per unit positive charge
- represent an electric field by means of field lines
- recall and use $E = V/d$ to calculate the field strength of the uniform field between charged parallel plates in terms of potential difference and separation
- calculate the forces on charges in uniform electric fields
- describe the effect of a uniform electric field on the motion of charged particles
- recall and use $E = Q/4\pi\epsilon_0 r^2$ for the field strength of a point charge in free space or air
- define potential at a point in terms of the work done in bringing unit positive charge from infinity to the point
- state that the field strength of the field at a point is numerically equal to the potential gradient at that point
- use the equation $V = Q/4\pi\epsilon_0 r$ for the potential in the field of a point charge.

Some effects of static electricity are familiar in everyday life. For example, a balloon rubbed on a jumper will stick to a wall, a TV screen that has been polished attracts dust, dry hair crackles (and may actually spark!) when brushed, and you may feel a shock when you touch the metal door-handle of a car on getting out after a journey in dry weather. All these are examples of insulated objects that have gained an electric charge by friction – that is, by being rubbed against other objects.

Insulators that are charged by friction will attract other objects. Some of the effects have been known for centuries. Greek scientists experimented with amber that was charged by rubbing it with fur. Our words *electron* and *electricity* come from the Greek word ηλεκτρον, which means amber. Today, electrostatics experiments are often carried out with plastic materials which are moisture-repellent and stay charged for longer.

Charging by friction can be hazardous. For example, a lorry which carries a bulk powder must be earthed before emptying its load. Otherwise, electric charge can build up on the tanker. This could lead to a spark from the tanker to earth, causing an explosion. Similarly, the pipes used for movement of highly flammable liquids (for example, petrol), are metal-clad. An aircraft moving through the air will also become charged. To prevent the first person touching the aircraft on landing from becoming seriously injured, the tyres are made to conduct, so that the aircraft loses its charge on touchdown.

There are two kinds of electric charge. Polythene becomes negatively charged when rubbed with wool, and cellulose acetate becomes positively charged, also when rubbed with wool. To understand why this happens, we need to go back to the model of the atom. Atoms consist of a positively charged nucleus with negatively charged electrons orbiting it. When the polythene is rubbed with wool, friction causes some electrons to be transferred from the wool to the polythene. So the polythene has a negative charge, and the wool is left with a positive charge. Cellulose acetate becomes positive because it loses some electrons to the wool when it is rubbed. Polythene and cellulose acetate are poorly conducting plastics, and the charges remain static on the surface.

Putting two charged polythene rods close to each other, or two charged acetate rods close to each other, shows that similar charges repel each other (Figure 11.1). Conversely, unlike charges attract. A charged polythene rod attracts a charged acetate rod.

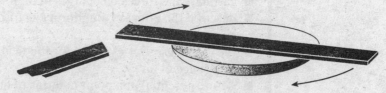

Figure 11.1 Like charges repel.

This is the basic law of force between electric charges:

Like charges repel, unlike charges attract.

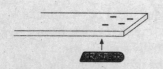

Figure 11.2 A charged rod can induce charges in an uncharged object.

Charged rods will also attract uncharged objects. A charged polythene rod will pick up small pieces of paper, for example. The presence of the negative charge on the rod causes a redistribution of the charges in the paper. Free electrons are repelled to the far side of the paper, so that the side nearest the rod is positive and is therefore attracted to the rod (Figure 11.2). The paper is said to be **charged by induction**. When the rod is removed, the free electrons will move back and cancel the positive charge.

Force between charges

In the late eighteenth century, the French scientist Charles Coulomb investigated the magnitude of the force between charges, and how this varies with the charges involved and the distance between them. He discovered the following rule.

> The force is proportional to the product of the charges and inversely proportional to the square of the distance between them. This is known as **Coulomb's law**.

Coulomb's experiments made use of small, charged insulated spheres. Strictly speaking, the law applies to point charges, but it can be used for charged spheres provided that their radii are small compared with their separation.

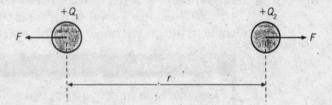

Figure 11.3

For point charges Q_1 and Q_2 a distance r apart (Figure 11.3), Coulomb's law gives the force F as

$$F \propto Q_1 Q_2 / r^2$$

or

$$F = \frac{k Q_1 Q_2}{r^2}$$

where k is a constant of proportionality, the value of which depends on the insulating medium around the charges and the system of units employed. In SI units, F is measured in newtons, Q in coulombs and r in metres. Then the constant k is given as

$$k = \frac{1}{4\pi\epsilon_0}$$

and so

$$F = \frac{Q_1 Q_2}{4\pi\epsilon_0 r^2}$$

when the charges are in a vacuum. The quantity ϵ_0 is called the **permittivity of free space** (or the permittivity of a vacuum).

The value of the permittivity of air is very close to that of a vacuum ($1.0005\epsilon_0$), so the equation can be used for the force between charges in a vacuum or in air.

The value of the permittivity of free space is given by

$$\epsilon_0 = 8.85 \times 10^{-12} \, C^2 \, N^{-1} m^{-2}$$

(We shall meet an alternative unit for ϵ_0 later, that is $F \, m^{-1}$.) This numerical value for ϵ_0 means that

$$k = \frac{1}{4\pi\epsilon_0} = 8.99 \times 10^9 \, C^{-2} \, N \, m^2$$

Coulomb's law is often referred to as an **inverse square** law of force, because the variation of force with distance r between the charges is proportional to $1/r^2$. We shall meet another important inverse square law of force when we come to the gravitational force between two point masses.

Example

Calculate the force between two point charges, each of $1.0 \times 10^{-9} \, C$, which are 4.0 cm apart in a vacuum.

Using $F = kQ_1Q_2/r^2$, $F = 9.0 \times 10^9 \times (1.0 \times 10^{-9})^2/(4.0 \times 10^{-2})^2$
$= \mathbf{5.6 \times 10^{-6} N}$

Now it's your turn

Calculate the force between two electrons which are 1.0×10^{-10} m apart in a vacuum.
(electron charge $e = -1.6 \times 10^{-19} \, C$)

Electric fields

Electric charges exert forces on each other when they are a distance apart. The idea of an **electric field** is used to explain this action at a distance. An electric field is a region of space where a stationary charge experiences a force.

Electric fields are invisible, but they can be represented by electric lines of force, just as magnetic fields can be represented by magnetic lines of force (see Chapter 12). The direction of the electric field is defined as the direction in which a positive charge would move if it were free to do so. So the lines of force can be drawn with arrows that go from positive to negative.

For any electric field:

■ the lines of force start on a positive charge, and end on a negative charge
■ the lines of force are smooth curves which never touch or cross
■ the strength of the electric field is indicated by the closeness of the lines: the closer they are, the stronger the field.

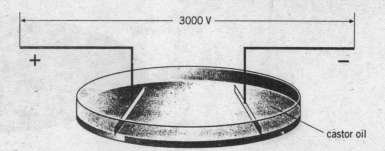

Figure 11.4 Apparatus for investigating electric field patterns

The apparatus in Figure 11.4 can be used to show electric field patterns. Semolina is sprinkled on a non-conducting oil and a high voltage supply is connected to the conducting plates. The semolina becomes charged by induction and lines up along the lines of force. Some electric field patterns are illustrated in Figure 11.5. The pattern for the charged conducting sphere (Figure 11.5c) is of particular importance. Although there are no electric field lines inside the conductor, the field lines appear to come from the centre. Thus, the charge on the surface of a conducting sphere appears as if it is all concentrated at the centre. This means that small conducting spheres may be used as an approximation to point charges.

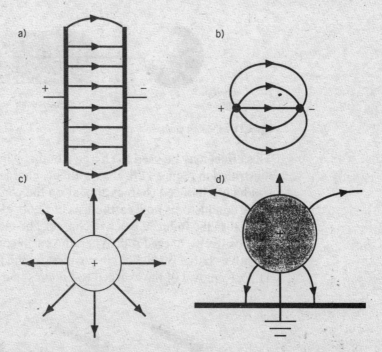

Figure 11.5 Some electric field patterns

Paint spraying makes use of some of the principles of electrostatics. It can be wasteful if the paint is not sprayed where it is needed. But if the spray is given a charge, and the object to be painted is given an opposite charge, the charged droplets follow the lines of force and end up on the surface. This is illustrated in Figure 11.6.

Figure 11.6 Paint spraying

Induced fields and charges

We have already stated that there is no resultant electric field in a conductor (unless a source of e.m.f. is connected across it). The reason for this is that electrons are free to move in the conductor. As soon as a charged body is placed near the conductor, the electric field of the charged body causes electrons in the conductor to move in the opposite direction to the field (because electrons have a negative charge). This is illustrated in Figure 11.7. These charges create a field in the opposite direction to the field due to the charged body. The induced charges will stop moving when the two fields are equal and opposite. Hence there is no resultant field in a conductor.

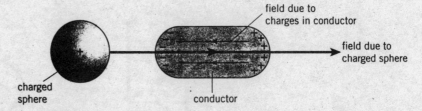

Figure 11.7 Induced charges

This effect may be used to charge a body by induction. The process is illustrated in Figure 11.8. If a positively charged rod is placed near an insulated conductor, induced charges appear on the conductor, as shown in Figure 11.8a. If the conductor is now earthed, as in 11.8b, electrons move from earth to neutralise the induced positive charge. The electrons are held in position by the positively charged rod. When the earth connection is removed, the negative charge is still held in position by the positively charged rod, as in 11.8c. Removal of the charged rod means that the electrons will distribute

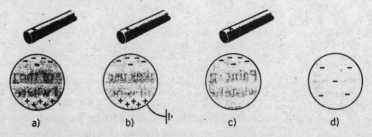

Figure 11.8 Charging by induction

themselves over the surface of the sphere, as in 11.8d. Note that if a positively charged rod is used, the final charge on the sphere is negative; if a negatively charged rod is used, the sphere becomes positively charged.

Electric field strength

> The **electric field strength** at a point is defined as the force per unit charge acting on a small positive charge placed at that point.

If the force experienced by a positive test charge $+Q$ placed in the field is F, the field strength E is given by

$$E = \frac{F}{Q}$$

(Be careful not to be confused by the symbol E for field strength. E may also be used for energy.) A unit of field strength can be deduced from this equation: force is measured in newtons and charge in coulombs, so the SI unit of field strength is N C^{-1}. We shall see later that there is another common SI unit for electric field strength, V m^{-1}. These two units are equivalent.

A uniform electric field

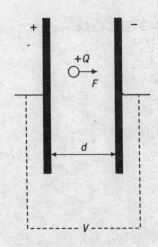

Figure 11.9

In Figure 11.5a, the electric field pattern between the parallel plates consists of parallel, equally spaced lines, except near the edges of the plates. This means that the field between charged parallel plates (as, for example, in a capacitor) is uniform.

Figure 11.9 illustrates parallel plates a distance d apart with a potential difference V between them. A charge $+Q$ in the uniform field between the plates has a force F acting on it. To move the charge towards the positive plate would require work to be done on the charge. Work is given by the product of force and distance. To move the charge from one plate to the other requires work W given by

$$W = Fd$$

From the definition of potential difference (page 179),

$$W = VQ$$

Thus $W = Fd = VQ$, or, rearranging,

$$F/Q = V/d$$

But F/Q is the force per unit charge, and this is the definition of electric field strength. Thus, for a uniform field, the field strength E is given by

$$E = \frac{V}{d}$$

This equation gives an alternative unit for electric field strength (remember that we have already derived N C^{-1} from the idea of force per unit charge). If we use the expression for a uniform field, the unit of potential difference is the volt and the unit of distance is the metre, so another unit for electric field strength is volts per metre (V m^{-1}). The two units N C^{-1} and V m^{-1} are equivalent.

Example

Two metal plates 5.0 cm apart have a potential difference of 1000 V between them. Calculate:
(a) the strength of the electric field between the plates,
(b) the force on a charge of 5.0 nC between the plates.

(a) From $E = V/d$, $E = 1000/5.0 \times 10^{-2} = $ **2.0×10^4 V m^{-1}**.
(b) From $F = EQ$, $F = 2.0 \times 10^4 \times 5.0 \times 10^{-9} = $ **1.0×10^{-4} N**.

Now it's your turn

Two metal plates 15 mm apart have a potential difference of 750 V between them. The force on a small charged sphere placed between the plates is 1.2×10^{-7} N. Calculate:
(a) the strength of the electric field between the plates,
(b) the charge on the sphere.

Field strength due to an isolated point charge

The electric field due to an isolated point charge is **radial** (see Figure 11.5c). We have to mention that the point charge is isolated. If any other object, charged or otherwise, is near it, the field would be distorted.

From Coulomb's law, the force on a test charge q a distance r from the isolated point charge Q is given by

$$F = \frac{Qq}{4\pi\epsilon_0 r^2}$$

The electric field E at the location of the test charge q is given by $E = F/q$. Thus, the electric field due to the isolated point charge is

$$E = \frac{Q}{4\pi\epsilon_0 r^2}$$

or

$$E = \frac{kQ}{r^2} \quad \text{where} \quad k = \frac{1}{4\pi\epsilon_0}$$

This is an inverse square law field of force because the force varies in proportion to $1/r^2$.

Example

In a simplified model of the hydrogen atom, the electron is at a distance of 5.3×10^{-11} m from the proton. The proton charge is $+1.6 \times 10^{-19}$ C. Calculate the electric field strength of the proton at this distance.

Assuming that the field is radial,
$E = kQ/r^2 = 9.0 \times 10^9 \times 1.6 \times 10^{-19}/(5.3 \times 10^{-11})^2$
$\qquad = \mathbf{5.1 \times 10^{11}\,N\,C^{-1}}$

Now it's your turn

A Van de Graaff generator has a dome of radius 15 cm. The dome carries a charge of 2.5×10^{-6} C. It can be assumed that this charge acts as if it were all concentrated at the centre of the spherical dome. Calculate the electric field strength at the dome's surface.

Electric potential energy and electric potential

We already know how useful the concept of energy is in dealing with mechanical problems. Energy is a conserved quantity and is thus a fundamental aspect of nature. We can also use the idea of energy in electricity.

We define electric potential energy in exactly the same way as we do for other types of potential energy. That is, the change in electric potential energy when a charge Q is moved between two points A and B in an electric field is the work done by the electric force in moving the charge from B back to A.

We defined electric field as the force per unit positive charge. Similarly, electric potential is defined as the potential energy per unit positive charge. The symbol for potential is V, and its unit is the volt. If a point charge Q has potential energy PE_A at a point A, the electric potential at that point is

$$V_A = PE_A/Q$$

We already know that only differences in potential energy are measurable. We need to specify a convenient reference point to act as a zero of potential energy and potential. In dealing with gravitational energy we often take the floor of the laboratory or the Earth's surface as zero, and measure mgh from one of these. Similarly, in electrical problems, it is often convenient to take earth potential as zero, especially if part of the circuit is earthed. But the 'official' definition of the zero of electric potential is the potential of a point an infinite distance away. This means that

> The **potential** at a point in an electric field is defined as the work done in bringing unit positive charge from infinity to the point.

There is an important link between the electric field at a point and the potential at that point.

> The field strength is equal to the negative of the potential gradient at that point.

Thus, if we have a graph showing how the potential changes in a field, the gradient of this graph at any point gives us the numerical value of the field strength at that point (see Figure 11.10). This link leads to an expression for the potential at points in the field of a point charge Q. The potential V is given by

$$V = \frac{kQ}{r} = \frac{Q}{4\pi\epsilon_0 r}$$

The potential very near a positive charge is large, and decreases towards zero the further away we move from the charge which is the source of the field. If the charge producing the field is negative, the potential is also negative and increases towards zero at large distances. Note that the variation of potential with distance is an inverse proportionality, and not the inverse square relationship that applies for the variation of field strength with distance.

Example

How much work must be done by an external force in moving a charge q of $+2.0\,\mu C$ from infinity to a point A, 0.40 m from a charge Q of $+30\,\mu C$?

The work is simply the change in electric potential energy. The potential energy at infinity is zero, so

$W = qV = qkQ/r = 2.0 \times 10^{-6} \times 9.0 \times 10^9 \times 30 \times 10^{-6}/0.40 = \mathbf{1.4\,J}$

Now it's your turn

Two $+30\,\mu C$ charges are placed on a straight line 0.40 m apart. A $+0.5\,\mu C$ charge is to be moved a distance of 0.10 m along the line from a point midway between the charges. How much work must be done?

In the case of a uniform electric field, such as that between two parallel plates (page 287) the relation between field strength and potential gradient is very simple. In Figure 11.9 the potential decreases uniformly as we move from the positive left-hand plate to the negative right-hand plate. If the left-hand plate is at a potential $+V$ and the right-hand plate is at a potential of zero, the graph of potential against distance is as shown in Figure 11.10a. It is a straight line, sloping downwards, with gradient $-V/d$. This gives the negative of the potential gradient as $+V/d$. As has been stated above, this is equal to the field strength E, a constant because the gradient is constant (Figure 11.10b). This is the same relation that was derived on page 287.

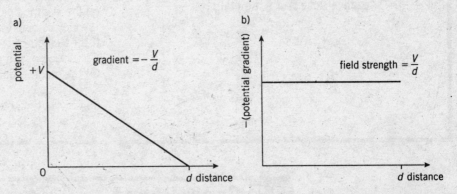

Figure 11.10 Graphs of a) potential and b) (the negative of the) potential gradient, for a uniform electric field

Section 11.1 Summary

- Insulators may be charged by friction.
- Like charges repel; unlike charges attract each other.
- When charged objects are placed near conductors, they cause a redistribution of charge in the conductor, thereby inducing charges.
- The force between two point charges is proportional to the product of the charges and inversely proportional to the square of the distance between them. This is Coulomb's law:
  $F = Q_1 Q_2/4\pi\epsilon_0 r^2$
- ϵ_0 is the permittivity of free space; its value is $8.85 \times 10^{-12} \, \mathrm{C^2 N^{-1} m^{-2}}$ ($\mathrm{F \, m^{-1}}$).
- An electric field is a region of space where a stationary charge experiences a force.
- Electric field strength is the force per unit positive charge: $E = F/Q$
- The electric field between parallel plates is uniform. The field strength is given by: $E = V/d$
- The electric field strength at a point in the field of an isolated point charge is given by: $E = Q/4\pi\epsilon_0 r^2$
- The electric potential at a point in an electric field is the work done in bringing unit positive charge from infinity to the point.
- The potential at a point in the field of an isolated point charge is given by: $V = Q/4\pi\epsilon_0 r$

Section 11.1 Questions

1 A glass rod rubbed with silk becomes positively charged. Explain what has happened to the glass. Explain also why the charged rod is able to attract small pieces of paper.

2 (a) Explain what is meant by an electric field.
(b) Sketch the electric field patterns
  (i) between two negatively charged particles,
  (ii) between a positive charge and a negatively charged flat metal plate.

3 In a simplified model, a uranium nucleus is a sphere of radius 8.0×10^{-15} m. The nucleus contains 92 protons (and rather more neutrons). The charge on a proton is 1.6×10^{-19} C. It can be assumed that the charge of these protons acts as if it were all concentrated at the centre of the nucleus. The nucleus releases an α-particle containing two protons (and two neutrons) at the surface of the nucleus. Calculate:
  (a) the electric field strength at the surface of the nucleus before emission of the α-particle,
  (b) the electric force on the α-particle at the surface of the nucleus,
  (c) the electric potential at the surface of the nucleus,
  (d) the electric potential energy of the α-particle when it is at the surface of the nucleus.

11.2 Capacitance

At the end of Section 11.2 you should be able to:
- define capacitance and the farad
- recall and solve problems using $C = Q/V$
- deduce from the area under a potential–charge graph the equation $W = \frac{1}{2}QV$ and hence $W = \frac{1}{2}CV^2$
- derive, using the formula $C = Q/V$, conservation of charge and the addition of p.d.s, formulae for capacitors in series and in parallel
- solve problems using formulae for capacitors in series and in parallel
- show an understanding of the function of capacitors in simple circuits.

Consider an isolated spherical conductor connected to a high voltage supply (Figure 11.11). it is found that, as the potential of the sphere is increased, the charge stored on the sphere also increases. The graph showing the variation of charge on the conductor with potential is shown in Figure 11.12. It can be seen that charge Q is related to potential V by

$$Q \propto V$$

Hence,

$$Q = CV$$

where C is a constant which depends on the size of the conductor. C is known as the capacitance of the conductor.

Capacitance is the ratio of charge to potential for a conductor.

$$C = \frac{Q}{V}$$

Another chance for confusion! The letter C is used as an abbreviation for the unit of charge, the coulomb; as an italic letter C it is used as the symbol for capacitance.

The unit of capacitance is the **farad** (symbol F). One farad is one coulomb per volt. The farad is an inconveniently large unit. In electronic circuits and laboratory experiments, the range of useful values of capacitance is from about 10^{-12} F (1 picofarad, or 1 pF) to 10^{-3} F (1 millifarad, or 1 mF). (See Table 1.1, page 3, for a list of decimal multiples and submultiples for use with units.)

Circuit components which store charge and therefore have capacitance are called **capacitors**.

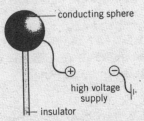

Figure 11.11 Charged spherical conductor

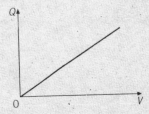

Figure 11.12 Relation between charge and potential

Example

Calculate the charge stored on an isolated conductor of capacitance 470 μF when it is at a potential of 20 V.

Using $C = Q/V$, we have $Q = CV = 470 \times 10^{-6} \times 20 = \mathbf{9.4 \times 10^{-3} C}$.

Now it's your turn

The charge on a certain isolated conductor is 2.5×10^{-2} C when its potential is 25 V. Calculate its capacitance.

Energy stored in a capacitor

When charging a capacitor, work is done by the battery to move charge on to the capacitor. Energy is transferred from the power supply and is stored as **electric potential energy** in the capacitor.

Camera flash units use a capacitor to store energy. The capacitor takes a few seconds to charge when connected to the battery in the camera. Then the energy is discharged very rapidly when the capacitor is connected to the flash-bulb to give a short but intense flash.

Since $Q = CV$, the charge stored on a capacitor is directly proportional to its potential (see Figure 11.12).

From the definition of potential, the work done (and therefore the energy transferred) is the product of the potential and the charge. That is,

$$W \text{ (and } E_p) = VQ$$

However, while more and more charge is transferred to the capacitor, the potential difference is increasing. Suppose the potential is V_0 when the charge stored is Q_0. When a further small amount of charge Δq is supplied, the energy transferred is given by

$$\Delta E_p = V_0 \Delta q$$

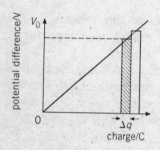

Figure 11.13 Graph of potential against charge for a capacitor

which is equal to the area of the shaded strip in Figure 11.13. Similarly, the energy transferred when a further charge Δq is transferred is given by the area of the next strip, and so on. If the amount of charge Δq is very small, the strips become very thin and their combined areas are just equal to the area between the line and the horizontal axis. Thus

The energy transferred from the battery when a capacitor is charged is given by the area under the graph line when charge (x-axis) is plotted against potential difference (y-axis).

Because the graph is a straight line through the origin, this area is just the area of the right-angled triangle formed by the line and the charge axis. Thus

$$E_p = \tfrac{1}{2} QV$$

This is the expression for the energy transferred from the battery in charging the capacitor. This is electric potential energy, and it is released when the capacitor is discharged. Since $C = Q/V$, this expression can be written in different forms.

$$E_p = \tfrac{1}{2} QV = \tfrac{1}{2} CV^2 = \frac{Q^2}{2C}$$

Example

Calculate the energy stored by a 470 μF capacitor at a potential of 20 V.

Using $E_p = \tfrac{1}{2} CV^2$, $E_p = \tfrac{1}{2} \times 470 \times 10^{-6} \times (20)^2 = \textbf{0.094 J}$.

Now it's your turn

A camera flash-lamp uses a 5000 μF capacitor which is charged by a 9 V battery. Calculate the energy transferred when the capacitor is fully discharged through the lamp.

Capacitors

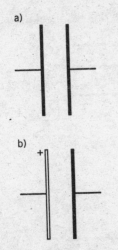

Figure 11.14 Circuit symbols for a) a capacitor and b) an electrolytic capacitor

The simplest capacitor in an electric circuit consists of two metal plates, with an air gap between them which acts as an insulator. This is called a parallel-plate capacitor. Figure 11.14a shows the circuit symbol for a capacitor. When the plates are connected to a battery, the battery transfers electrons from the plate connected to the positive terminal of the battery to the plate connected to the negative terminal. Thus the plates carry equal but opposite charges.

The **capacitance** of a capacitor is defined as the charge stored on one plate per unit potential difference between the plates.

The capacitance of an air-filled capacitor can be increased by putting an insulating material, such as mica or waxed paper, between the plates. The material between the plates is called the **dielectric**. In a type of capacitor known as an **electrolytic** capacitor the dielectric is deposited by an electrochemical reaction. These capacitors *must* be connected with the correct polarity for their plates, or they will be damaged. The circuit symbol for an electrolytic capacitor is shown in Figure 11.14b. Electrolytic capacitors are available with capacitances up to about 1 mF.

Factors affecting capacitance

As stated previously, the material used as a dielectric affects the capacitance of a capacitor. The other factors determining the capacitance are the area of the plates and the distance between them. Experiment shows that

The capacitance is directly proportional to the area A of the plates, and inversely proportional to the distance d between them.

Putting these two factors together gives

$$C \propto \frac{A}{d}$$

For a capacitor with air or a vacuum between the plates, the constant of proportionality is the permittivity of free space. Thus

$$C = \frac{\epsilon_0 A}{d}$$

Since C is measured in farads, A in square metres and d in metres, we can see that an alternative unit for ϵ_0 is farads per metre, F m^{-1} (see Section 11.1). We introduce a quantity called the **relative permittivity** ϵ_r of a dielectric to account for the fact that the use of a dielectric increases the capacitance.

The relative permittivity is defined as the capacitance of a parallel-plate capacitor with the dielectric between the plates divided by the capacitance of the same capacitor with a vacuum between the plates.

Relative permittivity ϵ_r is a ratio and has no units. Some values of relative permittivity are given in Table 11.1.

Table 11.1 Relative permittivity of different dielectric materials

material	relative permittivity ϵ_r
air	1.0005
polyethylene (polythene)	2.3
sulphur	4
paraffin oil	4.7
mica	6
barium titanate	1200

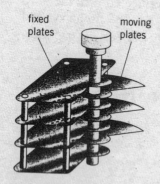

fixed plates   moving plates

Figure 11.15 Variable capacitor

Including the relative permittivity factor, the full expression for the capacitance of a parallel-plate capacitor is

$$C = \frac{\epsilon_0 \epsilon_r A}{d}$$

Variable capacitors (Figure 11.15) have one set of plates mounted on a spindle, so that the area of overlap can be changed. Varying the capacitance of a tuning circuit is one way of tuning in to different radio or TV stations.

Example

A parallel-plate, air-filled capacitor has square plates of side 30 cm that are a distance 1.0 mm apart. Calculate the capacitance of the capacitor.

Using $C = \epsilon_0 \epsilon_r A/d$, $C = 8.85 \times 10^{-12} \times 1 \times (30 \times 10^{-2})^2/1.0 \times 10^{-3}$
$$= 8.0 \times 10^{-10} \, \text{F}$$

Now it's your turn

A capacitor consists of two metal discs of diameter 15 cm separated by a sheet of polythene 0.25 mm thick. The relative permittivity of polythene is 2.3. Calculate the capacitance of the capacitor.

Capacitors in series and parallel

In Figure 11.16, the two capacitors of capacitance C_1 and C_2 are connected in series. We shall show that the combined capacitance C is given by

$$\frac{1}{C} = \frac{1}{C_1} + \frac{1}{C_2}$$

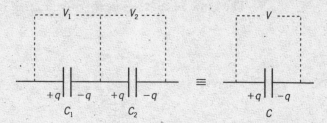

Figure 11.16 Capacitors in series

If the voltage across the equivalent capacitor is V and the charge stored on each plate is q, then $V = q/C$. The potential difference across the combination is the sum of the potential differences across the individual capacitors, $V = V_1 + V_2$, and each capacitor has charge q on each plate. A charge of $+q$ induced on one plate of one capacitor will induce a charge of $-q$ on the plate of the second capacitor. Since $V_1 = q/C_1$ and $V_2 = q/C_2$, then

$$q/C = q/C_1 + q/C_2$$

Dividing each side of the equation by q, we have

$$1/C = 1/C_1 + 1/C_2$$

A similar result applies for any number of capacitors connected in series.

The reciprocal of the combined capacitance equals the sum of the reciprocals of the individual capacitances in series.

Note that:

■ For two identical capacitors in series, the combined capacitance is equal to half of the value of each one.
■ For capacitors in series, the combined capacitance is always less than the value of the smallest individual capacitance.

In Figure 11.17, the two capacitors of capacitance C_1 and C_2 are connected in parallel. We shall show that the combined capacitance C is given by

$$C = C_1 + C_2$$

If the voltage across the equivalent capacitor is V and the charge stored on each plate is q, then $q = CV$. The total charge stored is the sum of the charges on the individual capacitors, $q = q_1 + q_2$, and there is the same potential difference V across each capacitor since they are connected in parallel. Since $q_1 = C_1V$ and $q_2 = C_2V$, then

$$CV = C_1V + C_2V$$

Dividing each side of the equation by V, we have

$$C = C_1 + C_2$$

The same result applies for any number of capacitors connected in parallel.

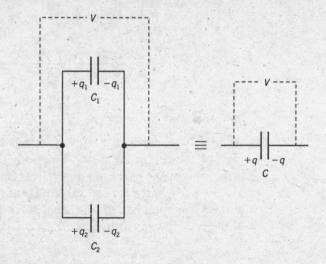

Figure 11.17 Capacitors in parallel

The combined capacitance equals the sum of all the individual capacitances in parallel.

Note that the equation for capacitors in *series* is similar to the equation for resistors in *parallel*, and the equation for capacitors in parallel is similar to the equation for resistors in series (see Chapter 7).

Examples

1 A $100\,\mu F$ capacitor in parallel with a $50\,\mu F$ capacitor is connected to a 12 V supply. Calculate:
   (a) the total capacitance,
   (b) the potential difference across each capacitor,
   (c) the charge stored on each capacitor.

   (a) Using the equation for capacitors in parallel,
   $C = C_1 + C_2 = 100 + 50 = \mathbf{150\,\mu F}$.
   (b) The potential difference across each capacitor is the same as the potential difference across the supply. This is **12 V**.
   (c) Using $Q = CV$, the charge stored on the $100\,\mu F$ capacitor is $100 \times 10^{-6} \times 12 = \mathbf{1.2 \times 10^{-3}\,C}$. The charge stored on the $50\,\mu F$ capacitor is $50 \times 10^{-6} \times 12 = \mathbf{6.0 \times 10^{-4}\,C}$.

2 A $100\,\mu F$ capacitor in series with a $50\,\mu F$ capacitor is connected to a 12 V supply. Calculate:
   (a) the combined capacitance,
   (b) the charge stored by each capacitor,
   (c) the potential difference across each capacitor.

(a) Using the equation for capacitors in series,
$1/C = 1/C_1 + 1/C_2 = 1/100 \times 10^{-6} + 1/50 \times 10^{-6} = 3 \times 10^4$.
Thus $C = 3.3 \times 10^{-5} = \mathbf{33\,\mu F}$.

(b) The charge stored by each capacitor is the same as the charge stored by the combination, so
$Q = CV = 33 \times 10^{-6} \times 12 = \mathbf{4.0 \times 10^{-4}\,C}$.

(c) Using $V = Q/C$, the potential difference across the $100\,\mu F$ capacitor is $4.0 \times 10^{-4}/100 \times 10^{-6} = \mathbf{4.0\,V}$. The potential difference across the $50\,\mu F$ capacitor is
$4.0 \times 10^{-4}/50 \times 10^{-6} = \mathbf{8.0\,V}$. Note that the two potential differences add up to the supply voltage.

Now it's your turn

(a) A $470\,\mu F$ capacitor is connected to a $20\,V$ supply. Calculate the charge stored on the capacitor.

(b) The capacitor in (a) is then disconnected from the supply and connected to an uncharged $470\,\mu F$ capacitor.

   (i) Explain why the capacitors are in parallel, rather than series, and why the total charge stored by the combination must be the same as the answer to (a).

   (ii) Calculate the capacitance of the combination.

   (iii) Calculate the potential difference across each capacitor.

   (iv) Calculate the charge stored on each capacitor.

Charging and discharging capacitors

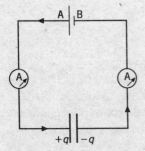

Figure 11.18 Charging a capacitor

The circuit in Figure 11.18 can be used to investigate what happens when a capacitor is connected to a battery. Start with the lead to battery terminal A connected, but that to B disconnected. When connection is first made to terminal B, both ammeters flick to the right and then return to zero, indicating a momentary pulse of current. If the process is repeated, nothing further happens. The capacitor has become charged.

During the initial connection, the ammeters have similar deflections. This tells us that the same charge has moved on to the left-hand plate as has been removed from the right-hand plate. The momentary current has caused the left-hand plate to acquire a charge of $+q$, while the right-hand plate has a charge of $-q$. The capacitor is now said to store a charge of q (although if we add up the charges on the left- and right-hand plates taking account of their sign, the net charge is zero). There can be no steady current because of the gap between the capacitor plates. The current stops when the potential difference across the capacitor is the same as the e.m.f. of the battery.

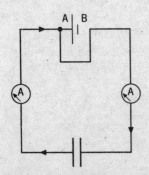

Figure 11.19 Discharging a capacitor

When the battery lead to terminal B is disconnected and joined to point A (Figure 11.19) so that the battery is no longer in circuit, both ammeters give similar momentary deflections to the left. This time a current in the opposite direction has moved the charge of $+q$ from the left-hand plate to cancel the charge of $-q$ on the right-hand plate. The capacitor has become discharged.

Remember that in metal wires the current is carried by free electrons. These move in the opposite direction to that of the conventional current (see Chapter 7). When the capacitor is charged, electrons move from the negative terminal of the battery to the right-hand plate of the capacitor, and from the left-hand plate to the positive terminal of the battery. When the capacitor is discharged, electrons flow from the negative right-hand plate of the capacitor to the positive left-hand plate.

The experiment described using the circuit in Figure 11.19 showed that there is a momentary current when a capacitor discharges. A resistor connected in series with the capacitor will reduce the current, so that the capacitor discharges more slowly.

The circuit shown in Figure 11.20 can be used to investigate more precisely how a capacitor discharges. When the two-way switch is connected to point A, the capacitor will charge up until the potential difference between its plates is equal to the e.m.f. V_0 of the supply. When the switch is moved to B, the capacitor will discharge through the resistor. When the switch makes contact with B, the current can be recorded at regular intervals of time as the capacitor discharges.

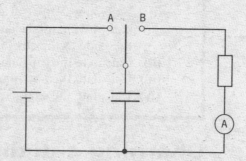

Figure 11.20 Circuit for investigating capacitor discharge

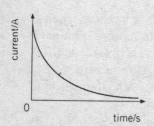

Figure 11.21 Graph of current against time for capacitor discharge

A graph of the discharge current against time is shown in Figure 11.21. The current is seen to change rapidly at first, and then more slowly. More detailed analysis shows that the curve is **exponential**. We shall meet exponential changes again when we deal with the decay of radioactive substances (Chapter 13).

All exponential decay curves have an equation of the form

$$x = x_0 e^{-kt}$$

where x is the quantity that is decaying (and x_0 is the value of x at $t = 0$), e to three decimal places is the number 2.718 (the root of natural logarithms) and k is a constant characteristic of the decay. A large value of k means that the decay is rapid, and a small value means a slow decay.

The solution for the discharge of a capacitor of capacitance C through a resistor of resistance R is of the form

$$Q = Q_0 e^{-t/CR}$$

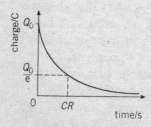

Figure 11.22 Graph of charge against time for capacitor discharge

This is the equation of the graph in Figure 11.22. Here Q_0 is the charge on the plates at time $t = 0$, and Q is the charge at time t.

The graphs of Figures 11.21 and 11.22 have exactly the same shape, and thus the equation for the discharge of a capacitor may be written

$$I = I_0 e^{-t/CR}$$

Furthermore, since for a capacitor Q is proportional to V, then the solution for the discharge can be written as

$$V = V_0 e^{-t/CR}$$

Time constant

As time progresses, the exponential curve in Figure 11.22 gets closer and closer to the time axis, but never actually meets it. Thus, it is not possible to quote a time for the capacitor to discharge completely.

However, the quantity CR in the decay equation may be used to give an indication of whether the decay is fast or slow, as shown in Figure 11.23.

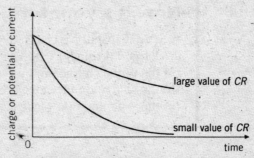

Figure 11.23 Decay curves for large and small time constants

CR is called the **time constant** of the capacitor–resistor circuit.

CR has the units of time, and is measured in seconds.

We can easily show that CR has units of time. From $C = Q/V$ and $R = V/I$, then $CR = Q/I$. Since charge Q is in coulombs and current I is in amperes, and one ampere is equal to one coulomb per second, Q/I is in seconds.

To find the charge Q on the capacitor plates after a time $t = CR$, we substitute in the exponential decay equation

$$Q = Q_0 e^{-CR/CR} = Q_0 e^{-1} = Q_0/e = Q_0/2.718$$

Thus

The time constant is the time for the charge to have decreased to $1/e$ (or $1/2.718$) of its initial charge.

In one time constant the charge stored by the capacitor drops to roughly one-third of its initial value. During the next time constant it will drop by the same ratio, to about one-ninth of the value at the beginning of the decay.

Example

A $500\,\mu F$ capacitor is connected to a 10 V supply, and is then discharged through a $100\,k\Omega$ resistor. Calculate:
(a) the initial charge stored by the capacitor,
(b) the initial discharge current,
(c) the value of the time constant,
(d) the charge on the plates after 100 s,
(e) the time at which the remaining charge is $2.5 \times 10^{-3} C$.

(a) From $Q = CV$, we have $Q = 500 \times 10^{-6} \times 10 = \mathbf{5.0 \times 10^{-3}\,C}$.
(b) From $I = V/R$, we have $I = 10/(100 \times 10^3) = \mathbf{1.0 \times 10^{-4}\,A}$.
(c) $CR = 500 \times 10^{-6} \times 100 \times 10^3 = \mathbf{50\,s}$.
(d) After 50 s, the charge on the plates is
$Q_0/e = 5 \times 10^{-3}/2.718 = 1.8 \times 10^{-3}$ C; after another 50 s, the charge is $1.8 \times 10^{-3}/2.718 = \mathbf{6.8 \times 10^{-4}\,C}$.
(e) Using $Q = Q_0 e^{-t/CR}$, $2.5 \times 10^{-3} = 5.0 \times 10^{-3} e^{-t/50}$, or $0.50 = e^{-t/50}$. Taking natural logarithms of both sides, $-0.693 = -t/50$, or $t = \mathbf{35\,s}$.

Now it's your turn

A $5.0\,\mu F$ capacitor is charged from a 12 V battery, and is then discharged through a $0.50\,M\Omega$ resistor. Calculate:
(a) the initial charge on the capacitor,
(b) the charge on the capacitor 2.0 s after the discharge starts,
(c) the potential difference across the capacitor at this time.

Section 11.2 Summary

- A capacitor stores charge. Its capacitance C is given by $C = Q/V$, where Q is the charge on the capacitor when there is a potential difference V between its plates.
- The unit of capacitance, the farad (F), is one coulomb per volt.
- The energy stored in a charged capacitor is given by:
$E = \frac{1}{2}QV = \frac{1}{2}CV^2 = \frac{1}{2}Q^2/C$
- The equivalent capacitance C of capacitors connected in series is given by: $1/C = 1/C_1 + 1/C_2 + \ldots$
- The equivalent capacitance C of capacitors connected in parallel is given by: $C = C_1 + C_2 + \ldots$
- When a charged capacitor discharges, the charge on the plates decays exponentially. The equation for the decay is: $Q = Q_0 e^{-t/CR}$
- The time constant of the circuit, given by CR, is the time for the charge to decay to $1/e$ of its initial value.

Section 11.2 Questions

1 A parallel-plate capacitor has rectangular plates of side 200 mm by 30 mm. The plates are separated by an air gap 0.50 mm thick. The permittivity of free space is 8.85×10^{-12} F m^{-1}.
  (a) Calculate the capacitance of the capacitor.
  (b) The capacitor is connected to a 12 V battery. Calculate:
    (i) the charge on each plate,
    (ii) the electric field between the plates.

2 Figure 11.24 shows an arrangement of capacitors.

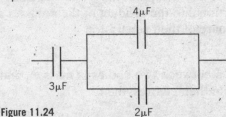

Figure 11.24

  (a) Calculate the capacitance of this arrangement.
  (b) The 4 μF capacitor is disconnected. Calculate the new capacitance.

3 A 10 μF capacitor is charged from a 9.0 V battery.
  (a) Calculate:
    (i) the electric potential energy stored by the capacitor,
    (ii) the charge stored by the capacitor.
  (b) The charged capacitor is discharged through a 150 kΩ resistor. Calculate:
    (i) the initial discharge current,
    (ii) the time constant,
    (iii) the time taken for the current to fall to 3.6×10^{-5} A.

11.3 Gravitational fields

At the end of Section 11.3 you should be able to:

- recall and use Newton's law of gravitation in the form $F = G(m_1 m_2)/r^2$
- show an understanding of the concept of a gravitational field as an example of a field of force and define gravitational field strength as force per unit mass
- derive, from Newton's law of gravitation and the definition of gravitational field strength, the equation $g = GM/r^2$ for the gravitational field strength of a point mass
- recall and solve problems using the equation $g = GM/r^2$ for the gravitational field strength of a point mass
- define potential at a point as the work done in bringing unit mass from infinity to the point
- solve problems using the equation $\phi = -GM/r$ for the potential in the field of a point mass
- recognise the analogy between certain qualitative and quantitative aspects of gravitational field and electric field
- analyse circular orbits in inverse square law fields by relating the gravitational force to the centripetal acceleration it causes
- show an understanding of geostationary orbits and their application.

Force between masses

We are familiar with the fact that the Earth's force of gravity is responsible for our weight, the force which pulls us towards the Earth. Isaac Newton concluded that the Earth's force of gravity is also responsible for keeping the Moon in orbit. His calculations showed that the Earth's force of gravity extends into space, but weakens with distance according to an inverse square law. That is, the Earth's force of gravity varies inversely with the square of the distance from the centre of the Earth. If you go two Earth radii from the Earth's surface, the force is a quarter of the force on the Earth's surface. This is part of **Newton's law of gravitation**.

> Newton's law of gravitation states that two point masses attract each other with a force that is proportional to the product of their masses and inversely proportional to the square of their separation.

Hence, if F is the force of attraction between two bodies of mass m_1 and m_2 respectively with distance r between their centres, then

$$F \propto m_1 m_2 / r^2$$

or

$$F = \frac{G m_1 m_2}{r_2}$$

where the constant of proportionality G is called the gravitational constant.

> The value of G is $6.67 \times 10^{-11} \, \text{N m}^2 \text{kg}^{-2}$.

Notice that this equation has a similar form to that for Coulomb's law between two charges (page 283). Both Newton's law and Coulomb's law are inverse square laws of force. We say that the equations are analogous, or that there is an analogy between this aspect of electric fields and gravitational fields. However, there are important differences.

■ The electric force acts on charges, whereas the gravitational force acts on masses.
■ The electric force can be attractive or repulsive, depending on the signs of the interacting charges, whereas two masses always attract each other.

It is possible to measure the gravitational constant G in a school laboratory, but the force of gravity between laboratory-sized masses is so small that it is not easy to obtain an accurate result.

Gravitational field strength

The **gravitational field strength** at a point is defined as the force per unit mass acting on a small mass placed at that point.

This continues the analogy between electric and gravitational fields. Remember that electric field strength is defined as the force per unit positive charge.

The force per unit mass is also a measure of the acceleration of free fall, by Newton's second law of motion $F = ma$ (See Chapter 5). Thus, the gravitational field strength is given by $F/m = g$, where F is in newtons and m is in kilograms. This means that the gravitational field strength at the Earth's surface is about $9.8\,\mathrm{N\,kg^{-1}}$. The unit $\mathrm{N\,kg^{-1}}$ is equivalent to the unit of acceleration, $\mathrm{m\,s^{-2}}$. (Remember that we had two equivalent units for electric field strength, $\mathrm{N\,C^{-1}}$ and $\mathrm{V\,m^{-1}}$. Although there is a clear *analogy* between $\mathrm{N\,kg^{-1}}$ and $\mathrm{N\,C^{-1}}$, there is *no direct link* between $\mathrm{m\,s^{-2}}$ and $\mathrm{V\,m^{-1}}$.)

Field strength due to an isolated point mass

We have already met the idea of a point mass in talking about the motion of a particle (Chapter 3). Of course, all the masses we come across in the laboratory have a finite size. But for calculations involving gravitational forces, it is fortunate that a spherical mass behaves as if it were a point mass at the centre of the sphere, with all the mass of the sphere concentrated at that point. This is very similar to the idea that the charge on a conducting sphere can be considered to be concentrated at the centre of the sphere.

The gravitational field outside a spherical mass is a radial one, just as the field associated with a point charge is radial. From Newton's law, the attractive force on a mass m caused by another mass M, with a distance of r between their centres, is given by

$$F = \frac{GMm}{r^2}$$

This means that the force per unit mass or gravitational field strength g is given by

$$g = \frac{F}{m} = \frac{GM}{r^2}$$

The field strengths due to masses that you find in a laboratory are tiny. For example, the field strength one metre away from an isolated mass of one kilogram is only $7 \times 10^{-11}\,\mathrm{N\,kg^{-1}}$. But field strengths due to the masses of objects such as the Earth or Moon are much larger. We already know that the field strength due to the Earth at the surface of the Earth is about $10\,\mathrm{N\,kg^{-1}}$. We can use this information to deduce information about the Earth, for example the mass of the Earth. Look at the Example that follows.

Example

The radius of the Earth is 6.4×10^6 m and the gravitational field strength at its surface is 9.8 N kg^{-1}.

(a) Assuming that the field is radial, calculate the mass of the Earth.

(b) The radius of the Moon's orbit about the Earth is 3.8×10^8 m. Calculate the strength of the Earth's gravitational field at this distance.

(c) The mass of the Moon is 7.4×10^{22} kg. Calculate the gravitational attraction between the Earth and the Moon.
(gravitational constant $G = 6.7 \times 10^{-11}$ N m^2 kg^{-2})

(a) Using $g = GM/r^2$, we have
$M = gr^2/G = 9.8 \times (6.4 \times 10^6)^2/6.7 \times 10^{-11} = \mathbf{6.0 \times 10^{24}\ kg}$

(b) Using $g = GM/r^2$, we have
$g = 6.7 \times 10^{-11} \times 6.0 \times 10^{24}/(3.8 \times 10^8)^2 = \mathbf{2.8 \times 10^{-3}\ N\ kg^{-1}}$

(c) Using $F = GMm/r^2$, we have
$F = 6.7 \times 10^{-11} \times 6.0 \times 10^{24} \times 7.4 \times 10^{22}/(3.8 \times 10^8)^2 = \mathbf{2.1 \times 10^{20}\ N}$

Now it's your turn

The mass of Jupiter is 1.9×10^{27} kg and its radius is 7.1×10^7 m. Calculate the gravitational field strength at the surface of Jupiter.
(gravitational constant $G = 6.7 \times 10^{-11}$ N m^2 kg^{-2})

Mass and weight

In Chapter 5, mass was said to be a measure of the inertia of a body to changes in velocity. Unless the body is travelling at speeds close to that of light, its mass is constant.

In a gravitational field, by definition, there is a force acting on the mass equal to the product of mass and gravitational field strength. This force is called the **weight**.

For an object of mass m in a gravitational field of strength g, the weight W is given by

weight = mass × gravitational field strength

or

$W = mg$

Although mass is invariant, weight depends on gravitational field strength. For example, a person of mass 60 kg has a weight of 600 N on Earth, but only 100 N on the Moon, although the mass is still 60 kg.

Gravitational potential energy and gravitational potential

In Section 11.1 we met the topic of electric potential energy and electric potential. There is a very strong analogy between electric and gravitational fields, and this is going to help us when we talk about gravitational potential energy and gravitational potential.

Gravitational potential at a point in a field is defined as the work done in bringing unit mass from infinity to the point.

The symbol for gravitational potential is ϕ and its unit is the joule per kilogram ($J\,kg^{-1}$). For a field produced by a point mass, the equation for the potential at a point in the field is

$$\phi = -\frac{GM}{r}$$

Note the minus sign. This is entirely consistent with the electrical case. We had a negative electric potential when the field was produced by a negative charge, so that the force between the negative field-producing charge and the positive test charge was attractive. Here the attractive gravitational force between field-producing mass and test mass also gives a negative potential.

Example

How much work is done by the gravitational field in moving a mass m of 2.0 kg from infinity to a point A, 0.40 m from a mass M of 30 kg?

The work which would have to be done by an external force is simply the change in gravitational potential energy. The potential energy at infinity is zero, so

$W = m\phi_A = -mGM/r = -2.0 \times 6.7 \times 10^{-11} \times 30/0.40 = -1.0 \times 10^{-8}\,J$

This is the work which would be done by an external force and is negative, so the work done by the field is positive and is equal to **$1.0 \times 10^{-8}\,J$**.

Note the similarity in the method of calculation with the electric potential energy calculation on page 290.

Now it's your turn

The Earth has mass 6.0×10^{24} kg and radius 6.4×10^6 m. Find the change in gravitational potential energy of a meteorite of mass 150 kg as it moves from an infinite distance to the Earth's surface. The meteorite starts from rest. With what speed does it strike the Earth? (gravitational constant $G = 6.67 \times 10^{-11}\,N\,m^2\,kg^{-2}$)

Planet and satellite orbits

Most planets in the solar system have orbits which are nearly circular. We now bring together the idea of a gravitational force and that of a centripetal force (see Chapter 10) to derive a relation between the period and the radius of the orbit of a planet describing a circular path about the Sun, or a satellite moving round the Earth or another planet.

Consider a planet of mass m in circular orbit about the Sun, of mass M. If the radius of the orbit is r, the gravitational force F_{grav} between Sun and planet is

$$F_{grav} = GMm/r^2$$

by Newton's law of gravitation. It is this force that provides the centripetal force as the planet moves in its orbit. The centripetal force F_{circ} is given by

$$F_{circ} = mv^2/r$$

where v is the linear speed of the planet. As has just been stated,

$$F_{grav} = F_{circ}$$

Thus

$$GMm/r^2 = mv^2/r$$

The period T of the planet in its orbit is the time required for the planet to travel a distance $2\pi r$. It is moving at speed v, so

$$v = 2\pi r/T$$

Putting this into the equation above, we have

$$GMm/r^2 = m(4\pi^2 r^2/T^2)/r$$

or, simplifying,

$$T^2 = (4\pi^2/GM)r^3$$

Another way of writing this is

$$\frac{T^2}{r^3} = \frac{4\pi^2}{GM}$$

Look at the right-hand side of this equation. The quantities π and G are constants. If we are considering the relation between T and r for planets in the solar system, then M is the same for each planet because it is the mass of the Sun.

This equation shows that:

> For planets or satellites describing circular orbits about the same central body, the square of the period is proportional to the cube of the radius of the orbit.

This relation is known as **Kepler's third law of planetary motion**. Johannes Kepler (1571–1630) analysed data collected by Tycho Brahe (1546–1601) on planetary observations. He showed that the observations fitted a law of the form $T^2 \propto r^3$. Fifty years later Newton showed that an inverse square law of gravitation, the idea of centripetal acceleration, and the second law of motion, gave an expression of exactly the same form. Newton cited Kepler's law in support of his law of gravitation. In fact, the orbits of the planets are not circular, but elliptical, a fact recognised by Kepler. The derivation is simpler for the case of a circular orbit.

Table 11.2 collects information about T and r for planets of the solar system. The last column shows that the value of T^2/r^3 is indeed a constant. Moreover, the value of T^2/r^3 agrees very well with the value of $4\pi^2/GM$, 2.97×10^{-25} yr^2m^{-3}.

Table 11.2

planet	T/(Earth years)	r/km	$T^2/r^3 (\text{yr}^2\text{m}^{-3})$
Mercury	0.241	57.9×10^6	2.99×10^{-25}
Venus	0.615	108.0×10^6	3.00×10^{-25}
Earth	1.00	150.0×10^6	2.96×10^{-25}
Mars	1.88	228.0×10^6	2.98×10^{-25}
Jupiter	11.9	778.0×10^6	3.01×10^{-25}
Saturn	29.5	1430.0×10^6	2.98×10^{-25}
Uranus	84.0	2870.0×10^6	2.98×10^{-25}
Neptune	165	4500.0×10^6	2.99×10^{-25}
Pluto	248	5900.0×10^6	2.99×10^{-25}
			(average 2.99×10^{-25})

Satellites are widely used in telecommunication. Many communication satellites are placed in what is called **geostationary orbit**. That is, they are in equatorial orbits with exactly the same period of rotation as the Earth, and move in the same direction as the Earth, so that they are always above the same point on the equator. Details of the orbit of such a satellite are worked out in the example which follows.

Example

For a geostationary satellite, calculate:
(a) the height above the Earth's surface,
(b) the speed in orbit.
(radius of Earth = 6.38×10^6 m; mass of Earth = 5.98×10^{24} kg)

(a) The period of the satellite is 24 hours = 8.64×10^4 s.
Using $T^2/r^3 = 4\pi^2/GM$, we have
$r^3 = 6.67 \times 10^{-11} \times 5.98 \times 10^{24} \times (8.64 \times 10^4)^2/4\pi^2$, giving
$r^3 = 7.54 \times 10^{22}$ m³. Taking the cube root, the radius r of the orbit is
4.23×10^7 m. The distance above the Earth's surface is
$(4.23 \times 10^7 - 6.4 \times 10^6) = \mathbf{3.59 \times 10^7\,m}$.
(b) Since $v = 2\pi r/T$, the speed is given by
$v = 2\pi \times 4.23 \times 10^7/8.64 \times 10^4 = \mathbf{3070\,m\,s^{-1}}$.

Now it's your turn

The radius of the Moon's orbit is 3.84×10^8 m, and its period is 27.4
days. Use Kepler's law to calculate the period of the orbit of a satellite
orbiting the Earth just above the Earth's surface.
(radius of Earth = 6.38×10^6 m)

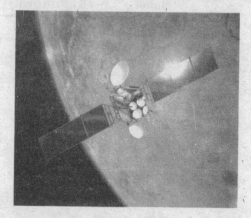

Figure 11.25 Communication satellite

Communication satellites are considered further in Section 16.3.

Weightlessness

Suppose that you are carrying out an experiment involving the use of a
newton balance in a lift. When the lift is stationary, an object of mass 10 kg,
suspended from the balance, will give a weight reading of $10g$ N. If the lift
accelerates upwards with acceleration $0.1g$, the reading on the balance
increases to $11g$ N. If the lift accelerates downwards with acceleration $0.1g$,
the apparent weight of the object decreases to $9g$ N. If, by an unfortunate
accident, the lift cable were to break and there were no safety restraints, the
lift would accelerate downwards with acceleration g. The reading on the

Newton balance would be zero. If during the fall you were to drop the pencil with which you are recording the balance readings, it would not fall to the floor of the lift, but remain stationary with respect to you. Both you and the pencil are in free fall. You are experiencing **weightlessness**. Figure 11.26 illustrates your predicament. It might be more correct to refer to this situation as *apparent* weightlessness, as you can only be truly weightless in the absence of a gravitational field – that is, at an infinite distance from the Earth or any other attracting object.

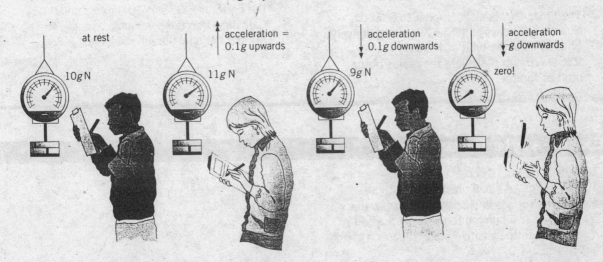

at rest   10g N

acceleration = 0.1g upwards   11g N

acceleration 0.1g downwards   9g N

acceleration g downwards   zero!

Figure 11.26 Lift experiment on weightlessness

A similar situation arises in a satellite orbiting the Earth. The force of gravity, which provides the centripetal force, is causing the body to fall out of its expected straight-line path. People and objects inside the satellite are experiencing a free fall situation and apparent weightlessness.

Section 11.3 summary

■ The attractive force between two point masses is proportional to the product of the masses and inversely proportional to the square of the distance between them. This is Newton's law of gravitation:
$F = Gm_1m_2/r^2$

■ A gravitational field is a region round a mass where another mass feels a force. Gravitational field strength g is the force per unit mass:
$g = F/m$
g is also the acceleration of free fall.

■ The gravitational field strength at a point in the gravitational field of a point mass M is: $g = GM/r^2$

■ The gravitational potential at a point in a gravitational field is the work done in bringing unit mass from infinity to the point.

■ The potential at a point in a field produced by a point mass is:
$\phi = -GM/r$

■ For a planet or satellite in circular orbit about a body, the square of the period is proportional to the cube of the radius of the orbit. This is Kepler's third law of planetary motion.

Section 11.3 Questions

1 The Earth's radius is about 6.4×10^6 m. The acceleration of free fall at the Earth's surface is 9.8 m s^{-2}. The gravitational constant G is 6.7×10^{-11} N m^2 kg^{-2}. Use this information to estimate the mean density of the Earth.

2 The times for Mars and Jupiter to orbit the Sun are 687 days and 4330 days respectively. The radius of the orbit of Mars is 228×10^6 km. Calculate the radius of the orbit of Jupiter.

3 The weight of a passenger in an aircraft on the runway is W. His weight when the aircraft is flying at an altitude of 16 km above the Earth's surface is W_a. The percentage change F in his weight is given by $F = 100(W_a - W)/W\%$. Taking the radius of the Earth as 6.378×10^3 km, calculate F. Calculate also the percentage change P in his gravitational potential energy.

Exam-style Questions

1 Two metal plates 10 cm apart are connected to a 1000 V supply. A small plastic ball with a conducting surface is suspended between the plates on an insulating thread. When the ball is made to touch the positive plate, it gains a charge of 5.0×10^{-9} C.
 (a) Calculate the strength of the electric field between the plates.
 (b) Explain why the ball shuttles to and fro between the plates.
 (c) Calculate the force on the ball when it is in the field.

2 In a simplified model of the hydrogen atom, the electron is separated from the proton by a distance r of 5.3×10^{-11} m. Use the following data to calculate the ratio of the electric force to the gravitational force at this distance. What would be the ratio if the charges were separated to a distance $2r$?

 Electron charge $e = -1.6 \times 10^{-19}$ C
 Electron mass $m_e = 9.1 \times 10^{-31}$ kg
 Proton mass $m_p = 1.7 \times 10^{-27}$ kg
 Permittivity of free space $\epsilon_0 = 8.9 \times 10^{-12}$ F m^{-1}
 Gravitational constant $G = 6.7 \times 10^{-11}$ N m^2 kg^{-2}

3 In Figure 11.27, there are two similar table-tennis balls, each of mass 2.5 g. The balls have equal charges and are suspended by insulating thread.

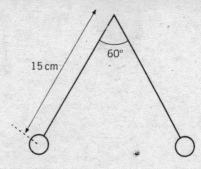

Figure 11.27

 (a) Calculate the tension in one of the threads.
 (b) Show that the electrostatic force between the two balls is 1.4×10^{-2} N.
 (c) Calculate the charge on each ball.

4 A capacitor is made from two strips of metal foil 50 mm wide and 50 m long, with paper of thickness 0.10 mm between them. The relative permittivity of the paper is 2.0.
 (a) Calculate the capacitance of the capacitor.
 (b) The strips are then rolled up into a cylinder, to make the capacitor easier to handle. However, a second sheet of paper must be applied on top of one of the metal sheets. Explain why this is necessary, and why the capacitance of the rolled-up sheets is greater than that in (a). (permittivity of free space $\epsilon_0 = 8.9 \times 10^{-12}$ F m^{-1})

5 Figure 11.28 illustrates the apparatus used by Cavendish in 1798 to find a value for the gravitational constant G. In a school experiment using similar apparatus, two lead spheres are attached to a light horizontal beam which is suspended by a wire. When a flask of mercury is brought close to each sphere, the gravitational attraction causes the beam to twist through a small angle. From measurements of the twisting (torsional) oscillations of the beam, a value can be found for the force producing a measured deflection. G can then be calculated if the large and small masses are known.

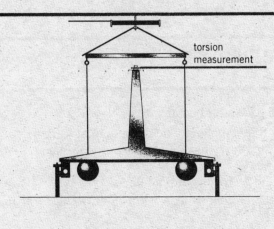

torsion measurement

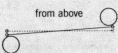

from above

Figure 11.28 Cavendish's experiment for G

(a) In such an experiment, one lead sphere has mass 6.22×10^{-3} kg and the mass of the mercury flask is 0.713 kg. Calculate the force between them when they are 72.0 mm apart. (gravitational constant $G = 6.67 \times 10^{-11}$ N m^2 kg^{-2})

(b) Comment on the size of the force.

6 Two point charges of $+2.4\,\mu$C and $-2.9\,\mu$C are placed at points A and B respectively in a vacuum, 0.15 m apart (Figure 11.29). It is required to find a point P at which the resultant electric field due to these two charges is zero.

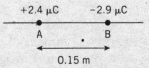

$+2.4\,\mu$C $-2.9\,\mu$C

A   •   B

0.15 m

Figure 11.29

(a) Explain why P cannot lie off the line joining the charges (extended if necessary).

(b) Decide whether P can lie to the left of A, between A and B, to the right of B, or in more than one of these regions. Give reasons for your answer.

(c) Hence deduce a location for P.

7 The gravitational field strength at the surface of the Moon is 1.62 N kg^{-1}. The radius of the Moon is 1740 km.

(a) Show that the mass of the Moon is 7.35×10^{22} kg.

(b) The Moon rotates about its axis (as well as moving in orbit about the Earth). In the future, scientists may wish to put a satellite into an orbit about the Moon such that the satellite remains stationary above one point on the Moon's surface.

  (i) Explain why this orbit must be an equatorial orbit.

  (ii) The period of rotation of the Moon about its axis is 27.4 days. Calculate the radius of the required orbit. ($G = 6.67 \times 10^{-11}$ N m^2 kg^{-2})

12.Electromagnetism

A study of electricity and of electromagnetic effects is one of the key areas for all students of physics. Not only are electromagnetic effects vital to modern living (imagine life without the electric motor) but they also give some insight into the pioneering research carried out by scientists such as Volta, Ampere, Ohm and Faraday. Their discoveries were exceptional because they were attempting to understand the invisible effects of electric currents and magnetic fields.

12.1 Magnetic fields

At the end of Section 12.1 you should be able to:
- show an understanding that a magnetic field is a field of force produced either by current-carrying conductors or by permanent magnets
- represent a magnetic field by means of field lines
- sketch flux patterns due to a long straight wire, a flat circular coil and a long solenoid
- show an understanding that the field due to a solenoid may be influenced by the presence of a ferrous core.

Some of the properties of magnets have been known for many centuries. The ancient Greeks discovered an iron ore called lodestone which, when hung from a thread, would come to rest always pointing in the same direction. This was the basis of the magnetic compass which has been in use since about 1500BC as a means of navigation.

The magnetic compass is dependent on the fact that a freely suspended magnet will come to rest pointing north–south. The ends of the magnet are said to be poles. The pole pointing to the north is referred to as the north-seeking pole (the north pole) and the other, the south-seeking pole (the south pole). It is now known that a compass behaves in this way because the Earth is itself a magnet.

Magnets exert forces on each other. These forces of either attraction or repulsion are used in many children's toys, in door catches and 'fridge magnets'. The effects of the forces may be summarised in the **law of magnets**.

Like poles repel.
Unlike poles attract.

The law of magnets implies that around any magnet, there is a region where a magnetic pole will experience a force. This region is known as a **magnetic field**. Magnetic fields are not visible but they may be represented by lines of magnetic force or **magnetic field lines**, or magnetic 'flux'. A simple way of imaging magnetic field lines is to think of one such line as the direction in which a free magnetic north pole would move if placed in the field. Magnetic field lines may be plotted using a small compass (a plotting compass) or by the use of iron filings and a compass (Figure 12.1).

Figure 12.1 The iron filings line up with the magnetic field of the bar magnet which is under the sheet of paper. A plotting compass will give the direction of the field.

The magnetic field lines of a bar magnet are shown in Figure 12.2. Effects due to the Earth's magnetic field have not been included since the Earth's field is relatively weak and would be of importance only at some distance from the magnet. It is important to realise that, although the magnetic field has been drawn in two dimensions, the actual magnetic field is three-dimensional.

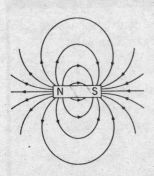

Figure 12.2 Magnetic field pattern of a bar magnet

For any magnetic field:

- the magnetic field lines start at a north pole and end at a south pole
- the magnetic field lines are smooth curves which never touch or cross
- the strength of the magnetic field is indicated by the distance between the lines – closer lines mean a stronger field.

It can be seen that these properties are very similar to those for electric field lines (page 284).

Figure 12.3 illustrates the magnetic field pattern between the north pole of one magnet and the south pole of another. This pattern is similar to that produced between the poles of a horseshoe magnet.

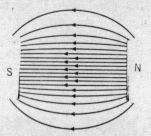

Figure 12.3 Magnetic field pattern between a north and south pole

Figure 12.4 shows the magnetic field between the north poles of two magnets. The magnetic field due to one magnet opposes that due to the other. The field lines cannot cross and consequently there is a point X, known as a **neutral point**, where there is no resultant magnetic field because the two fields are equal in magnitude but opposite in direction.

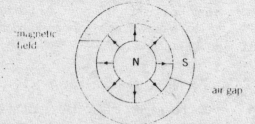

Figure 12.4 Magnetic field pattern between two north poles

Example

A circular magnet is made with its north pole at the centre, separated from the surrounding circular south pole by an air gap. Draw the magnetic field lines in the gap.

Figure 12.5

Now it's your turn

1 Draw a diagram to illustrate the magnetic field between the south poles of two magnets.

2 Two bar magnets are placed on a horizontal surface (Figure 12.6).

Figure 12.6

Draw the two magnets, and sketch the magnetic field lines around them. On your diagram, mark the position of any neutral points.

Magnetic effect of an electric current

The earliest discovery of the magnetic effect of an electric current was made in 1820 when Oersted, a Danish physicist, noticed that a compass was deflected when brought near to a wire carrying an electric current. It is now known that all electric currents produce magnetic fields. The size and shape of the magnetic field depends on the size of the current and the shape (configuration) of the conductor through which the current is travelling.

The magnetic field due to a long straight wire may be plotted using the apparatus illustrated in Figure 12.8. Note that the current must be quite large (about 5 A). Iron filings are sprinkled on to the horizontal board and a plotting compass is used to determine the direction of the field.

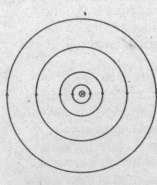

Figure 12.7 Hans Christian Oersted

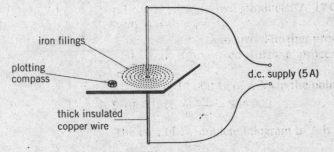

Figure 12.8 Apparatus to plot the magnetic field due to a long wire

Figure 12.9 shows the field pattern due to a long straight current-carrying wire. The lines are concentric circles centred on the middle of the wire. The separation of the lines increases with distance from the wire, indicating that the field is decreasing in strength. The direction of the magnetic field may be found using the **right-hand rule** (see Figure 12.10).

Imagine holding the conductor in the right hand with the thumb pointing in the direction of the current. The direction of the fingers gives the direction of the magnetic field.

Figure 12.9 Magnetic field pattern due to a long straight wire

Similar apparatus to that in Figure 12.8 may be used to investigate the shapes of the magnetic field due to a flat coil and to a solenoid (a long coil).

magnetic
field

current

Figure 12.10 The right-hand rule

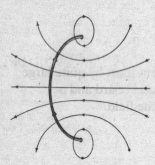

Figure 12.11 Magnetic field pattern due to a flat coil

Figure 12.11 illustrates the magnetic field pattern due to a flat coil. The field has been drawn in a plane normal to the coil and through its centre.

A solenoid may be thought to be made up of many flat coils placed side-by-side. The magnetic field pattern of a long solenoid (that is, a coil which is long in comparison with its diameter) is shown in Figure 12.12. The field lines are parallel and equally spaced over the centre section of the solenoid, indicating that the field is uniform. The field lines spread out towards the ends.

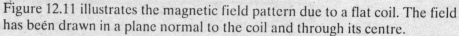

Figure 12.12 Magnetic field pattern due to a solenoid

The strength of the magnetic field at each end is one half that at the centre. The direction of the magnetic field in a flat coil and in a solenoid may be found using the **right-hand grip rule**, as illustrated in Figure 12.13.

Grasp the coil or solenoid in the right hand with the fingers pointing in the direction of the current. The thumb gives the direction of the magnetic field.

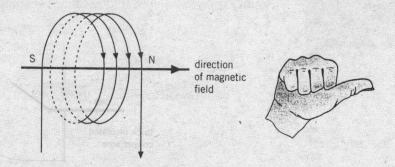

S   N   direction
of magnetic
field

Figure 12.13 The right-hand grip rule

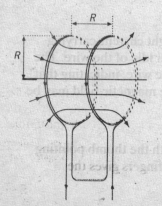

Figure 12.14 Magnetic field in Helmholtz coils

The magnetic north end of the coil or solenoid is the end from which the lines of magnetic force are emerging. Note the similarities and, more importantly, the differences between this rule and the right-hand rule for the long straight wire (Figure 12.10).

Uniform magnetic fields are of importance in the study of charged particles. A uniform field is produced in a solenoid but this field is inside the solenoid and consequently, it may be difficult to make observations and to take measurements. This problem is overcome by using **Helmholtz coils**. These are two identical flat coils, with the same current in each. The coils are positioned so that their planes are parallel and separated by a distance equal to the radius of either coil. The coils and their resultant magnetic field are illustrated in Figure 12.14.

Example

Two long straight wires, of circular cross-section, are each carrying the same current directly away from you, down into the page. Draw the magnetic field due to the two current-carrying wires.

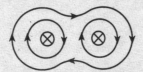

Figure 12.15

Now it's your turn

1 Draw magnetic field patterns, one in each case to represent
   (a) a uniform field,
   (b) a field which is decreasing in strength in the direction of the field,
   (c) a field which is increasing in strength along the direction of the field.

2 Draw a diagram of the magnetic field due to two long straight wires when the currents in the two wires are in opposite directions.

Electromagnets and their uses

The strength of the magnetic field due to a flat coil or a solenoid may be increased by winding the coil on a bar of soft iron. The bar is said to be the **core** of the coil. The iron is referred to as being 'soft' because it can be magnetised and demagnetised easily. With such a core, the strength of the magnetic field may be increased by up to 1000 times. With ferrous alloys (iron alloyed with cobalt or nickel), the field may be 10^4 times stronger. Magnets such as these are called **electromagnets**. Electromagnets have many uses because, unlike a permanent magnet, the magnetic field can be switched off by switching off the current in the coil.

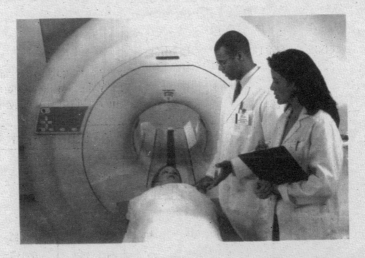

Figure 12.16 Magnetic imaging of body structures requires the use of a very large uniform magnetic field.

Section 12.1 Summary

- A magnetic field is a region of space where a magnet will experience forces.
- Magnetic field lines never touch or cross.
- The separation of magnetic field lines indicates the strength of the magnetic field.
- An electric current gives rise to a magnetic field, the strength and direction of which depends on the size of the current and the shape of the current-carrying conductor.
- The direction of the field due to a straight wire is given by the right-hand rule.
- The direction of the field in a solenoid is given by the right-hand grip rule.
- The field of a solenoid may be increased in strength by a ferrous core; this is the principle of an electromagnet.

12.2 Forces and fields

At the end of Section 12.2 you should be able to:

- show an appreciation that a force might act on a current-carrying conductor placed in a magnetic field
- recall and solve problems using the equation $F = BIl \sin \theta$, with directions as interpreted by Fleming's left-hand rule
- define magnetic flux density and the tesla
- show an understanding of how the force on a current-carrying conductor can be used to measure the flux density of a magnetic field, using a current balance
- explain the forces between current-carrying conductors and predict the direction of the forces
- predict the direction of the force on a charge moving in a magnetic field
- recall and solve problems using $F = Bqv \sin \theta$
- describe and compare the forces on mass, charge and current in gravitational, electric and magnetic fields, as appropriate
- show an appreciation of the main principles of the determination of e by Millikan's experiment
- summarise and interpret the experimental evidence for quantisation of charge
- explain the main principles of one method for the determination of v and e/m for electrons
- describe and analyse qualitatively the deflection of beams of charged particles by uniform electric and magnetic fields
- explain how electric and magnetic fields can be used in velocity selection.

In Section 12.1, we saw that the plotting of lines of magnetic force gives the direction and shape of the magnetic field. Also, the distance between the lines indicates the strength of the field. However, the strength of the magnetic field was not defined. Physics is the science of measurement and consequently, the topic would not be complete without defining and measuring magnetic field strength. Magnetic field strength is defined through a study of the **motor effect**.

The motor effect

The interaction of the magnetic fields produced by two magnets causes forces of attraction or repulsion between the two. If a conductor is placed between magnet poles and a current is passed through the conductor, the magnetic fields of the current-carrying conductor and the magnet may interact, causing forces between them. This is illustrated in Figure 12.17.

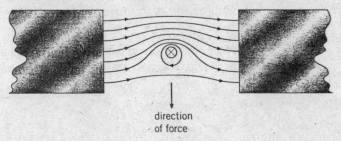

direction
of force

Figure 12.17 The interacting magnetic fields of a current-carrying conductor and the poles of a magnet

The existence of the force may be demonstrated with the apparatus shown in Figure 12.18. The strip of aluminium foil is held loosely between the poles of a horseshoe magnet so that the foil is at right-angles to the magnetic field. When the current is switched on, the foil jumps and becomes taut, showing that a force is acting on it. The direction of the force, known as the **electromagnetic force**, depends on the directions of the magnetic field and of the current. This phenomenon, when a current-carrying conductor is at an angle to a magnetic field, is called the **motor effect**.

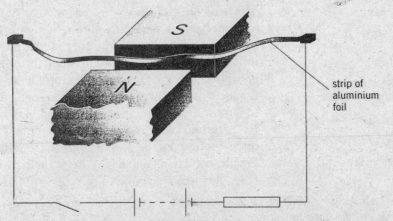

strip of
aluminium
foil

Figure 12.18 Demonstrating the motor effect acting on a piece of aluminium

The direction of the force relative to the directions of the current and the magnetic field may be predicted using **Fleming's left-hand rule**. This is illustrated in Figure 12.19.

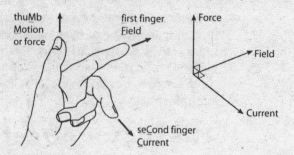

Figure 12.19 Fleming's left-hand rule

If the first two fingers and thumb of the left hand are held at right angles to one another with the **F**irst finger in the direction of the **F**ield and the se**C**ond finger in the direction of the **C**urrent, then the thu**M**b gives the direction of the force or **M**otion.

Note that, if the conductor is held fixed, motion will not be seen but nevertheless, there will be an electromagnetic force.

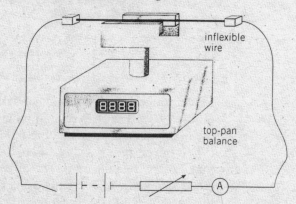

Figure 12.20 Apparatus to investigate the magnitude of the electromagnetic force

The magnitude of the electromagnetic force may be investigated using the 'current balance' apparatus shown in Figure 12.20. A horseshoe magnet is made by placing two flat magnadur magnets on a U-shaped iron core. A length of inflexible wire is firmly fixed between the poles of the magnet. When the current is switched on, the reading on the top-pan balance changes. This change in reading is a measure of the electromagnetic force. Variation of the current leads to the conclusion that the electromagnetic force F is proportional to the current I. By using magnadur magnets of different lengths (but of similar strengths), the force F is found to be proportional to the length of wire l in the magnetic field. Finally, by varying the angle θ between the wire and the direction of the magnetic field, the force F is found to be proportional to $\sin \theta$. Hence the expression

$$F \propto Il \sin \theta$$

is derived. The expression can be rewritten as

$$F = BIl \sin \theta$$

where B is a constant.

The constant B depends on the strength of the magnet and, if stronger magnets are used, the constant has a greater value. The equation can, therefore, be used as the defining equation for magnetic field strength.

The equation can be rewritten as

$$\frac{F}{l} = BI \sin \theta$$

So,

> For a long straight conductor carrying unit current at right-angles to a uniform magnetic field, the magnetic field strength B is numerically equal to the force per unit length of the conductor.

Magnetic field strength is usually referred to as **magnetic flux density**. In SI units, this is measured in **tesla (T)**. An alternative name for this unit is **weber per square metre (Wb m^{-2})**.

> One tesla is the uniform magnetic flux density which, acting normally to a long straight wire carrying a current of 1 ampere, causes a force per unit length of 1 N m^{-1} on the conductor.

Since force is measured in newtons, length in metres and current in amperes, it can be derived from the defining equation for the tesla that the tesla may also be expressed as $\text{N m}^{-1} \text{A}^{-1}$. The unit involves force which is a vector quantity and thus magnetic flux density is also a vector.

When using the equation $F = BIl \sin \theta$, it is sometimes helpful to think of the term $B \sin \theta$ as being the component of the magnetic flux density which is at right-angles (or normal) to the conductor (see Figure 12.21).

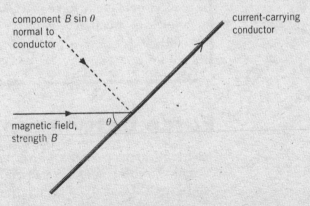

component $B \sin \theta$ normal to conductor

current-carrying conductor

magnetic field, strength B

θ

Figure 12.21 $B \sin \theta$ is the component of the magnetic field which is normal to the conductor.

The tesla is a large unit for the measurement of flux density. A strong magnet may have a flux density between its poles of a few teslas. The magnetic flux density due to the Earth in the UK is about $44\,\mu T$ at an angle of $66°$ to the horizontal.

Example

The horizontal component of the Earth's magnetic flux density is $1.8 \times 10^{-5}\,T$. The current in a horizontal cable is 150 A. Calculate, for this cable:

(a) the maximum force per unit length,
(b) the minimum force per unit length.

In each case, state the angle between the cable and the magnetic field.

(a) force per unit length $= \dfrac{F}{l} = BI\sin\theta$

Force per unit length is a maximum when $\theta = 90°$ and $\sin\theta = 1$.
Force per unit length $= 1.8 \times 10^{-5} \times 150 = 2.7 \times 10^{-3}\,N\,m^{-1}$
Maximum force per unit length $= \mathbf{2.7 \times 10^{-3}\,N\,m^{-1}}$ when the cable is at right-angles to the field.

(b) Force per unit length is a minimum when $\theta = 0$ and $\sin\theta = 0$.
Minimum force per unit length $= \mathbf{0}$ when the cable is along the direction of the field.

Now it's your turn

1 The effective length of the filament in a light bulb is 3.1 cm and, for normal brightness, the current in the filament is 0.25 A. Calculate the maximum electromagnetic force on the filament when in the Earth's field of flux density $44\,\mu T$.

2 A straight conductor carrying a current of 6.5 A is situated in a uniform magnetic field of flux density 4.3 mT. Calculate the electromagnetic force per unit length of the conductor when the angle between the conductor and the field is:
(a) $90°$,
(b) $45°$.

Force between parallel conductors

A current-carrying conductor has a magnetic field around it. If a second current-carrying conductor is placed parallel to the first, this second conductor will be in the magnetic field of the first and, by the motor effect, will experience a force.

By similar reasoning, the first conductor will also experience a force. By Newton's third law (see Chapter 5), these two forces will be equal in magnitude and opposite in direction. The effect can be demonstrated using the apparatus in Figure 12.22. It can be seen that, if the currents are in the same direction, the pieces of foil move towards one another (the pinch effect). If the currents are in opposite directions, the pieces of foil move apart. An explanation for the effect can be found by reference to Figure 12.23.

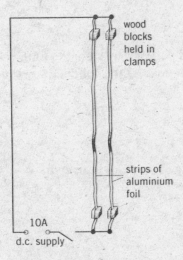

Figure 12.22 Apparatus to demonstrate the force between parallel current-carrying conductors

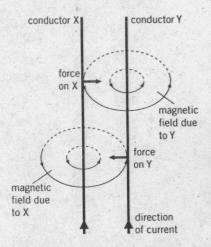

Figure 12.23 Diagram to illustrate the force between parallel current-carrying conductors

The current in conductor X causes a magnetic field and the field lines are concentric circles. These field lines will be at right angles to conductor Y and so, using Fleming's left-hand rule, there will be a force on Y in the direction of X. Using similar reasoning, the force on X due to the magnetic field of Y is towards Y. Reversing the direction of the current in one conductor will reverse the directions of the two forces and thus, when the currents are in opposite directions, the conductors tend to move apart.

The force per unit length on each of the conductors depends on the magnitude of the current in each conductor and also their separation. Since force can be expressed in terms of base units, then it is possible to use the effect to define the ampere in terms of SI units. The definition may be given as follows.

Consider two long straight parallel conductors of negligible area of cross-section, situated one metre apart in a vacuum. Then, if the current in each conductor is 1 ampere, the force per metre length acting on each conductor is 2.0×10^{-7}N.

It may not be necessary to learn the definition of the ampere, but it is important to realise how the ampere is defined in terms of SI units and that the definition does not involve the property of a substance.

Force on a moving charged particle in a magnetic field

An electric current is charge in motion. Since charge is always associated with particles, then the current in a conductor is a movement of charged particles. If a current-carrying conductor is placed in a magnetic field, it may experience a force depending on the angle between the field and the conductor. The force arises from the force on the individual moving charged particles in the conductor. It has been shown that a conductor of length l carrying a current I at an angle θ to a uniform magnetic field of flux density B experiences a force F given by

$$F = BIl \sin \theta$$

If there are n charged particles in a length l of the conductor, each carrying a charge q, which pass a point in the conductor in time t, then the current in the conductor is given by

$$I = \frac{nq}{t}$$

and the speed of charged particles is given by $v = \dfrac{l}{t}$. Hence,

$$F = B\frac{nq}{t}l \sin \theta$$

and

$$F = Bnqv \sin \theta$$

This force is the force on n charged particles. Therefore,

> The force on a particle of charge q moving at speed v at an angle θ to a uniform magnetic field of flux density B is given by
>
> $$F = Bqv \sin \theta$$

The direction of the force will be given by Fleming's left-hand rule. However, it must be remembered that the second finger is used to indicate the direction of the *conventional* current. If the particles are positively charged, then the second finger is placed in the same direction as the velocity. However, if the particles are negatively charged (e.g. electrons), the finger must point in the opposite direction to the velocity.

Consider a positively charged particle of mass m carrying charge q and moving with velocity v as shown in Figure 12.24.

path of particle,
charge $+q$
mass m
velocity v

uniform magnetic
field B out of page

Figure 12.24

The particle enters a uniform magnetic field of flux density B which is normal to the direction of motion of the particle. As the particle enters the field, it will experience a force normal to its direction. This force will not change the speed of the particle but it will change its direction of motion. As the particle moves through the field, the force will remain constant since the speed has not changed and it will always be normal to the direction of motion. The particle will, therefore, move in an arc of a circle (see Chapter 9).

The force $F = Bqv \sin \theta$ (in this case, $\sin \theta = 1$), provides the centripetal force for the circular motion. Hence,

centripetal force $= mv^2/r = Bqv$

Re-arranging,

radius of path $= r = \dfrac{mv}{Bq}$

The importance of this equation is that, if the speed of the particle and the radius of its path are known, then the specific charge, i.e. the ratio of charge to mass, can be found. Then, if the charge on the particle is known, its mass may be calculated. The technique is also used in nuclear research to try to identify some of the fundamental particles. The tracks of charged particles are made visible in a bubble chamber (Figure 12.25). Analysing these tracks gives information as to the sign of the charge on the particle and its specific charge.

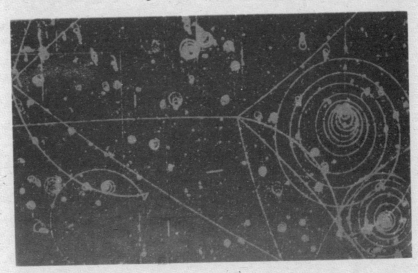

Figure 12.25 Tracks of particles produced in a bubble chamber

Specific charge of the electron

We will now look at two important experiments, both involving forces on charges in fields, which led to information on fundamental properties of the electron, its charge and its mass.

The electron charge

The principle of a method to measure the electron charge is very simple.
It involves the balancing of the gravitational force on a tiny, charged sphere by an electric force in the opposite direction, so that the sphere remains stationary.

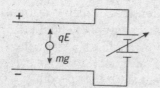

Figure 12.26 Principle of Millikan's experiment for *e*

In Millikan's original experiment, the sphere was an oil droplet, which became charged by friction as it passed out of the jet of a spray. If you do this experiment in a school laboratory now, you will probably use a very small polystyrene sphere.

Millikan's arrangement is sketched in Figure 12.26. The charged droplets are sprayed between two parallel plates connected to a source of variable potential difference, so that the electric field between them can be changed. The tendency is for a droplet to fall under its own weight, *mg*. The observer concentrates on one droplet and adjusts the electric field until the weight is exactly balanced by the electric force, *qE*, in the opposite direction. This gives the equilibrium condition $qE = qV/d = mg$. Millikan found that the charge *q* was always an integral multiple of a fundamental quantity of charge *e* equal to 1.6×10^{-19} C. This fundamental charge was ascribed to the charge of the electron. The experimental result that the charges on the droplets seems to be only in integral multiples of *e* means that charge is quantised, or exists only in discrete amounts.

Example

A negatively charged polystyrene sphere of mass 3.3×10^{-15} kg is held at rest between two parallel plates separated by 5.0 mm when the potential difference between them is 170 V. How many excess electrons are on the sphere?

Apply the equation for the equilibrium condition, $qV/d = mg$. This gives $q = 3.3 \times 10^{-15} \times 9.8 \times 5.0 \times 10^{-3}/170 = 9.5 \times 10^{-19}$ C. Taking the electron charge *e* as 1.6×10^{-19} C, *q* corresponds to 5.9 excess electrons. Because you can't have 0.9 of an electron, the answer must be **6 excess electrons**. The discrepancy must be due to experimental inaccuracy or to rounding errors; we have been working to only two significant figures.

Now it's your turn

The charged sphere in the example above suddenly loses one of its excess electrons. The potential difference between the plates remains the same. Describe the initial motion of the sphere.

The electron mass

How do you find the mass of an electron? This cannot be done directly but the ratio of charge to mass *e/m* can be found and, knowing the charge *e* on the electron, the mass *m* can then be calculated.

We have seen that a particle of mass *m* and charge *q* moving with speed *v* at right angles to a uniform magnetic field of flux density *B* experiences a force F_B given by

$$F_B = Bqv$$

The direction of this force is given by Fleming's left-hand rule and is always normal to the velocity. The fact that we have a constant force which is normal to the direction of motion is the condition necessary for circular motion. The force F_B provides the centripetal force.

The motion can be represented by the equation

$Bqv = mv^2/r$

Re-arranging the terms,

$q/m = v/Br$

The ratio *e/m* for an electron, sometimes referred to as the specific charge, may be determined using a fine-beam tube, as shown in Figure 12.27. The path of electrons is made visible by having low-pressure gas in the tube and thus the radius of the orbit may be measured. By accelerating the electrons through a known potential difference, their speed on entry into the region of the magnetic field may be calculated (see Section 7.1). The magnetic field is provided by a pair of current-carrying coils (Helmholtz coils, page 318).

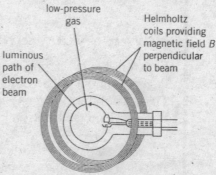

Figure 12.27 Fine-beam tube

It is of interest to rotate the tube slightly, so that the velocity of the electrons is not normal to the magnetic field. In this case, the path of the electrons is seen to be a helix (rather like the coils of a spring). The component of the velocity normal to the field gives rise to circular motion. However, there is also a component of velocity along the direction of the field. There is no force on the electrons resulting from this component of velocity. Consequently, the electrons execute circular motion *and* move in a direction normal to the plane of the circle. The circle is 'pulled out' into a helix (see Figure 12.28). The helical path is an important aspect of the focusing of electron beams by magnetic fields in an electron microscope (Figure 12.29).

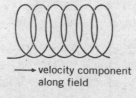

→ velocity component along field

Figure 12.28 Electrons moving in a helix

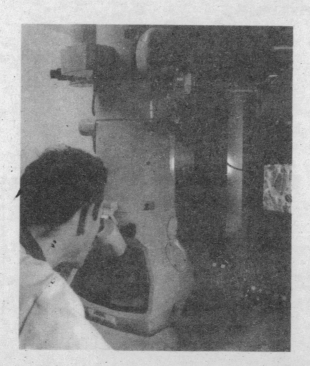

Figure 12.29 Laboratory technician using an electron microscope

Example

Electrons are accelerated to a speed of $8.4 \times 10^6 \text{m s}^{-1}$. They then pass into a region of uniform magnetic flux of flux density 0.50 mT. The path of the electrons in the field is a circle with a radius 9.6 cm. Calculate:
(a) the specific charge of the electron,
(b) the mass of the electron, assuming the charge on the electron is 1.6×10^{-19} C.

(a) $e/m = v/Br$
$= 8.4 \times 10^6/(0.50 \times 10^{-3} \times 0.096)$
$= \mathbf{1.8 \times 10^{11} C\,kg^{-1}}$
(b) $e/m = 1.8 \times 10^{11} = 1.6 \times 10^{-19}/m$
$m = \mathbf{9.1 \times 10^{-31} kg}$

Now it's your turn

Electrons are accelerated through a potential difference of 220 V. They then pass into a region of uniform magnetic flux of flux density 0.54 mT. The path of the electrons is normal to the magnetic field. Given that the charge on the electron is 1.6×10^{-19} C and its mass is 9.1×10^{-31} kg, calculate:
(a) the speed of the accelerated electron,
(b) the radius of the circular path in the magnetic field.

Velocity selection of charged particles

We have seen that when particles of mass m and charge $+q$ enter an electric field of field strength E, there is a force F_E on the particle given by

$F_E = qE$

If the velocity of the particles before entry into the field is v at right angles to the field lines (see Figure 12.30), the particles will follow a parabolic path as they pass through the field.

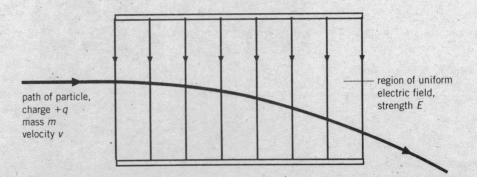

path of particle, charge $+q$ mass m velocity v

region of uniform electric field, strength E

Figure 12.30

Now suppose that a uniform magnetic field acts in the same region as the electric field. If this field acts downwards into the plane of the page, then, by Fleming's left-hand rule (see page 322), a force will act on the charged particle in the direction opposite to the force due to the electric field. The magnitude F_B of this force is given by

$$F_B = Bqv$$

where B is the flux density of the magnetic field.

If the magnitude of one of the two fields is adjusted, then a situation can arise where the two forces, F_E and F_B, are equal in magnitude but opposite in direction. Thus

$$Bqv = qE$$

and

$$v = E/B$$

For the value of the velocity given by E/B, the particles will not be deflected, as shown in Figure 12.31. Particles with any other velocities will be deflected. If a parallel beam of particles enters the field then all the particles passing undeviated through the slit will have the same velocity. Note that the mass does not come into the equation for the speed and so, particles with a different mass (but the same charge) will all pass undeviated through the region of the fields if they satisfy the condition $v = E/B$.

The arrangement shown in Figure 12.31 is known as a velocity selector. Velocity selectors are very important in the study of ions. Frequently, the production of the ions gives rise to different speeds but to carry out investigations on the ions, the ions must have one speed only.

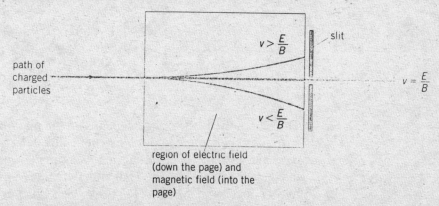

Figure 12.31 Velocity selector

Example

It is required to select charged ions which have a speed of $4.2 \times 10^6 \, \text{m s}^{-1}$. The electric field strength in the velocity selector is $3.2 \times 10^4 \, \text{V m}^{-1}$. Calculate the magnetic field strength required.

$v = E/B$
$B = 3.2 \times 10^4 / 4.2 \times 10^6$
$\quad = \mathbf{7.6 \times 10^{-3} \, T}$

Now it's your turn

Singly charged ions pass undeviated through a velocity selector. The electric field strength in the selector is $3.6 \times 10^4\,\text{V m}^{-1}$ and the magnetic flux density is 8.5 mT. Calculate the selected speed of the ions.

The Hall effect

Consider a thin slice of a conductor which is normal to a magnetic field, as illustrated in Figure 12.32. When there is a current in the conductor in the direction shown, charge carriers (electrons in a metal) will be moving at right angles to the magnetic field. They will experience a force which will tend to make them move to one side of the conductor. A potential difference known as the Hall voltage V_H will develop across the conductor. The Hall voltage does not increase indefinitely but reaches a constant value when the force on the charge carrier due to the magnetic field is equal to the force due to the electric field set up as a result of the Hall voltage.

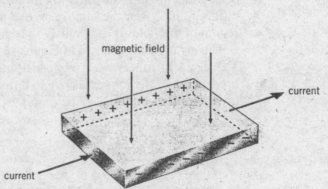

Figure 12.32 The Hall effect

The size of the Hall voltage depends on the material of the conductor, the current in the sample and the magnetic flux density. If the current is kept constant, then the Hall voltage across a sample will be proportional to the magnetic flux density. The Hall effect provides a means by which flux densities may be measured, using a Hall probe (see Chapter 2). In general, the material used for a Hall probe is a semiconductor. This gives a Hall voltage many thousands of times greater than that for a metal under the same conditions.

Comparing the effects of fields

We can summarise the effects of the different sorts of fields on masses, charges and current-carrying conductors. Although, as we have seen, there are close analogies between gravitational and electric fields, there are some ways in which they behave very differently.

Start with the effect of a gravitational field on a mass. Because masses always attract each other, a mass placed in a gravitational field will always move in the direction of the field, from a position of higher potential to a lower potential. For a field produced by a point mass, the field strength obeys an inverse square law relationship, and the potential obeys a reciprocal relationship with distance from the source of the field.

Electric fields are like gravitational fields in that, for a field produced by a point charge, the field strength is also given by the inverse square law, and the potential by a reciprocal relationship. However, we can have both positive charges and negative charges. A positive electric charge moves in the direction of the field, from a position of higher potential to a lower potential (just like a mass in a gravitational field). But a negative charge does just the opposite, against the direction of the field and from a low potential to a high potential.

What about electric charges in a magnetic field? A stationary charge is unaffected, but a moving charge experiences a force F given by $F = Bqv \sin \theta$. The direction of the force is given by Fleming's left-hand rule (for positive charges).

Finally, a current-carrying conductor in a magnetic field does not experience a force if the conductor is parallel to the field direction, but for all other directions it experiences a force given by $F = BIl \sin \theta$. The direction of the force is again given by Fleming's left-hand rule.

Example

A charged particle has mass 6.7×10^{-27} kg and charge $+3.2 \times 10^{-19}$ C. It is travelling at speed 2.5×10^8 m s^{-1} when it enters a region of space where there is a uniform magnetic field of flux density 1.6 T at right angles to its direction of motion. Calculate:
(a) the force on the particle due to the magnetic field,
(b) the radius of its orbit in the field.

(a) Force $= Bqv \sin \theta$
$\qquad = 1.6 \times 3.2 \times 10^{-19} \times 2.5 \times 10^8 \times 1$
$\qquad = \mathbf{1.3 \times 10^{-10}}$ **N**
(b) Centripetal force is provided by the electromagnetic force
$$mv^2/r = Bqv$$
$$6.7 \times 10^{-27} \times (2.5 \times 10^8)^2/r = 1.3 \times 10^{-13}$$
$$r = \mathbf{3.2\,m}$$

Now it's your turn

1 An electron of mass 9.1×10^{-31} kg and charge -1.6×10^{-19} C is travelling at a speed of 7.0×10^7 m s^{-1} when it enters a region of space where there is a uniform magnetic field of flux density 4.5 mT at right angles to its direction of motion. Calculate:
   (a) the force on the electron due to the magnetic field,
   (b) the radius of its orbit in the field.

2 An electron and an α-particle travelling at the same speed both enter the same region of uniform magnetic flux which is at right angles to their direction of motion. State and explain any differences between the two paths in the field.

Section 12.2 Summary

- There is a force on a current-carrying conductor whenever it is at an angle to a magnetic field.
- The direction of the force is given by Fleming's left-hand rule – place the first two fingers and thumb of the left hand at right-angles to each other, first finger in the direction of the magnetic field, second finger in the direction of the current, then the thumb gives the direction of the force.
- Magnetic flux density (field strength) is measured in tesla (T). $1\,T = 1\,Wb\,m^{-2}$
- The magnitude of the force F on a conductor of length L carrying a current I at an angle θ to a magnetic field of flux density B is given by the expression: $F = BIl \sin \theta$
- The force between parallel current-carrying conductors provides a means by which the ampere may be defined in terms of SI units.
- The force F on a particle with charge q moving at speed v at an angle θ to a magnetic field of flux density B is given by the expression: $F = Bqv \sin \theta$
- The path of a charged particle, moving at constant speed in a plane at right-angles to a uniform magnetic field, is circular.

Section 12.2 Questions

1 A stiff straight wire has a mass per unit length of $45\,g\,m^{-1}$. The wire is laid on a horizontal bench and a student passes a current through it to try to make it lift off the bench. The horizontal component of the Earth's magnetic field is $18\,\mu T$ in a direction from south to north and the acceleration of free fall is $10\,m\,s^{-2}$.

(a) (i) State the direction in which the wire should be laid on the bench.

   (ii) Calculate the minimum current required.

(b) Suggest whether the student is likely to be successful with his experiment.

2 The magnetic flux density B at a distance r from a long straight wire carrying a current I is given by the expression

$$B = 2.0 \times 10^{-7} \times \frac{I}{r}$$

where r is in metres and I is in amps.

(a) Calculate:
  (i) the magnetic flux density at a point distance $2.0\,cm$ from a wire carrying a current of $15\,A$,
  (ii) the force per unit length on a second wire, also carrying a current of $15\,A$, which is parallel to, and $2.0\,cm$ from, the first wire.

(b) Suggest why the force between two wires is demonstrated in the laboratory using aluminium foil rather than copper wires.

3 A small horseshoe magnet is placed on a balance and a stiff wire is clamped in the space between its poles. The length of wire between the poles is $5.0\,cm$. When a current of $3.5\,A$ is passed through the wire, the reading on the balance increases by $0.027\,N$.

(a) Calculate the magnetic flux density between the poles of the magnet.

(b) State three assumptions which you have made in your calculation.

4 Figure 12.33 shows the track of a particle in a
bubble chamber as it passes through a thin
sheet of metal foil. A uniform magnetic field
was applied into the plane of the paper.
State with a reason:
(a) in which direction the particle was
moving,
(b) whether the particle is positively or
negatively charged.

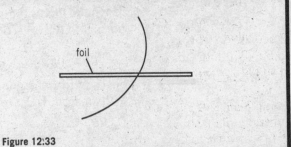

Figure 12:33

12.3 Electromagnetic induction

At the end of Section 12.3 you should be able to:
- define magnetic flux and the weber
- recall and solve problems using $\Phi = BA$
- define magnetic flux linkage
- infer from appropriate experiments on electromagnetic induction:
  - that a changing magnetic flux can induce an e.m.f. in a circuit
  - that the direction of the induced e.m.f. opposes the change producing it
  - the factors affecting the magnitude of the induced e.m.f.
- recall and solve problems using Faraday's law of electromagnetic induction and Lenz's law
- explain simple applications of electromagnetic induction
- show an understanding of the principle of operation of a simple iron-cored transformer and recall and solve problems using $N_s/N_p = V_s/V_p = I_p/I_s$ for an ideal transformer
- show an appreciation of the scientific and economic advantages of alternating current and of high voltages for the transmission of electrical energy.

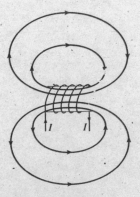

Figure 12.34 Continuous magnetic field lines

Magnetic flux

Figures 12.9, 12.11 and 12.12 illustrate the pattern of the magnetic field in the region of various conductors. However, the patterns are not all complete. All magnetic field lines should be continuous, as illustrated in Figure 12.34. Early experimenters thought that there was a flow of something along these lines and this gave rise to the idea of a magnetic flux, since 'flux' means 'flow'. Magnetic flux density (field strength) varies and this is shown by the spacing of the lines. Magnetic flux density may be considered as the number of lines of magnetic force per unit area of an area at right-angles to the lines. Magnetic flux Φ can be thought of as the total number of lines.

Magnetic flux is the product of the magnetic flux density and the area normal to the lines of flux.

For a uniform magnetic field of flux density B which makes an angle θ with an area A, the magnetic flux Φ is given by the expression $\Phi = BA \sin \theta$.

The unit of magnetic flux is the **weber** (Wb). One weber is equal to one tesla metre-squared i.e. $T\, m^2$.

The concept of magnetic flux is used frequently when studying electromagnetic induction.

Electromagnetic induction

The link between electric current and magnetic field was discovered by Oersted (1820). In 1831, Henry in the United States and Faraday in England demonstrated that an e.m.f. could be induced by a magnetic field. The effect was called **electromagnetic induction**.

Electromagnetic induction is now easy to demonstrate in the laboratory because sensitive meters are available. Figure 12.35 illustrates apparatus which may be used for this purpose. The galvanometer detects very small currents but it is important to realise that what is being detected are small electromotive forces (e.m.f.s). The current arises because there is a complete circuit incorporating an e.m.f. The following observations can be made.

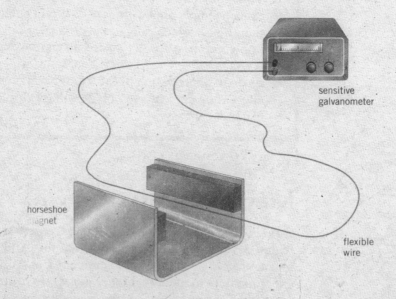

sensitive galvanometer

horseshoe magnet

flexible wire

Figure 12.35 Apparatus to demonstrate electromagnetic induction

■ An e.m.f. is induced when
  – the wire is moved through the magnetic field, across the face of the pole-pieces
  – the magnet is moved so that the wire passes across the face of the pole-pieces.
■ An e.m.f. is *not* induced when
  – the wire is held stationary between the pole-pieces
  – the magnet is moved so that the pole-pieces move along the length of the wire
  – the wire moves lengthways so that it does not change its position between the poles of the magnet.

These observations lead to the conclusion that an e.m.f. is induced whenever lines of magnetic flux are cut. The cutting may be caused by a movement of either the wire or of the magnet. The magnitude of the e.m.f. is also observed to vary.

■ The magnitude of the e.m.f.
  – increases as the speed at which the wire is moved increases
  – increases as the speed at which the magnet is moved increases
  – increases, if the wire is made into a loop with several turns (see Figure 12.36)
  – increases as the number of turns on the loop increases.

Figure 12.36 Wire wound to form a loop of several turns

It can be concluded that the magnitude of the induced e.m.f. depends on the rate at which magnetic flux lines are cut. The rate may be changed by varying the rate at which the flux lines are cut by a single wire or by using different numbers of turns of wire. The two factors are taken into account by using the term **magnetic flux linkage**. Change in magnetic flux linkage $\Delta(N\Phi)$ is equal to the product of the change in magnetic flux $\Delta\Phi$ and the number of turns N of a conductor involved in the change in flux.

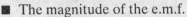

change in magnetic flux linkage $\Delta(N\Phi) = N\Delta\Phi$

The experimental observations are summarised in **Faraday's law of electromagnetic induction**:

The e.m.f. induced is proportional to the rate of change of magnetic flux linkage.

The experimental observations made with the apparatus of Figures 12.35 and 12.36 have been concerned with the magnitude of the e.m.f. However, it is noticed that the direction of the induced e.m.f. changes and that the direction is dependent on the direction in which the magnetic flux lines are being cut. The direction of the induced e.m.f. or current in a wire moving through a magnetic field at right angles to the fields may be determined using **Fleming's right-hand rule**. This rule is illustrated in Figure 12.37.

If the first two fingers and thumb of the right hand are held at right angles to one another, the **F**irst finger in the direction of the magnetic **F**ield and the thu**M**b in the direction of **M**otion, then the se**C**ond finger gives the direction of the induced e.m.f. or **C**urrent.

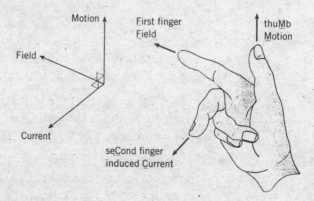

Figure 12.37 Fleming's right-hand rule

An explanation for the direction of the induced e.m.f. can be found by reference to the motor effect and conservation of energy.

Figure 12.38 shows a wire being moved downwards through a magnetic field. Since the wire is in the form of a continuous loop, the induced e.m.f. gives rise to a current, and the direction of this current can be found using Fleming's right-hand rule. This current is at right-angles to the magnetic flux and, by the motor effect, there will be a force on the wire. Using Fleming's left-hand rule (see page 322), the force is upwards when the wire is moving downwards. Reversing the direction of motion of the wire causes a current in the opposite direction and hence the electromagnetic force would once again oppose the motion. This conclusion is not surprising when conservation of energy is considered. An electric current is a form of energy and this energy must have been converted from some other form. Movement of the wire against the electromagnetic force means that work has been done in overcoming this force and it is this work which is seen as electrical energy. Anyone who has ridden a bicycle with a dynamo will realise that work has to be done to light the lamp!

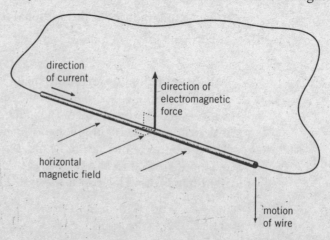

Figure 12.38

This application of conservation of energy is summarised in **Lenz's law**:

The direction of the induced e.m.f. is such as to cause effects to oppose the change producing it.

Faraday's law of electromagnetic induction and Lenz's law may be summarised using the equation

$$E = -\frac{d(N\Phi)}{dt}$$

(see the Maths Note below) where E is the e.m.f. induced by a rate of change of flux linkage of $d(N\Phi)/dt$. The minus sign indicates that the induced e.m.f. causes effects to oppose the change producing it.

Maths Note

The shorthand way of expressing the rate of change of a quantity x with time t is

$$\text{rate of change of } x \text{ with } t = \frac{dx}{dt}$$

This represents a mathematical operation known as *differentiation*. It is achieved by finding the gradient of the graph of x against t.

You will come across this notation here, in connection with the rate of change of magnetic flux linkage, and in one of the equations for radioactive decay (page 362).

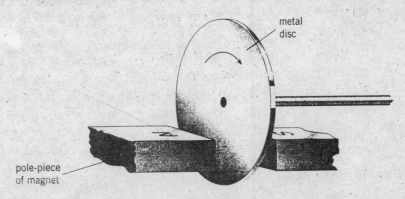

metal
disc

pole-piece
of magnet

Figure 12.39 Apparatus to demonstrate eddy current damping

The conversion of mechanical energy to electrical energy may be shown by spinning a metal disc in a magnetic field, as illustrated in Figure 12.39. The disc is seen to slow down much more rapidly with the magnet in place than when it has been removed. As the disc spins, it cuts through the flux lines of the magnet. This cutting becomes more obvious if a radius of the disc is considered. As the radius rotates, it will cut flux. An e.m.f. will be induced in the disc but, because the rate of cutting of flux varies from one part of the disc to another, the e.m.f. will have different magnitudes in different regions. The disc is metal and therefore electrons will move between regions with different e.m.f. values. Currents are induced in the disc. Since these currents vary in magnitude and direction there are called **eddy currents**. The eddy currents cause heating in the disc and the dissipation of the energy of rotation of the disc is referred to as **eddy current damping**. If the permanent magnet in Figure 12.39 is replaced by an electromagnet, the spinning disc will be slowed down whenever there is a current in the electromagnet. This is the principle behind electromagnetic braking. The advantage over conventional brakes is that there is no physical contact with the spinning disc. This makes such brakes very useful for trains travelling at high speeds. However, the disadvantage is that, as the disc slows down, the induced eddy currents will be smaller and therefore the braking will be less efficient. This system would be useless as the parking brake on a car!

Example

The uniform flux density between the poles of a magnet is 0.080 T. A small coil of area of cross-section 6.5 cm² has 250 turns and is placed with its plane normal to the magnetic field. The coil is withdrawn from the field in a time of 0.26 s.

Determine:
(a) the magnetic flux through the coil when it is between the poles of the magnet,
(b) the change in magnetic flux linkage when the coil is removed from the field,
(c) the average e.m.f. induced in the coil whilst it is being withdrawn.

(a) *magnetic flux* $\Phi = BA \sin\theta$
$$= 0.080 \times 6.5 \times 10^{-4}$$
$$= \mathbf{5.2 \times 10^{-5}\,Wb}$$

(b) *change in flux linkage* $= (N\Phi)_{\text{FINAL}} - (N\Phi)_{\text{INITIAL}}$
$$= 0 - (250 \times 5.2 \times 10^{-5})$$
$$= \mathbf{-1.3 \times 10^{-2}\,Wb}$$
(the sign indicates that the flux linkage is decreasing)

(c) *induced e.m.f.* $= \dfrac{change\ in\ flux\ linkage}{time\ taken}$
$$= (1.3 \times 10^{-2})/0.26$$
$$= \mathbf{0.050\ V}$$

Now it's your turn

1 An aircraft has a wingspan of 17 m and is flying horizontally in a northerly direction at a speed of 94 m s⁻¹. The vertical component of the Earth's magnetic field is 40 μT in a downward direction.

(a) Calculate:
  (i) the area swept out per second by the wings,
  (ii) the magnetic flux cut per second by the wings,
  (iii) the e.m.f. induced between the wingtips.
(b) State which wing-tip will be at the higher potential.

2 A current-carrying solenoid produces a uniform magnetic flux of density 4.6×10^{-2} T along its axis. A small circular coil of radius 1.2 cm has 350 turns of wire and is placed on the axis of the solenoid with its plane normal to the axis. Calculate the average e.m.f. induced in the coil when the current in the solenoid is switched off in a time of 85 ms.

3 A metal disc is made to spin at 15 revolutions per second about an axis through its centre normal to the plane of the disc. The disc has radius 24 cm and spins in a uniform magnetic field of flux density 0.15 T, parallel to the axis of rotation.
Calculate:
(i) the area swept out each second by a radius of the disc,
(ii) the flux swept out each second by a radius of the disc,
(iii) the e.m.f. induced in the disc.

e.m.f. induced between two coils

A current-carrying solenoid or coil is known to have a magnetic field. If the current in the coil is switched off, there will be a change in flux in the region around the coil. Consider the apparatus illustrated in Figure 12.40.

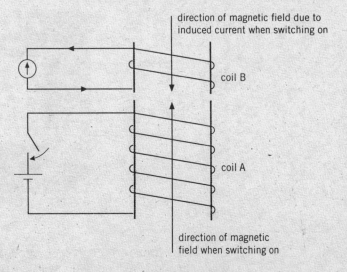

Figure 12.40

As the current in coil A is being switched on, the magnetic field in this coil grows. The magnetic field links with the turns on coil B and, as a result, there is a change in flux linkage in coil B and an e.m.f. is induced in this coil.

Coil B forms part of a complete circuit and hence there is a current in coil B. The direction of this current can be determined using Lenz's law.

The change which brought about the induction of a current was a *growth* in the magnetic flux in coil A. The induced current in coil B will give rise to a magnetic field in coil B and this field will, by Lenz's law, try to oppose the growth of the field in coil A. Consequently, since the field in coil A is vertically upwards (the right-hand grip rule), the field in coil B will be vertically downwards and the induced current will be in an anticlockwise direction through the meter. When the current in coil A is switched off, the magnetic field in coil A will *decay*. The magnetic field in coil B due to the induced current must try to prevent this decay and hence it will be vertically upwards. The induced current has changed in direction.

The magnitude of the induced e.m.f. can be increased by inserting a soft iron core into the coil (but be careful not to damage the meter as any induced e.m.f. will be very much greater) or by increasing the number of turns on the coils or by switching a larger current in coil A.

It is important to realise that an e.m.f. is induced only when the magnetic flux in coil A is changing; that is, when the current in coil A is changing. A steady current in coil A will not give rise to an induced e.m.f. An e.m.f. may be induced continuously in coil B if an alternating current is provided for coil A. This is the principle of the **transformer**.

A simple transformer is illustrated in Figure 12.41. Two coils of insulated wire are wound on to a laminated soft-iron core. An alternating e.m.f. is applied across one coil (the primary coil) and an e.m.f. is induced in the secondary coil. The ratio of the output e.m.f. to the applied e.m.f. is dependent on the ratio of the number of turns on the two coils. The applied e.m.f. and the induced e.m.f. across the secondary coil have the same frequency but are out of phase by π rad (180°).

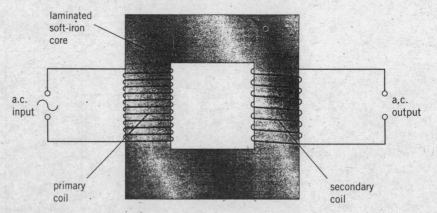

Figure 12.41 Simple transformer

Consider a transformer with N_p turns and N_s turns on the primary and secondary coils respectively, with currents in the coils of I_p and I_s when the e.m.f.s across the coils and V_p and V_s.

For a transformer which is 100% efficient (an **ideal** transformer),

input power = output power
$$V_p I_p = V_s I_s$$

and

$$\frac{N_s}{N_p} = \frac{V_s}{V_p} = \frac{I_p}{I_s}$$

When V_s is greater than V_p, there are more turns on the secondary coil than the primary and the transformer is said to be a **step-up transformer**. Conversely, if V_s is less than V_p, the secondary coil has fewer turns than the primary and the transformer is said to be a **step-down transformer**.

In practice, the transformer will not be 100% efficient due to power losses. Some sources of these losses are:

■ loss of magnetic flux between the primary and secondary coils
■ resistive heating in the primary and secondary coils
■ heating of the core due to eddy currents
■ heating of the core due to repeated magnetisation and demagnetisation.

The use of soft iron reduces heating due to magnetisation and demagnetisation of the core. However, since iron (an electrical conductor) is used as the material for the core so that the magnetic flux linkage is large, eddy currents cannot be prevented. Laminating the core, i.e. building up the core from thin strips of soft iron which are electrically insulated from one another, reduces eddy current losses.

Figure 12.42 Transformers are produced in many shapes and sizes.

Example

A transformer is to be used with an alternating supply of 240 V to power a lamp rated as 12 V, 3 A. The secondary coil of the transformer has 80 turns. Assuming that the transformer is ideal, calculate, for the primary coil:
(a) the number of turns,
(b) the current.

(a) For an ideal transformer,

$$V_s/V_p = N_s/N_p$$
$$12/240 = 80/N_p$$
$$N_p = \textbf{1600 turns}$$

(b) For an ideal transformer,

$$V_s/V_p = I_p/I_s$$
$$12/240 = I_p/3$$
$$I_p = \textbf{0.15 A}$$

Now it's your turn

1 The primary coil of an ideal transformer has 1200 turns and is connected to a 240 V alternating supply. The transformer is to be used to step down the voltage to 9.0 V.

(a) Calculate the number of turns on the secondary coil.
(b) An appliance, rated as 9.0 V, 1.5 A is connected to the secondary coil of the transformer. Calculate the current in the primary coil.

2 An ideal transformer is to be used to step up a voltage from 220 V to supply a current of 15 mA at 6.6 kV. The primary coil has 250 turns. Calculate:

(a) the ratio of the number of turns on the secondary coil to that on the primary coil,

(b) the number of turns on the secondary coil,

(c) the current in the primary coil.

Transmission of electrical energy

Transformers play an essential part in the transmission of electrical energy. Generating stations are often situated a long way from the cities they serve. Hydroelectric plants must be placed at the dam providing the water supply, and nuclear generators need a plentiful supply of cooling water; there is also a safety factor to consider. Fossil-fuel plants are also placed some distance from towns, mainly because of air pollution concerns and because land is cheaper away from towns. This means that, whatever the means of generating the electricity, it is likely to need to be transmitted over long distances. There is always a power loss in transmission lines, simply due to the I^2R effect (see page 170). However, it turns out that this loss can be reduced if the power is transmitted at a high voltage. Look at the following Example.

Example

A small town, several km from an electricity generating station, requires 120 kW of power on average. The total resistance of the transmission lines is 0.40 Ω. Calculate the power loss if the transmission is made at a voltage of **(a)** 240 V, **(b)** 24 kV.

First calculate the current I in the transmission lines in each case. In (a), $I = P/V = 1.2 \times 10^5/240 = 500$ A. In (b), the same calculation gives $I = 5.0$ A.

Now find the power loss from $P = I^2R$. In (a), it is $(500)^2 \times 0.40 = \textbf{100 kW}$, or more than 80% of the power required by the town. In (b), the power loss is $5.0^2 \times 0.40 = \textbf{10 W}$, less than one-hundredth of 1% of the power requirement.

The advantage of transmitting power at the higher voltage is now obvious. The advantage of using alternating current rather than direct current is that a.c. is so efficiently stepped up to a high voltage for transmission using a transformer. Transformers used in transmission networks have very high efficiencies, often over 98%.

Section 12.3 Summary

- Magnetic field is a general term used for a region of space where a current-carrying conductor experiences an electromagnetic force.
- Magnetic flux density (magnetic field strength) measures the strength of a magnetic field and has a magnitude and direction at every point in the field.
- Magnetic flux is the product of flux density and area normal to the flux.
- The direction of the induced current in a conductor moving through a magnetic field is given by Fleming's right-hand rule. That is:
  If the first two fingers and thumb of the right hand are held at right-angles to each other, the first finger in the direction of the magnetic field and the thumb in the direction of motion, then the second finger gives the direction of the induced e.m.f. or current.
- Faraday's law of electromagnetic induction states that the e.m.f. induced is proportional to the rate of change of magnetic flux linkage.
- Lenz's law states that the direction of the induced e.m.f. is such as to cause effects to oppose the change producing it.
- Faraday's law of electromagnetic induction and Lenz's law may be summarised using the equation:

$$E = -d(N\Phi)/dt$$

where E is the e.m.f. induced by a rate of change of flux linkage of $d(N\Phi)/dt$. The sign indicates the relative direction of the e.m.f. and the change in flux linkage.

- In a transformer, an alternating current in one coil induces an alternating e.m.f. in a second coil. The ratio of the output (secondary) e.m.f. to the applied (primary) e.m.f. equals the ratio of the number of turns on the secondary coil to the number of turns on the primary coil:

$$V_s/V_p = N_s/N_p$$

Section 12.3 Questions

1 A coil is constructed by winding 400 turns of wire on to a cylindrical iron core. The mean radius of the coil is 3.0 cm. It is found that the flux density B in the core due to a current I in the coil is given by the expression

$B = 2.2\ I$

where B is in tesla.
(a) Calculate, for a current of 0.64 A in the coil:
  (i)   the magnetic flux density in the core,
  (ii) the magnetic flux in the core,
  (iii) the flux linkage of the coil.

(b) The current in the coil is switched off in a time of 0.011 s. Calculate the e.m.f. induced and state where this e.m.f. will be observed.
(c) Hence suggest why, when switching off a large electromagnet, the current is reduced gradually rather than switched off suddenly.

2 A flat coil contains 250 turns of insulated wire and has a mean radius of 1.5 cm. The coil is placed in a region of uniform magnetic flux of flux density 85 mT such that there is an angle θ between the plane of the coil and the flux lines. The coil is withdrawn from the magnetic field in a time of 0.30 s.

Calculate the average e.m.f. induced in the coil for an angle θ of:
(a) zero,
(b) 90°,
(c) 35°.

3 An ideal transformer is to be used to step down a 220 V alternating supply to 7.5 V.

(a) Calculate the ratio of the number of turns on the secondary coil to that on the primary coil.
(b) Calculate the number of turns on the primary coil given that there are 140 turns on the secondary coil.
(c) Suggest why the wire forming the secondary coil would, in this case, be thicker than that for the primary coil.

Exam-style Questions

1 A long straight current-carrying wire is held in a north-south direction vertically above a small compass. The vertical distance between the compass and the wire is 4.6 cm.

The horizontal component of the Earth's field is 18 µT and the magnetic flux density B due to a current I in a long straight wire is given by the expression

$$B = 2.0 \times 10^{-7} \times \frac{I}{r}$$

where B is measured in tesla and r is the distance from the wire in metres.
(a) State the direction in which the compass points when there is no current in the wire.
(b) A steady current of 6.0 A is switched on in the wire.
  (i) State the direction of the magnetic field of the wire in the region of the compass.
  (ii) Calculate the magnetic flux density due to the current in the wire at the position of the compass.
  (iii) Determine the angle through which the compass will rotate when the current in the wire is switched on.
(c) The compass is replaced by a long straight wire so that this wire is parallel to and 4.5 cm from the first wire. This second wire carries a current of 7.7 A. Calculate the force per unit length between the two wires.

2 The magnetic flux density B at the centre of a long solenoid carrying a current I is given by the expression

$$B = 2.5\,I$$

where B is in millitesla and I is in amps. The cross-section of the solenoid has a diameter of 2.5 cm. A length of wire is wound tightly round the centre of the solenoid to form a small coil of 50 turns, as illustrated in Figure 12.43.
(a) Calculate, for a current of 1.8 A in the solenoid:
  (i) the flux density in the solenoid,
  (ii) the magnetic flux linkage of the small coil.

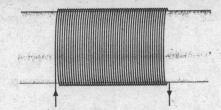

Figure 12.43

(b) The current is reversed in direction in 0.031 s. Calculate the average e.m.f. induced in the small coil.

3 Two coils A and B are placed as shown in Figure 12.44.

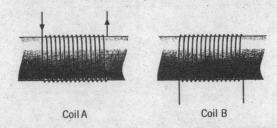

Coil A   Coil B

Figure 12.44

The current in coil A is switched on and then off so that the variation with time t of the current I is as shown in Figure 12.45.

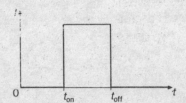

Figure 12.45

Sketch a graph to show the variation with time t of the e.m.f. E induced in coil B.

13. Nuclear physics

In the early 1900s, Rutherford showed that the atom consists of a very small nucleus containing the majority of the atom's mass, surrounded by orbiting electrons. In this chapter, we shall look at some of the physics of the nucleus. We start by looking at the structure of the nucleus. The decay of unstable nuclei leads to radioactive emissions. We shall investigate the properties of these emissions. Nuclei may also be transformed by particle impact. We shall look at the processes of fission and fusion, with particular reference to energy changes.

13.1 Atomic structure

At the end of Section 13.1 you should be able to:

- describe a simple model for the nuclear atom to include protons, neutrons and orbital electrons
- distinguish between nucleon number (mass number) and proton number (atomic number)
- show an understanding that an element can exist in various isotopic forms, each with a different number of neutrons
- use the usual notation for the representation of nuclides.

The atoms of all elements are made up of three particles called **protons**, **neutrons**, and **electrons**. The protons and neutrons are at the centre or nucleus of the atom. The electrons orbit the nucleus.

We shall see later that the diameter of the nucleus is only about 1/10 000 of the diameter of an atom.

Figure 13.1 illustrates very simple models of a helium atom and a lithium atom.

The protons and neutrons both have a mass of about one atomic mass unit ($1 \, u = 1.66 \times 10^{-27} \, kg$). By comparison, the mass of an electron is very small, about 1/2000 of 1 u. The vast majority of the mass of the atom is therefore in the nucleus.

The basic properties of the proton, neutron and electron are summarised in Table 13.1.

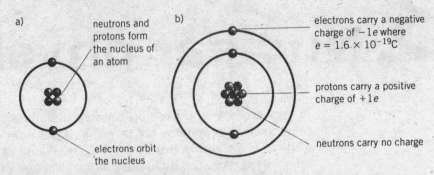

Figure 13.1 Structures of a) a helium atom and b) a lithium atom

Table 13.1

	approximate mass	charge	position
proton	u	$+e$	in nucleus
neutron	u	0	in nucleus
electron	u/2000	$-e$	orbiting nucleus

Atoms and ions

Atoms are uncharged because they contain equal numbers of protons and electrons. If an atom loses one or more electrons so that it does not contain an equal number of protons and electrons, it becomes charged and is called an **ion**.

For example, if a sodium atom loses one of its electrons it becomes a positive sodium ion.

$$\text{Na} \quad \rightarrow \quad \text{Na}^+ \quad + \quad \text{e}^-$$
sodium atom   sodium ion   electron

If an atom gains an electron, it becomes a negative ion.

Proton number and nucleon number

The number of protons in the nucleus of an atom is called the **proton number** (or **atomic number**) Z.

The number of protons together with the number of neutrons in the nucleus is called the **nucleon number** (or **mass number**) A.

A **nucleon** is the name given to either a proton or a neutron. The difference between the nucleon number (A) and the proton number (Z) gives the number of neutrons in the nucleus. Because atoms are electrically neutral, the proton number is equal to the number of electrons orbiting the nucleus.

Nuclear representation

The nucleus of one form of sodium contains 11 protons and 12 neutrons. Therefore its proton number Z is 11 and the nucleon number A is $11 + 12 = 23$. This nucleus can be shown as $^{23}_{11}$Na. In general, a nucleus can be represented as

$$^{\text{nucleon number}}_{\text{proton number}} X$$

where X is the chemical symbol.

Example

An oxygen nucleus is represented by $^{16}_{8}$O. Describe its atomic structure.

The nucleus has a proton number of 8 and a nucleon number of 16. Thus, its nucleus contains **8 protons** and $16 - 8 = $ **8 neutrons**. There are also **8 electrons** (equal to the number of protons) orbiting the nucleus.

Now it's your turn

Write down the proton number and the nucleon number for the potassium nucleus $^{40}_{19}$K. Deduce the number of neutrons in the nucleus.

Isotopes

Sometimes atoms of the same element have different numbers of neutrons in their nuclei. The most abundant form of chlorine contains 17 protons and 18 neutrons in its nucleus, giving it a nucleon number of $17 + 18 = 35$. This is often called chlorine-35. Another form of chlorine contains 17 protons and 20 neutrons in the nucleus, giving it a nucleon number of 37. This is chlorine-37. Chlorine-35 and chlorine-37 are said to be **isotopes** of chlorine.

Isotopes are different forms of the same element which have the same numbers of protons but different numbers of neutrons in their nuclei.

Some elements have many isotopes, but others have very few. For hydrogen, the most common isotope is hydrogen-1. Its nucleus is a single proton. Hydrogen-2 is called deuterium; its nucleus contains one proton and one neutron. Hydrogen-3, with one proton and two neutrons, is called tritium.

Note that the term isotope is also used to describe *nuclei* with the same proton number (that is, nuclei of the same element) but with different nucleon numbers. You may also come across the term **nuclide**.

A nuclide is one type of nucleus with a particular nucleon number and a particular proton number.

- An atom consists of a nucleus containing protons and neutrons surrounded by orbiting electrons.
- Most of the mass of an atom is contained in its nucleus.
- Neutral atoms contain equal numbers of protons and electrons.
- Atoms which have gained or lost electrons are charged, and are called ions.
- The nucleon number A of a nucleus is the number of nucleons (protons and neutrons) in the nucleus.
- The proton number Z of a nucleus is the number of protons in the nucleus; hence the number of neutrons in the nucleus is $A - Z$.
- A nucleus (chemical symbol X) may be represented by:

$$^{\text{nucleon number}}_{\text{proton number}} X$$

- Isotopes are different forms of the same element (that is, with the same proton number) but with a different nucleon number.

13.2 α-particles, β-particles and γ-radiation

At the end of Sections 13.2–13.4 you should be able to:

- show an understanding of the nature of α-, β- and γ-radiations
- appreciate that nucleon number, proton number, and energy and mass are all conserved in nuclear processes
- represent simple nuclear reactions by nuclear equations of the form $^{14}_{7}N + ^{4}_{2}He \rightarrow ^{17}_{8}O + ^{1}_{1}H$
- show an appreciation of the spontaneous and random nature of nuclear decay
- infer the random nature of radioactive decay from the fluctuations in count-rate.

Some elements have nuclei which are unstable. In order to become more stable, they emit particles and/or electromagnetic radiation. The nuclei are said to be **radioactive**, and the emission is called **radioactivity**. The emissions are invisible to the eye, but their tracks were first made visible in a device called a cloud chamber. The photograph shows tracks created by one type of emission, α-particles.

Examination of the nature and properties of the emitted particles or radiation shows that the emissions are of three different types. These are **α-particles** (alpha-particles), **β-particles** (beta-particles) and **γ-radiation** (gamma radiation). All three emissions orginate from the nucleus.

Figure 13.2 Tracks of α-particles

α-particles

Like a helium atom, an α-particle contains two protons and two neutrons, and hence carries a charge of $+2e$. α-particles travel at speeds of up to about $10^7 \, \text{m s}^{-1}$ (about 5% of the speed of light). α-particle emission is the least penetrating of the three types of emission. It can pass through very thin

> An α-particle is identical to the nucleus of a helium atom.

paper, but is unable to penetrate thin card. Its range in air is a few centimetres. Because α-particles are charged, they can be deflected by electric and magnetic fields.

As α-particles travel through matter, they interact with nearby atoms causing them to lose one or more electrons. The ionised atom and the dislodged electron are called an ion pair. The production of an ion pair requires the separation of unlike charges, and this process requires energy. α-particles have a relatively large mass and charge, and consequently they are efficient ionisers. They may produce as many as 10^5 ion pairs for every centimetre of air through which they travel. Thus they lose energy relatively quickly, and have low penetrating power.

When the nucleus of an atom emits an α-particle, it is said to undergo α-decay. The nucleus loses two protons and two neutrons in this emission.

> In α-decay the proton number of the nucleus decreases by two, and the nucleon number decreases by four.

Each element has a particular proton number, and therefore α-decay causes one element to change into another. (This process is sometimes called **transmutation**.) The original nuclide is called the **parent** nuclide, and the new one the **daughter** nuclide.

For example, uranium-234 (the parent nuclide) may emit an α-particle. The daughter nuclide is thorium-230. In addition, energy is released. This emission is represented by the nuclear equation

$$^{234}_{92}U \rightarrow \, ^{230}_{90}Th + \, ^4_2He + energy$$

β-particles

β-particles are fast-moving electrons.

β-particles have speeds in excess of 99% of the speed of light. These particles have half the charge and very much less mass than α-particles. Consequently, they are much less efficient than α-particles in producing ion pairs. They are thus far more penetrating than α-particles, being able to travel up to about a metre in air. They can penetrate card and sheets of aluminium up to a few millimetres thick. Their charge means that they are affected by electric and magnetic fields. However, there are important differences between the behaviour of α- and β-particles in these fields. β-particles carry a negative charge, and are thus deflected in the opposite direction to the positively charged α-particles. Because the mass of a β-particle is much less than that of an α-particle, β-particles experience a much larger deflection when moving at the same speed as α-particles.

It was stated in Section 13.1 that the nucleus contains protons and neutrons. What, then, is the origin of β-particle emission? The β-particles certainly come from the nucleus, not from the electrons outside the nucleus. What happens is that just prior to β-emission a neutron in the nucleus forms a proton and an electron, in order to change the ratio of protons to neutrons in the nucleus. This makes the daughter nucleus more stable.

In fact, *free* neutrons are known to decay like this:

$$^1_0n \rightarrow \, ^1_1p + \, ^0_{-1}e + energy$$

A similar process happens in the nucleus. In β-decay the electron is emitted from the nucleus as a β-particle. This leaves the nucleus with the same number of nucleons as before, but with one extra proton and one fewer neutrons.

In β-decay a daughter nuclide is formed with the proton number increased by one, but with the same nucleon number.

For example, strontium-90 (the parent nuclide) may decay with the emission of a β-particle to form the daughter nuclide yttrium-90. As in the case of α-emission, energy is evolved. The decay is represented by the nuclear equation

$$^{90}_{38}\text{Sr} \rightarrow {}^{90}_{39}\text{Y} + {}^{0}_{-1}\text{e} + \text{energy}$$

γ-radiation

γ-radiation is part of the electromagnetic spectrum with wavelengths between 10^{-11} m and 10^{-13} m.

Since γ-radiation has no charge, its ionising power is much less than that of either α- or β-particles. γ-radiation penetrates almost unlimited thicknesses of air, several metres of concrete or several centimetres of lead.

α- and β-particles are emitted by unstable nuclei which have excess energy. The emission of these particles results in changes in the ratio of protons to neutrons, but the nuclei may still have excess energy. The nucleus may return to its unexcited (or ground) state by emitting energy in the form of γ-radiation.

In γ-emission no particles are emitted and there is therefore no change to the proton number or nucleon number of the parent nuclide.

For example, when uranium-238 decays by emitting an α-particle, the resulting nucleus of thorium-234 contains excess energy (it is in an excited state) and emits a photon of γ-radiation to return to the ground state. This process is represented by the nuclear equation

$$^{234}_{90}\text{Th}^* \rightarrow {}^{234}_{90}\text{Th} + {}^{0}_{0}\gamma$$

The * next to the symbol Th on the left-hand side of the equation shows that the thorium nucleus is in an excited state.

Note that in all radioactive decay equations (and, in fact, in all equations representing nuclear reactions) nucleon number and proton number are conserved. The sum of the numbers at the top of the symbols on the left-hand side of the equation (the sum of the nucleon numbers) is equal to the sum of the nucleon numbers on the right-hand side. Similarly, the sum of the numbers at the bottom of the symbols on the left-hand side (the sum of the proton numbers) is equal to the sum of the proton numbers on the right-hand side. Energy and mass, taken together, are also conserved in all nuclear processes.

Summary of the properties of radioactive emissions

Table 13.2 summarises the properties of α-particles, β-particles and γ-radiation.

Table 13.2

property	α-particle	β-particle	γ-radiation
mass	4 u	about u/2000	0
charge	$+2e$	$-e$	0
nature	helium nucleus (2 protons + 2 neutrons)	electron	short-wavelength electromagnetic waves
speed	up to $0.05c$	more than $0.99c$	c
penetrating power	few cm of air	few mm of aluminium	few cm of lead
relative ionising power	10^4	10^2	1
affects photographic film?	yes	yes	yes
deflected by electric, magnetic fields?	yes, see Figure 13.3	yes, see Figure 13.3	no

Figure 13.3 illustrates a hypothetical demonstration of the effect of a magnetic field on the three types of emission. The direction of the magnetic field is perpendicularly into the page. γ-radiation is uncharged, and is not deflected by the magnetic field. Because α- and β-particles have opposite charges, they are deflected in opposite directions. Note that the deviation of the α-particles is less than that of the β-particles, and that α-particles all deviate by the same amount. β-particles by contrast have a range of deflections indicating that they have a range of energies.

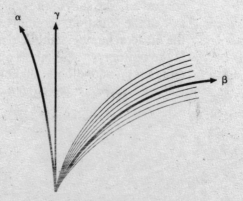

Figure 13.3 Deflection of α- and β-particles and γ-rays by a magnetic field into the page

13.3 Radioactive decay series

The daughter nuclide of a radioactive decay may itself be unstable and so may emit radiation to give another different nuclide. This sequence of radioactive decay from parent nuclide through succeeding daughter nuclides is called a **radioactive decay series**. The series ends when a stable nuclide is reached.

Part of such a radioactive decay series, the uranium series, is shown in Table 13.3.

Table 13.3 Part of the decay series of uranium-238

decay	radiation emitted
$^{238}_{92}U \rightarrow\ ^{234}_{90}Th + ^{4}_{2}He + ^{0}_{0}\gamma$	α, γ
$^{234}_{90}Th \rightarrow\ ^{234}_{91}Pa + ^{0}_{-1}e + ^{0}_{0}\gamma$	β, γ
$^{234}_{91}Pa \rightarrow\ ^{234}_{92}U + ^{0}_{-1}e + ^{0}_{0}\gamma$	β, γ
$^{234}_{92}U \rightarrow\ ^{230}_{90}Th + ^{4}_{2}He + ^{0}_{0}\gamma$	α, γ
$^{230}_{90}Th \rightarrow\ ^{226}_{88}Ra + ^{4}_{2}He + ^{0}_{0}\gamma$	α, γ
$^{226}_{88}Ra \rightarrow\ ^{222}_{86}Rn + ^{4}_{2}He$	α
$^{222}_{86}Rn \rightarrow\ ^{218}_{84}Po + ^{4}_{2}He$	α

13.4 Detecting radioactivity

Some of the methods used to detect radioactive emissions are based on the ionising properties of the particles or radiation.

The Geiger counter

Figure 13.4 illustrates a Geiger-Müller tube with a scaler connected to it. When radiation enters the window, it creates ion pairs in the gas in the tube. These charged particles, and particularly the electrons, are accelerated by the potential difference between the central wire anode and the cylindrical cathode. These accelerated particles then cause further ionisation. The result of this continuous process is described as an **avalanche effect**. That is, the entry of one particle into the tube and the production of one ion pair results in very large numbers of electrons and ions arriving at the anode and cathode respectively. This gives a pulse of charge which is amplified and counted by the scaler or ratemeter. (A scaler measures the total count of pulses in the tube during the time that the scaler is operating. A ratemeter continuously monitors the number of counts per second.) Once the pulse has been registered, the charges are removed from the gas in readiness for further radiation entering the tube.

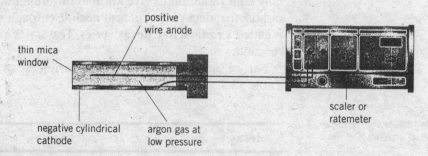

Figure 13.4 Geiger-Müller tube and scaler

Photographic plates

When a radioactive emission strikes a photographic film, the film reacts as if it had been exposed to a small amount of visible light. When the film is developed, fogging or blackening is seen. This fogging can be used to detect not only the presence of radioactivity, but also the intensity of the radiation.

Figure 13.5 Film badge dosimeter

Figure 13.5 shows a film badge dosimeter. It contains a piece of photographic film which becomes fogged when exposed to radiation. Workers who are at risk from radiation wear such badges to gauge the type and dose of radiation to which they have been exposed. The radiation passes through different filters before reaching the film. Consequently the type of radiation, as well as the quantity, can be assessed.

The scintillation counter

Early workers with radioactive materials used glass screens coated with zinc sulphide to detect radiation. When radiation is incident on the zinc sulphide, it emits a tiny pulse of light called a **scintillation**. The rate at which these pulses are emitted indicates the intensity of the radiation.

The early researchers worked in darkened rooms, observing the zinc sulphide screen by eye through a microscope and counting the number of flashes of light occurring in a certain time. Now a scintillation counter is used (Figure 13.6).

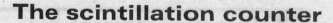

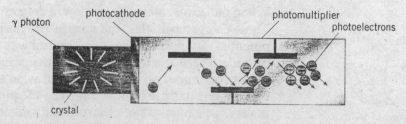

Figure 13.6 Scintillation counter

Often a scintillator crystal is used instead of a zinc sulphide screen. The crystal is mounted close to a device known as a photomultiplier, a vacuum-tube device which uses the principle of photoelectric emission (see Chapter 9). Flashes of light cause the emission of photoelectrons from the negative electrode of the photomultiplier. The photoelectric current is amplified inside the tube. The output electrode is connected to a scaler or ratemeter, as with the Geiger-Müller tube.

Background radiation

Radioactivity is a natural phenomenon. Rocks such as granite contain small amounts of radioactive nuclides, some foods we eat emit radiation, and even our bodies are naturally radioactive. Although the atmosphere provides life on Earth with some shielding, there is, nevertheless, some radiation from outer space (cosmic radiation). In addition to this natural radioactivity, we are exposed to radiation from man-made sources. These are found in medicine, in fallout from nuclear explosions, and in leaks from nuclear power stations. The sum of all this radiation is known as **background radiation**. Figure 13.7 indicates the relative proportions of background radiation coming from various sources.

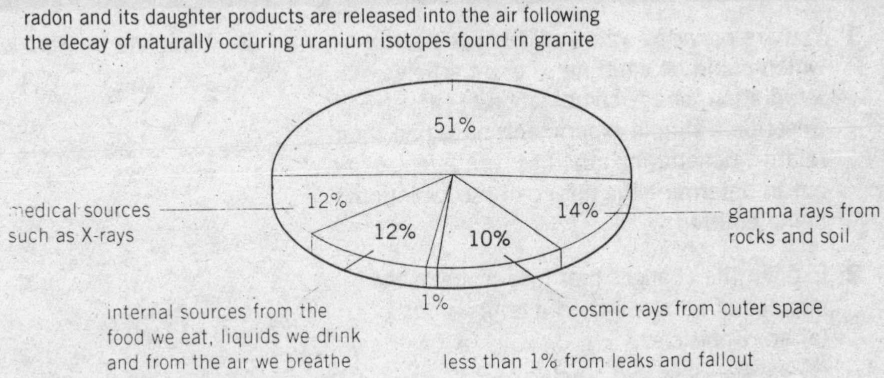

radon and its daughter products are released into the air following the decay of naturally occuring uranium isotopes found in granite

Figure 13.7 Sources of background radiation

In carrying out experiments with radioactive sources, it is important to take account of background radiation. In order to determine the count-rate due to the radioactive source, the background count-rate must be subtracted from the total measured count-rate. Allowance for background radiation gives the corrected count-rate.

The spontaneous and random nature of radioactive decay

Detection of the count-rate of radioactive sources shows that the emission of radiation is both *spontaneous* and *random*. It is a spontaneous process because it is not affected by any external factors, such as temperature or pressure. Decay is random in that it is not possible to predict which nucleus in a sample will decay next. There is, however, a constant probability (or chance) that a nucleus will decay in any fixed period of time. We will look at this in more detail in Section 13.5.

Sections 13.2–13.4 Summary

- An α-particle is a helium nucleus (two protons + two neutrons).
- A β-particle is a fast-moving electron.
- γ-radiation consists of short-wavelength electromagnetic waves.
- Nuclear notation: $^{\text{nucleon number}}_{\text{proton number}}\text{X}$, where X is the symbol for the chemical element.
- In nuclear notation the emissions are represented as: α-particle ^4_2He; β-particle $^0_{-1}\text{e}$; γ-radiation $^0_0\gamma$
- α-emission reduces the nucleon number of the parent nucleus by 4, and reduces the proton number by 2.
- β-emission causes no change to the nucleon number of the parent nucleus, and increases the proton number by 1.
- γ-emission causes no change to nucleon number or proton number of the parent nucleus.
- Radioactive decay is a spontaneous, random process.

Sections 13.2–13.4 Questions

1 You are provided with a radioactive source which could be emitting α- or β-particles, or γ-radiation, or any combination of these. Describe a simple experiment, based on their relative penetrating qualities, you might carry out to determine the nature of the radiation(s) being emitted.

2 Explain the changes that take place to the nucleus of an atom when it emits:
(a) an α-particle,
(b) a β-particle,
(c) γ-radiation.

3 Complete the following radioactive series.
$$^a_b\text{X} \rightarrow \,^?_?\text{Y} + \,^4_2\text{He}$$
$$^?_?\text{Y} \rightarrow \,^?_?\text{Z} + \,^0_{-1}\text{e}$$
$$^?_?\text{Z} \rightarrow \,^?_?\text{Z} + \,^?_??$$

13.5 Random decay

At the end of Sections 13.5–13.6 you should be able to:

- define half-life
- define the terms activity and decay constant and recall and solve problems using $A = \lambda N$
- infer and sketch the exponential nature of radioactive decay and solve problems using the relationship $x = x_0 \exp(-\lambda t)$ where x could represent activity, number of undecayed particles or received count-rate
- solve problems using the relation $\lambda = 0.693/t_{\frac{1}{2}}$

We now look at some of the consequences of the random nature of radioactive decay. If six dice are thrown simultaneously (Figure 13.8), it is likely that one of them will show a six. If twelve dice are thrown, it is likely that two of them will show a six, and so on. While it is possible to predict the likely number of sixes that will be thrown, it is impossible to say which of the dice will actually show a six. We describe this situation by saying that the throwing of a six is a **random process**.

Figure 13.8 The dice experiment

In an experiment similar to the one just described, some students throw a large number of dice (say 6000). Each time a six is thrown, that die is removed. The results for the number of dice remaining after each throw are shown in Table 13.4.

Table 13.4 Results of dice-throwing experiment

number of throws	number of dice remaining	number of dice removed
0	6000	
1	5000	1000
2	4173	827
3	3477	696
4	2897	580
5	2414	483
6	2012	402
7	1677	335
8	1397	280

Figure 13.9 is a graph of the number of dice remaining against the number of throws. This kind of graph is called a **decay curve**. The rate at which dice are removed is not linear, but there is a pattern. After between 3 and 4 throws, the number of dice remaining has halved. Reading values from the graph shows that approximately 3.8 throws would be required to halve the number of dice. After another 3.8 throws the number has halved again, and so on.

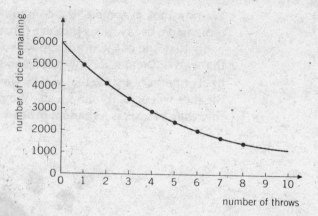

Figure 13.9

We have met this type of curve before, in talking about the decay of current in a circuit containing a capacitor and a resistor (Chapter 11).

We can apply the dice experiment to model radioactive decay. The 6000 dice represent radioactive nuclei. To score a six represents radioactive emission. All dice scoring six are removed, because once a nucleus has undergone radioactive decay, it is no longer available for further decay. Thus, we can describe how rapidly a sample of radioactive material will decay.

A graph of the number of undecayed nuclei in a sample against time has the typical decay curve shape shown in Figure 13.10. It is not possible to state how long the entire sample will take to decay (its 'life'). However, after 3 minutes the number of undecayed nuclei in the sample has halved. After a further 3 minutes, the number of undecayed atoms has halved again. We describe this situation by saying that this radioactive nuclide has a **half-life** $t_{\frac{1}{2}}$ of 3 minutes.

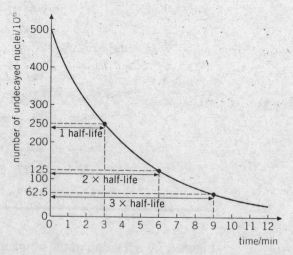

Figure 13.10 Radioactive decay curve

The half-life of a radioactive nuclide is the time taken for the number of undecayed nuclei to be reduced to half its original number.

The half-lives of different nuclides have a very wide range of values. Examples of some radioactive nuclides and their half-lives are given in Table 13.5.

Table 13.5 Examples of half-lives

radioactive nuclide	half-life
uranium-238	4.5×10^9 years
radium-226	1.6×10^3 years
radon-222	3.8 days
francium-221	4.8 minutes
astatine-217	0.03 seconds

We shall see later that half-life may also be expressed in terms of the activity of the material.

Note that half-life is similar to the time constant, which we met in Chapter 11. Both half-life and the time constant relate to the time taken for a certain fraction of the quantity concerned to decay. For the half-life, that fraction is 1/2. For the time constant, the fraction is 1/e, or about 1/2.7.

If you measure the count-rate from a nuclide with a very long half-life, you might expect to obtain a constant value. In fact, the reading of the ratemeter fluctuates about an average value. This demonstrates the random nature of radioactive decay.

Example

The half-life of francium-221 is 4.8 minutes. Calculate the fraction of a sample of francium-221 remaining undecayed after a time of 14.4 minutes.

The half-life of francium-221 is 4.8 min, so after 4.8 min half of the sample will remain undecayed. After two half-lives (9.6 min), $0.5 \times 0.5 = 0.25$ of the sample will remain undecayed. After three half-lives (14.4 min), $0.5 \times 0.25 = 0.125$ will remain undecayed. So the fraction remaining undecayed is **0.125** or **1/8**.

Now it's your turn

Using the half-life values given in Table 13.5, calculate:
(a) the fraction of a sample of uranium-238 remaining undecayed after 9.0×10^9 years,
(b) the fraction of a sample of astatine remaining undecayed after 0.30 s,
(c) the fraction of a sample of radium-226 that has decayed after 3200 years,
(d) the fraction of a sample of radon-222 that has decayed after 15.2 days.

13.6 Mathematical descriptions of radioactive decay

Activity and decay constant

As we saw in the dice experiment, increasing the number of dice increases the number of sixes that appear with each throw. Similarly, if we investigate the decay of a sample of radioactive material we find that the greater the number of radioactive nuclei in the sample the greater the rate of decay. This can be described mathematically as $dN/dt \propto N$. This gives

$$\frac{dN}{dt} = -\lambda N$$

N is the number of undecayed atoms in the sample. dN/dt is the rate at which the number of nuclei in the sample is changing, so $-dN/dt$ represents the rate of decay. $-dN/dt$ is known as the **activity** A of the source, and is measured in **becquerels**.

The activity of a radioactive source is the number of nuclear decays produced per unit time in the source. Activity is measured in becquerels (Bq), where 1 becquerel is 1 decay per second.

$$1 \, \text{Bq} = 1 \, \text{s}^{-1}$$

Combining $A = -dN/dt$ and $dN/dt = -\lambda N$, we have

$$A = \lambda N$$

λ is a constant of proportionality known as the **decay constant**. It has the units s^{-1}, yr^{-1} and so on.

λ is defined as the probability per unit time that a nucleus will undergo decay.

This is an important equation because it relates a quantity we can measure ($-dN/dt$, the rate of decay or activity) to a quantity which cannot in practice be determined (N, the number of undecayed nuclei). We shall see later that the decay constant λ is directly related to the half-life $t_{\frac{1}{2}}$, which can be obtained by experiment. This opens the way to calculating the number of undecayed nuclei in a sample. Trying to count nuclei when they are decaying is similar to counting sheep in a field while some are escaping through a gap in the hedge!

Example

Calculate the number of phosphorus-32 nuclei in a sample which has an activity of 5.0×10^6 Bq.
(decay constant of phosphorus-32 = $5.6 \times 10^{-7} s^{-1}$)

From $dN/dt = -\lambda N$, $N = (-dN/dt)/\lambda = -5.0 \times 10^6/5.6 \times 10^{-7}$
$= -8.9 \times 10^{12}$. The minus sign in this answer arises because dN/dt is the rate of decay. The quantity measured by a ratemeter is the rate of decay, and so should be negative, but it is always displayed as a positive quantity. Similarly, activities are always quoted as positive. So don't be worried about discarding the minus sign here! The number of phosphorus-32 nuclei = $\mathbf{8.9 \times 10^{12}}$.

Note: In this type of calculation, because the activity is measured in becquerel (s^{-1}), the decay constant λ must be measured in consistent units. If λ had been quoted as 4.2 day^{-1}, it would have been necessary to convert to s^{-1}.

Now it's your turn

Calculate the activity of the following samples of radioactive materials:
(a) 6.7×10^{21} atoms of strontium-90,
  (decay constant of strontium-90 = 8.3×10^{-10} s^{-1})
(b) 2.0 mg of uranium-238. 0.238 kg of uranium-238 contains 6.0×10^{23} atoms.
  (decay constant of uranium-238 = $5.0 \times 10^{-13} s^{-1}$)

To solve the equation $dN/dt = -\lambda N$ requires mathematics beyond the scope of A/AS Physics. However, it is important to know the solution, in order to find the variation with time of the number of nuclei remaining in the sample. The solution is

$$N = N_0 e^{-\lambda t} \quad \text{or} \quad N = N_0 \exp(-\lambda t)$$

where N_0 is the initial number of undecayed nuclei in the sample, and N is the number of undecayed nuclei at time t. This equation is of exactly the same form as the one used to describe the decay of current in a circuit containing a capacitor and a resistor. Note that the decay constant λ is analogous to the reciprocal of the time constant, that is to $1/CR$.

The equation represents an **exponential** decay. The decay curve of N against t is as shown in Figure 13.10. Since activity A is proportional to N ($A = \lambda N$), the curve of A against t is the same shape, and we can write

$$A = A_0 e^{-\lambda t}$$

Example

A sample of phosphorus-32 contains 8.6×10^{12} nuclei at time $t = 0$. The decay constant of phosphorus-32 is 4.8×10^{-2} day^{-1}. Calculate the number of undecayed phosphorus-32 nuclei in the sample after 10 days.

From $N = N_0\, e^{-\lambda t}$, we have $N = 8.6 \times 10^{12} \times e^{-0.048 \times 10}$, so $N = \mathbf{5.3 \times 10^{12}}$. (Again, it is important to measure λ and t in consistent units. Here λ is in day^{-1} and t is in days, so there is no problem.)

Now it's your turn

A sample of phosphorus-32 contains 8.6×10^{12} nuclei at time $t = 0$. The decay constant of phosphorus-32 is 4.8×10^{-2} day^{-1}. Calculate the number of undecayed phosphorus-32 nuclei in the sample after:
(a) 20 days,
(b) 40 days.

Decay constant and half-life

Using the equation $N = N_0\, e^{-\lambda t}$, we can derive an equation which relates the half-life to the decay constant. For any radioactive nuclide, the number of undecayed nuclei after one half-life is, by the definition of half-life, equal to $N_0/2$, where N_0 is the original number of undecayed nuclei. Using the radioactive decay equation

$$N = N_0\, e^{-\lambda t}$$

we have, at time $t = t_{\frac{1}{2}}$

$$N = N_0\, e^{-\lambda t_{\frac{1}{2}}}$$

and, dividing each side of the equation by N_0,

$$\frac{N}{N_0} = \frac{1}{2} = e^{-\lambda t_{\frac{1}{2}}}$$

or

$$2 = e^{\lambda t_{\frac{1}{2}}}$$

Taking natural logarithms of both sides,

$$\ln 2 = \lambda t_{\frac{1}{2}}$$

So that

$$t_{\frac{1}{2}} = \frac{\ln 2}{\lambda}$$

or

$$t_{\frac{1}{2}} = \frac{0.693}{\lambda}$$

Example

Calculate the half-life of radium-226, which has a decay constant of $1.42 \times 10^{-11}\,\text{s}^{-1}$.

Using $t_{\frac{1}{2}} = 0.693/\lambda$, we have $t_{\frac{1}{2}} = 0.693/1.42 \times 10^{-11} = \mathbf{4.88 \times 10^{10}\,\text{s}}$.

Now it's your turn

1 Calculate the half-lives of the following radioactive nuclides:
  (a) bismuth-214, which has a decay constant of $4.3 \times 10^3\,\text{s}^{-1}$,
  (b) carbon-14, which has a decay constant of $4.1 \times 10^{-12}\,\text{s}^{-1}$.

2 Calculate the decay constants of the following radioactive nuclides:
  (a) helium-5, which has a half-life of $6.0 \times 10^{-20}\,\text{s}$,
  (b) sodium-24, which has a half-life of 15 h.

Sections 13.5–13.6 Summary

- The half-life $t_{\frac{1}{2}}$ of a radioactive nuclide is the time taken for the number of undecayed nuclei to be reduced to half the original number.
- The activity of a radioactive source is the number of decays per unit time. The unit of activity is the becquerel (Bq); 1 becquerel = 1 decay per second.
- The activity dN/dt of a source is related to the number N of undecayed nuclei by the equation:

  $$dN/dt = -\lambda N$$

  where λ is the decay constant.
- The decay constant is defined as the probability of decay per unit time of a nucleus.
- The number N of undecayed nuclei in a radioactive sample at time t is given by the equation:

  $$N = N_0\,e^{-\lambda t}$$

  where N_0 is the number of undecayed nuclei at time $t = 0$.
- The half-life $t_{\frac{1}{2}}$ and the decay constant λ are related by the equation:

  $$t_{\frac{1}{2}} = 0.693/\lambda$$

Sections 13.5–13.6 Questions

1 The table below shows the variation with time t of the activity of a sample of a radioactive nuclide X. The average background count during the experiment was 36 min^{-1}.

(a) Plot a graph to show the variation with time of the corrected count-rate.
(b) Use the graph to determine the half-life of the nuclide X.

t/hour	0	1	2	3	4	5	6	7	8	9	10
activity/min^{-1}	854	752	688	576	544	486	448	396	362	334	284

2 A radioactive source with a half-life of 10 days has an initial activity of 2.0×10^{10} Bq. Calculate the activity of this source after 30 days.

3 The half-lives of two radioactive nuclides X and Y are 40 s and 70 s respectively. Initially, samples of each contain 2.0×10^7 undecayed nuclei.
(a) Calculate the initial activity of each sample.
(b) Draw, on the same axes, two lines to represent the decay of these samples.

4 Calculate the mass of caesium-137 that has an activity of 4.0×10^5 Bq. 0.137 kg of caesium-137 contains 6.0×10^{23} atoms. The half-life of caesium-137 is 30 years.

5 (a) The activity of a radioactive source X falls from 5.0×10^{10} Bq to 1.0×10^{10} Bq in 5.0 hours. Calculate the half-life.
(b) The activity of a certain mass of carbon-14 is 5.00×10^9 Bq. The half-life of carbon-14 is 5570 years. Calculate the number of carbon-14 nuclei in the sample.

13.7 Mass defect

At the end of Sections 13.7–13.11 you should be able to:
- show an appreciation of the association between energy and mass as represented by $E = mc^2$ and recall and solve problems using this relationship
- sketch the variation of binding energy per nucleon with nucleon number
- explain the relevance of binding energy per nucleon to nuclear fusion and to nuclear fission.

At a nuclear level, the masses we deal with are so small that it would be very clumsy to measure them in kilograms. Instead, we measure the masses of nuclei and nucleons in **atomic mass units (u)**.

One atomic mass unit (1 u) is defined as being equal to one-twelfth of the mass of a carbon-12 atom. 1 u is equal to 1.66×10^{27} kg.

Using this scale of measurement, to six decimal places, we have

$$\text{proton mass } m_p = 1.007276 \, \text{u}$$

$$\text{neutron mass } m_n = 1.008665 \, \text{u}$$

$$\text{electron mass } m_e = 0.000549 \, \text{u}$$

Because all atoms and nuclei are made up of protons, neutrons and electrons, we should be able to use these figures to calculate the mass of any atom or nucleus. For example, the mass of a helium-4 nucleus, consisting of two protons and two neutrons, should be

$(2 \times 1.007276) + (2 \times 1.008665) = 4.031882\,u$

However, the actual mass of a helium nucleus is 4.001508 u.

The difference between the expected mass and the actual mass of a nucleus is called the **mass defect** of the nucleus. In the case of the helium-4 nucleus, the mass defect is 4.031882 – 4.001508 = 0.030374 u.

The mass defect of a nucleus is the difference between the total mass of the separate nucleons and the combined mass of the nucleus.

Example

Calculate the mass defect for a carbon-14 ($^{14}_{6}C$) nucleus. The measured mass is 14.003240 u.

The nucleus contains 6 protons and 8 neutrons, of total mass $(6 \times 1.007276) + (8 \times 1.008665) = 14.112976\,u$. The mass defect is 14.112976 – 14.003240 = **0.109736 u**.

Now it's your turn

Calculate the mass defect for a nitrogen-14 ($^{14}_{6}N$) nucleus. The measured mass is 14.003070 u.

13.8 Mass–energy equivalence

In 1905, Albert Einstein proposed that there is an equivalence between mass and energy. The relationship between energy E and mass m is

$E = mc^2$

where c is the speed of light. E is measured in joules, m in kilograms and c in metres per second.

Using this relation, we can calculate that 1.0 kg of matter is equivalent to $1.0 \times (3.0 \times 10^8)^2 = 9.0 \times 10^{16}\,J$.

The mass defect of the helium nucleus, calculated previously as 0.030374 u, is equivalent to $0.030374 \times 1.66 \times 10^{-27} \times (3.00 \times 10^8)^2 = 4.54 \times 10^{-12}\,J$. (Note that the mass in u must be converted to kg by multiplying by 1.66×10^{-27}.)

The joule is an inconveniently large unit to use for nuclear calculations. A more convenient energy unit is the mega electronvolt (MeV), as many energy changes that take place in the nucleus are of the order of several MeV. One mega electronvolt is the energy gained by one electron when it is accelerated through a potential difference of one million volts.

Since *electrical energy = charge × potential difference*, and the electron charge is 1.60×10^{-19} C,

$$1\,\text{MeV} = 1.60 \times 10^{-19} \times 1.00 \times 10^6$$

or

$$1\,\text{MeV} = 1.60 \times 10^{-13}\,\text{J}$$

The energy equivalent of the mass defect of the helium nucleus is thus $4.54 \times 10^{-12}/1.60 \times 10^{-13} = 28.4\,\text{MeV}$.

If mass is measured in u and energy in MeV,

1 u is the equivalent of 931 MeV

13.9 Binding energy

Within the nucleus there are strong forces which bind the protons and neutrons together. To completely separate all these nucleons requires energy. This energy is referred to as the **binding energy** of the nucleus. Stable nuclei, those which have little or no tendency to disintegrate, have large binding energies. Less stable nuclei have smaller binding energies.

Similarly, when protons and neutrons are joined together to form a nucleus, this binding energy must be released. The binding energy is the energy equivalent of the mass defect.

We have seen that the binding energy of the helium-4 nucleus is 28.4 MeV. 28.4 MeV of energy is required to separate, to infinity, the two protons and two neutrons of this nucleus.

Binding energy is the energy equivalent of the mass defect of a nucleus. It is the energy required to separate to infinity all the nucleons of a nucleus.

Example

Calculate the binding energy, in MeV, of a carbon-14 nucleus with a mass defect of 0.109736 u.

Using the equivalence 1 u = 931 MeV, 0.109736 u is equivalent to 102 MeV. Since the binding energy is the energy equivalent of the mass defect, the binding energy = **102 MeV**.

Now it's your turn

Calculate the binding energy, in MeV, of a nitrogen-14 nucleus with a mass defect of 0.108517 u.

Stability of nuclei

A stable nucleus is one which has a very low probability of decay. Less stable nuclei are more likely to disintegrate. A useful measure of stability is the **binding energy per nucleon** of the nucleons in the nucleus.

> Binding energy per nucleon is defined as the total energy needed to completely separate all the nucleons in a nucleus divided by the number of nucleons in the nucleus.

Figure 13.11 shows the variation with nucleon number of the binding energy per nucleon for different nuclides. The most stable nuclides are those with the highest binding energy per nucleon; that is, those near point B on the graph. Iron is one of these stable nuclides. Typically, very stable nuclides have binding energies per nucleon of about 8 MeV.

Light nuclei, between A and B on the graph, may combine or fuse to form larger nuclei with larger binding energies per nucleon. This process is called **nuclear fusion**. For the process to take place, conditions of very high temperature and pressure are required, such as in stars like the Sun.

Heavy nuclei, between B and C on the graph, when bombarded with neutrons, may break into two smaller nuclei, again with larger binding energy per nucleon values. This process is called **nuclear fission**.

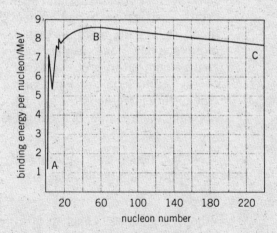

Figure 13.11 Binding energy per nucleon, against nucleon number

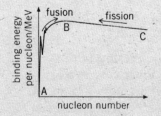

Figure 13.12

Figure 13.12 highlights nuclides which may undergo fusion (blue part of curve) or fission (red part of curve) in order to increase their binding energy per nucleon.

When nuclear fusion or fission takes place, the nucleon numbers of the nuclei involved change. A higher binding energy per nucleon is achieved, and this is accompanied by a *release* of energy. This release of energy during fission reactions is how the present generation of nuclear power stations produce electricity.

Example

The binding energy of a helium-4 nucleus is 28.4 MeV. Calculate the binding energy per nucleon.

The helium-4 nucleus has 4 nucleons. The binding energy per nucleon is thus 28.4/4 = **7.1 MeV per nucleon**.

Now it's your turn

The binding energy of a carbon-14 nucleus is 102 MeV. Calculate the binding energy per nucleon.

13.10 Nuclear fission

Within the nucleus of an atom the nucleons experience both attractive and repulsive forces. The attractive force is called the **strong nuclear force**. This acts like a 'nuclear glue' to hold the nucleons together. The repulsive forces are the electric (Coulomb-law) forces between the positively charged protons. Gravitational forces of attraction exist, but are negligible in comparison to the other forces. Stable nuclei have much larger attractive forces than repulsive forces. Stable nuclides generally have approximately the same number of neutrons and protons in the nucleus; that is, the neutron-to-proton ratio is close to one. In heavy nuclei such as uranium and plutonium there are far more neutrons than protons, giving a neutron-to-proton ratio of more than one. For example, uranium-235 has 92 protons and 143 neutrons, giving a neutron-to-proton ratio of 1.55. This leads to a much lower binding energy per nucleon compared with iron, and such nuclides are less stable. Any further increase in the number of neutrons in such nuclei is likely to cause the nucleus to undergo **nuclear fission**.

> Nuclear fission is the splitting of a heavy nucleus into two lighter nuclei of approximately the same mass.

When a uranium-235 nucleus absorbs a neutron, it becomes unstable and splits into two lighter, more stable nuclei. The most likely nuclear reaction is

$$^{235}_{92}\text{U} + {}^{1}_{0}\text{n} \rightarrow {}^{236}_{92}\text{U} \rightarrow {}^{141}_{56}\text{Ba} + {}^{92}_{36}\text{Kr} + 3{}^{1}_{0}\text{n} + \text{energy}$$

This process is called *induced* nuclear fission, because it is started by the capture of a neutron by the uranium nucleus.

Each of the fission reactions described by this equation results in the release of three neutrons. Other possible fission reactions release two or three neutrons. If these neutrons are absorbed by other uranium-235 nuclei, these too may become unstable and undergo fission, thereby releasing even more neutrons. The reaction is described as being a **chain reaction** which is accelerating. This is illustrated in Figure 13.13. If this type of reaction continues uncontrolled, a great deal of energy is released in a short time, and a nuclear explosion results.

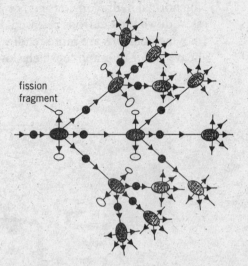

fission fragment

Figure 13.13 Accelerating chain reaction

If the number of neutrons which take part in the chain reaction is controlled so that the number of fissions per unit time is constant, rather than increasing, the rate of release of energy can be controlled. This situation is illustrated in Figure 13.14. These conditions apply in the reactor of a modern nuclear power station, where some of the neutrons released in fission reactions are absorbed by control rods in order to limit the rate of fission reactions.

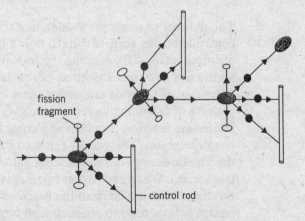

fission fragment

control rod

Figure 13.14 Controlled chain reaction. Two of the neutrons produced by fission are absorbed by control rods. The third neutron induces further nuclear fission.

Nuclear fission reactor

There are several possible fission reactions which may occur in a nuclear reactor. Most are represented by the general equation

$$^{235}_{92}\text{U} + {}^{1}_{0}\text{n} \rightarrow {}^{236}_{92}\text{U} \rightarrow 2 \text{ new nuclides} + 2 \text{ or } 3 \text{ neutrons} + \text{energy}$$

The nuclides formed in this reaction are called **fission fragments**.

Figure 13.15 illustrates a pressurised water-cooled reactor (PWR). Inside the reactor, the uranium-235 is placed in long fuel rods which are surrounded by heavy water (water in which deuterium atoms, hydrogen-2, take the place of the normal hydrogen-1 atoms) or graphite. The role of these materials is to slow down the neutrons released during fission. This is important because slow-moving neutrons are more readily absorbed by uranium-235. Materials such as heavy water or graphite which slow down the neutrons are called **moderators**.

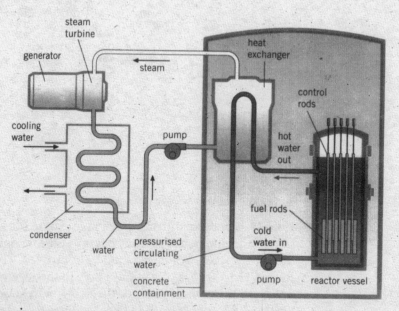

Figure 13.15 Pressurised water-cooled reactor

The number of neutrons available to trigger further fission reactions is controlled using **control rods** of boron or cadmium. These substances readily absorb neutrons. By pushing the rods further into the reactor core, the number of neutrons absorbed increases and the rate of fission reactions decreases. If the rods are pulled out a little, fewer neutrons are absorbed and the rate of fissioning increases. Ideally, the rods should be adjusted so that the fission reaction proceeds at a constant rate; that is, it neither accelerates nor decelerates. The energy released during fission reactions is seen mainly as the kinetic energy of the resulting lighter nuclei and neutrons (the fission fragments). When the fission fragments are slowed down, the heat energy produced is removed from the reactor by a circulating coolant. This heat energy is then used to produce high temperature steam in a heat exchanger. The steam is used to drive turbine generators.

There are two main types of nuclear reactor used for the commercial generation of electricity. These are the pressurised water-cooled reactor (PWR) and the advanced gas-cooled reactor (AGR).

In the PWR, water circulates through the reactor vessel at a high pressure, typically 150 atmospheres. At such pressure the water does not boil, even though its temperature may be as high as 300 °C. This super-heated water passes into a heat exchanger through a series of pipes surrounded by water at much lower pressure. The low-pressure water absorbs heat energy and changes into steam, which then drives the turbines and electricity generators.

In AGR reactors, helium gas at a temperature of about 600 °C is the coolant. Again, energy is transferred from the core to water to a heat exchanger in order to produce the steam to drive the turbines.

13.11 Nuclear fusion

Most of the energy on Earth comes from the Sun, where it is produced by nuclear fusion reactions. Light nuclei, such as isotopes of hydrogen, join together to produce heavier, more stable nuclei, and in doing so release energy. One of these fusion reactions is

$$^2_1H + {}^2_1H \rightarrow {}^3_2He + {}^1_0n + \text{energy}$$

From the binding energy per nucleon curve (Figure 13.11) we see that the binding energy per nucleon for light nuclei, such as hydrogen, is low. But if two light nuclei are made to fuse together, they may form a new heavier nucleus which has a higher binding energy per nucleon. It will be more stable than the two lighter nuclei from which it was formed. Because of this difference in stability, a fusion reaction such as this will release energy.

Although fusion reactions are the source of solar energy, we are at present unable to duplicate this reaction in a controlled manner on Earth. This is because the nuclei involved in fusion have to be brought very close together, and conditions of extremely high temperature and pressure, similar to those found at the centre of the Sun, are required. Reactions requiring these conditions are called **thermonuclear reactions**. Some fusion reactions involving hydrogen isotopes have been made to work in the Joint European Torus (JET), although not yet in a controlled, sustainable manner. In 2006, an international consortium agreed to undertake the ITER (International Tokamak Engineering Research) project, which is designed to produce up to 500 MW of fusion power sustained for over 400 seconds by the fusion of a ^2H–^3H mixture. Construction on a site in southern France will take 10 years.

Figure 13.16 Most of our energy on Earth comes from the Sun. It is produced by nuclear fusion reactions. Light nuclei such as isotopes of hydrogen join together to produce heavier, more stable nuclei. This photo shows eruptions from the Sun's surface caused by particularly energetic fusion reactions.

Sections 13.7–13.11 Summary

- The mass defect of a nucleus is the difference between the total mass of the separate nucleons and the mass of the nucleus.
- Einstein's mass–energy equivalence relation: $E = mc^2$
- The binding energy of a nucleus is the energy needed to separate completely all its constituent nucleons.
- The binding energy per nucleon is a measure of the stability of a nucleus: a high binding energy per nucleon means the nucleus is stable.
- Nuclear fission is the splitting of a heavy nucleus into two smaller, lighter nuclei of approximately equal mass.
- Nuclear fusion is the joining together of light nuclei to form a larger, heavier nucleus.

Sections 13.7–13.11 Questions

1 Calculate the mass defect, the binding energy of the nucleus, and the binding energy per nucleon for the following nuclei:
  (a) hydrogen-3, $^3_1$H; nuclear mass 3.01605 u,
  (b) zirconium-97, $^{97}_{40}$Zr; nuclear mass 97.09801 u,
  (c) radon-222, $^{222}_{86}$Rn; nuclear mass 222.01754 u.
  (proton mass = 1.00728 u; neutron mass = 1.00867 u)

2 One possible reaction taking place in the core of a reactor is

$$^{235}_{92}U + {}^1_0n \rightarrow {}^{95}_{42}Mo + {}^{139}_{57}La + 2{}^1_0n + 7{}^{\,0}_{-1}e$$

For this reaction, calculate:
  (a) the mass on each side of the equation,
  (b) the change in mass after fission has taken place,
  (c) the energy released per fission of uranium-235,
  (d) the energy available from the complete fission of 1.00 g of uranium-235,
  (e) the mass of uranium-235 used by a 500 MW nuclear power station in one hour, assuming 30% efficiency.
  (masses: $^{235}_{92}$U, 235.123 u; $^{95}_{42}$Mo, 94.945 u; $^{139}_{57}$La, 138.955 u; proton, 1.007 u; neutron, 1.009 u. 0.235 kg of uranium-235 contains 6.0×10^{23} atoms.)

3 Two fusion reactions which take place in the Sun are described below.
  (a) A hydrogen-2 (deuterium) nucleus absorbs a proton to form a helium-3 nucleus.
  (b) Two helium-3 nuclei fuse to form a helium-4 nucleus plus two free protons.
  For each reaction, write down the appropriate nuclear equation and calculate the energy released.
  (masses: $^2_1$H, 2.01410 u; $^3_2$He, 3.01605 u; $^4_2$He, 4.00260 u; $^1_1$p, 1.00728 u; $^1_0$n, 1.00867u)

13.12 Probing matter

At the end of Section 13.12 you should be able to:
■ infer from the results of the α-particle scattering experiment the existence and small size of the nucleus.

Figure 13.17 Ion microscope photograph of iridium

Figure 13.17 shows a photograph taken with an ion microscope, a device which makes use of the de Broglie wavelength of gas ions (see Chapter 9). It shows a sample of iridium at a magnification of about five million. The positions of individual iridium atoms can be seen.

Photographs like this reinforce the idea that all matter is made of very small particles we call atoms. Advances in science at the end of the nineteenth and the beginning of the twentieth century led most physicists to believe that atoms themselves are made from even smaller particles, some of which have positive or negative charges. Unfortunately even the most powerful microscopes cannot show us the internal structure of the atom. Many theories were put forward about the structure of the atom, but it was a series of experiments carried out by Ernest Rutherford and his colleagues around 1910 that led to the birth of the model we now know as the **nuclear atom**.

Probing matter using α-particles

In 1911, Rutherford and two of his associates, Geiger and Marsden, fired a beam of α-particles at a very thin piece of gold foil. A zinc sulphide detector was moved around the foil to see the directions in which the α-particles travelled after striking the foil (Figure 13.18).

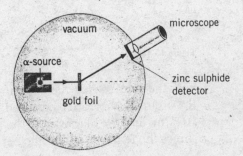

Figure 13.18 α-scattering experiment

They discovered that:

- the vast majority of the α-particles passed through the foil with very little or no deviation from their original path
- a small number of particles were deviated through an angle of more than about 10°
- an extremely small number of particles (one in ten thousand) were deflected through an angle greater than 90°.

From these observations, the following conclusions could be drawn.

- The majority of the mass of an atom is concentrated in a very small volume at the centre of the atom. Most α-particles would therefore pass through the foil undeviated.
- The centre (or nucleus) of an atom is positively charged. α-particles, which are also positively charged, passing close to the nucleus will experience a repulsive force causing them to deviate.
- Only α-particles that pass very close to the nucleus, almost striking it head-on, will experience large enough repulsive forces to cause them to deviate through angles greater than 90°. The fact that so few particles did so confirms that the nucleus is very small, and that most of the atom is empty space.
- Because atoms are neutral, the atom must contain negative particles. These travel around, or orbit, the nucleus.

Figure 13.19 shows some of the possible trajectories of the α-particles. Using the nuclear model of the atom and equations to describe the force between charged particles, Rutherford calculated the fraction of α-particles that he would expect to be deviated through various angles. The calculations agreed

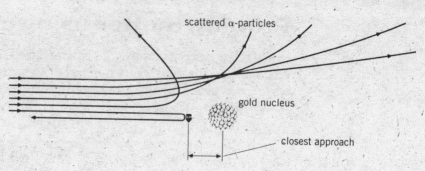

scattered α-particles

gold nucleus

closest approach

Figure 13.19

with the results from the experiment. This confirmed the nuclear model of the atom. Rutherford calculated that the diameter of the nucleus is about 10^{-15} m, and the diameter of the whole atom about 10^{-10} m. Figure 13.20 shows the features of the nuclear model of a nitrogen atom.

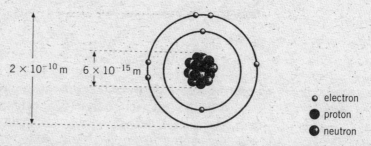

2×10^{-10} m 6×10^{-15} m

○ electron
● proton
◑ neutron

Figure 13.20 The diameter of a nitrogen atom is more than 30 000 times bigger than the diameter of its nucleus.

Some years later the α-particle scattering experiment was repeated using α-particles with higher energies. Some discrepancies between the experimental results and Rutherford's scattering formula were observed. These seemed to be occurring because the high-energy α-particles were passing very close to the nucleus, and were experiencing not only the repulsive electrostatic force but also a strong attractive force which appears to act over only a very short range. This became known as the **strong nuclear force**. This is the force that holds the nucleus together.

Probing matter using electrons

Electrons are not affected by the strong nuclear force. It was suggested that they might therefore be a more effective tool with which to investigate the structure of the atom. As we saw in Chapter 9, moving electrons have a wavelike property, and can be diffracted.

If a beam of electrons is directed at a sample of powdered crystal and the electron wavelength is comparable with the interatomic spacing in the crystal, the electron waves are scattered from planes of atoms in the tiny crystals, creating a diffraction pattern (Figure 13.21). The fact that a diffraction pattern is obtained confirms the regular arrangement of the atoms in a crystalline solid. Measurements of the angles at which strong scattering is obtained can be used to calculate the distances between planes of atoms.

Figure 13.21 Electron diffraction pattern of a sample of pure titanium

If the energy of the electron beam is increased, the wavelength decreases. Eventually the electron wavelength may be of the same order of magnitude as the diameter of the nucleus. Probing the nucleus with high-energy electrons rather than α-particles gives a further insight into the dimensions of the nucleus, and also gives information about the distribution of charge in the nucleus itself.

Section 13.12 Summary

- The Rutherford α-particle experiment confirmed the nuclear model of the atom: the atom consists of a small, heavy, positively-charged nucleus, surrounded by negatively-charged electrons in orbit about the nucleus.
- The diameter of the nucleus is about 10^{-15} m; the diameter of the atom is about 10^{-10} m.
- Electron diffraction gives evidence for the regular arrangement of atoms in crystals, and allows the measurement of the distance between planes of atoms in solids.

Exam-style Questions

1 Cobalt-60 is often used as a radioactive source in medical physics. It has a half-life of 5.25 years. Calculate how long after a new sample is delivered the activity will have decreased to one-eighth of its value when delivered.

2 Iodine-131 is a beta-emitter with half-life of 194 hours. It decays to xenon (Xe). The decay is represented by

$$^{131}_{53}I \rightarrow ^{P}_{Q}Xe + \beta$$

(a) Obtain the values of P and Q in the decay equation.
(b) A sample of iodine-131 of mass 0.20 g is prepared in a reactor.
   (i) Given that 0.131 kg of iodine-131 contains 6.02×10^{23} atoms, calculate the number of iodine-131 atoms contained in the sample at the time of preparation.
   (ii) Calculate the mass of iodine-131 which will have decayed 48 hours after preparation.

3 The half-life of a radioactive isotope of sodium used in medicine is 15 hours.
(a) Determine the decay constant for this nuclide.
(b) A small volume of a solution containing this nuclide has an activity of 1.2×10^4 disintegrations per minute when it is injected into the bloodstream of a patient. After 30 hours, the activity of 1.0 cm^3 of blood taken from the patient is 0.50 disintegrations per minute. Estimate the volume of blood in the patient. Assume that the solution is uniformly diluted in the blood, that it is not taken up by the body tissues, and that there is no loss by excretion.

4 A fusion reaction that takes place in the Sun is represented by the equation

$$^{2}_{1}H + ^{2}_{1}H \rightarrow ^{3}_{2}He + ^{1}_{0}n + energy$$

(a) Calculate the energy released in this reaction. (masses: $^{2}_{1}H$, 2.0141 u; $^{3}_{2}He$, 3.0160 u; $^{1}_{0}n$, 1.0087 u)
(b) Suggest why it is difficult to achieve controlled reactions of this type.

5 A stationary radium nucleus ($^{224}_{88}$Ra) of mass 224 u spontaneously emits an α-particle. The α-particle is emitted with an energy of 9.2×10^{-13} J, and the reaction gives rise to a nucleus of radon (Rn).

(a) Write down a nuclear equation to represent α-decay of the radium nucleus.

(b) Show that the speed with which the α-particle is ejected from the radium nucleus is 1.7×10^7 m s^{-1}.

(c) Calculate the speed of the radon nucleus on emission of the α-particle. Explain how the principle of conservation of momentum is applied in your calculation.

6 When an α-particle travels through air, it loses energy by ionisation of air molecules. For every air molecule ionised, approximately 5.6×10^{-18} J of energy is lost by the α-particle.

(a) Suggest a typical value for the range of an α-particle in air. Hence estimate the number of air molecules ionised per millimetre of the path of the α-particle, given that the α-particle has initial energy 9.2×10^{-13} J.

(b) It has been discovered that the number of ionisations per unit length of the path of an α-particle suddenly increases just before the α-particle stops. State, with a reason, the effect that this observation will have on your estimate.

14. Electronic sensors

Electronic sensors have become such everyday items that we hardly appreciate them – except when they fail to work! Some sensors are simple, such as those detecting temperature in a thermostat or detecting light levels in an automatic switch for a lamp. Other sensors may be more complex and used to control systems. For example, in the braking system of a car an electronic circuit may be used to sense whether a wheel is skidding. If it is, then the circuit controls the brakes not only to stop the skid but also to prevent further skidding.

An electronic sensor may be thought to consist of three parts: a sensing device, a processing unit and an output device, as illustrated in Figure 14.1.

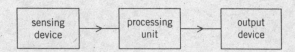

Figure 14.1 Block diagram for an electronic sensor

14.1 Sensing devices

At the end of Section 14.1 you should be able to:
- show an understanding that an electronic sensor consists of a sensing device and a circuit that provides an output voltage
- show an understanding of the change in resistance with light intensity of a light-dependent resistor (LDR)
- sketch the temperature characteristic of a negative temperature coefficient thermistor
- show an understanding of the action of a piezo-electric transducer and its application in a simple microphone
- describe the structure of a metal wire strain gauge
- relate extension of a strain gauge to change in resistance of the gauge.

The **sensing device** is the first stage of any electronic sensor and is the means by which whatever is to be detected or monitored is converted into an electrical property. This electrical property is frequently resistance. A physical property of the sensing device changes with a change in whatever it is monitoring.

Sensing devices are available that can be used to detect many different types of change – for example, changes in light intensity, temperature, sound level, humidity, pressure, strain, or magnetic field. Some of the more common sensing devices will be considered.

The light-dependent resistor (LDR)

a)

A light-dependent resistor (LDR) consists of two metal grids that intersect each other. The space between the grids is filled with a semiconductor material, for example, cadmium sulphide doped with copper.

When light is incident on the semiconductor material, the number of electrons in the semiconductor that are free to conduct increases. The higher the intensity of light on the LDR, the greater the number of electrons that can move freely. Hence, as the light intensity increases, the resistance of the LDR decreases. Figure 14.3 shows the variation with light intensity of the resistance of a typical LDR.

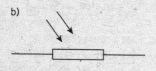

b)

Figure 14.2 An LDR and its symbol

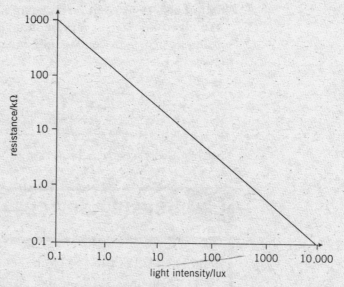

Figure 14.3 Resistance v. light intensity for an LDR

Note: Both light intensity, measured in lux, and resistance, measured in ohms, are plotted on logarithmic scales in Figure 14.3. The graph is a straight line *but* this does not mean that resistance is inversely proportional to intensity. Data relating to light intensity and resistance for a typical LDR are shown in Table 14.1.

Table 14.1 Typical LDR data

light level	illumination/lux	LDR resistance/Ω
moonlight	0.1	1×10^6
normal room lighting	450	900
sunlight	28000	100

The lux is a unit that is used to measure the light power incident per unit area of a surface.

The thermistor

The resistance of most metals increases to a certain extent with rise in temperature. Negative temperature coefficient devices, often referred to as **thermistors**, are made from semiconductor material, usually the oxides of metals. The resistance of thermistors decreases significantly with rise in temperature. Thermistors are manufactured in various shapes and sizes, including rods, discs and beads. Figure 14.4a shows an example of a disc and bead thermistor.

Data relating to the temperature and resistance of a typical bead thermistor are shown in Table 14.2. The variation with temperature of a typical thermistor is shown in Figure 14.5. This variation is non-linear and is, in fact, approximately exponential over a limited range of temperature.

a)

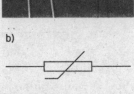

b)

Figure 14.4 Thermistors and their symbol

Table 14.2 Typical thermistor data

temperature/°C	thermistor resistance/Ω
1	3700
10	2500
20	1800
30	1300
40	900
50	660

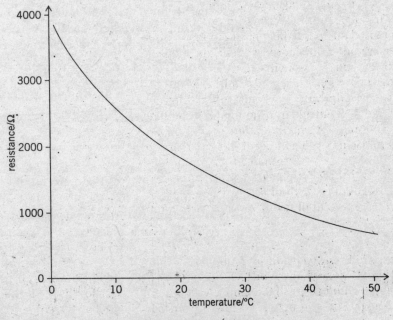

Figure 14.5 Resistance v. temperature for a thermistor

The piezo-electric transducer

A **transducer** is any device that converts energy from one form into another.

Piezo-electric crystals such as quartz have a complex ionic structure (see Section 15.2). In the normal unstressed state of the crystal, the centre of charge of the positive ions coincides with the centre of charge of the negative ions. However, when pressure is applied to the crystal, the crystal changes shape by a small amount and the centres of the positive and the negative charge no longer coincide. A voltage is generated across the crystal. This is known as the **piezo-electric effect**.

The magnitude of the voltage that is generated depends on the magnitude of the pressure on the crystal, and its polarity depends on whether the crystal is compressed or expanded – that is, whether the pressure applied is greater or less than the ambient pressure.

A sound wave consists of a series of compressions and rarefactions. If a sound wave is incident on a piezo-electric crystal, then a voltage will be produced across the crystal that varies in a similar way to the variation in pressure of the sound wave. To detect the voltages, opposite faces of the crystal are coated with a metal and electrical connections are made to these metal films. Since the voltages generated are small, they are amplified. The crystal and its amplifier may be used as a simple microphone for converting sound signals into electrical signals. The symbol for a microphone is shown in Figure 14.6.

Figure 14.6 Symbol for a microphone

The metal wire strain gauge

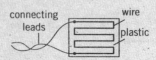

Figure 14.7 Metal wire strain gauge

In engineering, it is frequently necessary to test for the strains experienced in different parts of structures. This can be achieved using a metal wire **strain gauge**. A strain gauge consists of a length of very fine metal wire sealed into a small rectangle of thin plastic, as shown in Figure 14.7.

In use, the strain gauge is attached firmly to the structure to be monitored. When the plastic is stretched, the fine wire will also be stretched. The wire is strained, causing its length to increase and its cross-sectional area to decrease. An increase in length and a decrease in cross-sectional area both result in an increase in the electrical resistance of the wire.

The resistance R of a uniform wire is related to its length L and area of cross-section A by the expression

$$R = \frac{\rho L}{A}$$

where ρ is the resistivity of the material of the wire (see page 173).

If the wire increases in length by only a small amount ΔL, then the change in the cross-sectional area may be assumed to be negligible. The new resistance of the wire is given by

$$(R + \Delta R) = \frac{\rho(L + \Delta L)}{A}$$

By subtracting these two expressions, the change in resistance ΔR is given by

$$\Delta R = \frac{\rho \Delta L}{A}$$

Since it is assumed that A is constant, and ρ is also a constant, then

$$\Delta R \propto \Delta L$$

Since the strain is proportional to the extension ΔL, then the strain is also proportional to the change in resistance.

Examples

1 Explain what is meant by a negative temperature coefficient thermistor.

It is an electrical device whose resistance decreases as its temperature rises.

2 Suggest why the resistance of a strain gauge changes when the object on to which it is fixed is strained.

The strain gauge consists of a grid of fine metal wire in a plastic envelope. When the envelope is stretched, the length of the wire increases and its diameter decreases (slightly). Increase in length and decrease in diameter both cause the resistance to increase.

Now it's your turn

1 Draw a sketch graph to show the variation with temperature θ of the resistance R of a thermistor. Mark typical values on the axes of your graph.

2 (a) State what is meant by the piezo-electric effect.
   (b) Describe how a quartz crystal may be used to provide an electrical signal that is dependent on the sound pressure incident on it.

Section 14.1 Summary

- Sensing devices are used to detect changes in the environment or the properties of a body.
- In general, the change being monitored causes a change in an electrical property of the sensor.
- Frequently used sensors include the light-dependent resistor (LDR), the thermistor, the piezo-electric transducer and the strain gauge.
- The resistance of an LDR decreases as the light intensity on it increases.
- The resistance of a negative temperature coefficient thermistor decreases with increasing temperature.
- A piezo-electric transducer converts variations in pressure (such as due to a sound wave) into variations in voltage.
- The strain on a metal wire strain gauge is proportional to its change in resistance.

Section 14.1 Questions

1 Suggest suitable sensing devices to detect changes in
   (a) sunlight,
   (b) the width of a crack in a wall,
   (c) the weight of water in a tank,
   (d) the temperature of the oil in an engine.

2 Describe how a sensor may be used to count the number of mice that pass through a small hole in a wall.

14.2 Processing units

At the end of Section 14.2 you should be able to:
- show an understanding that the output from sensing devices can be registered as a voltage
- recall the main properties of an ideal operational amplifier (op-amp)
- deduce, from the properties of an ideal operational amplifier, the use of an operational amplifier as a comparator
- show an understanding of the effects of negative feedback on the gain of an operational amplifier
- recall the circuit diagrams for both the inverting and the non-inverting amplifier for single signal input
- show an understanding of the virtual earth approximation and derive an expression for the gain of inverting amplifiers
- recall and use expressions for the voltage gain of inverting and non-inverting amplifiers.

The change in the physical property of a sensing device must, in general, be processed in some way before the change can be displayed or measured. This processing is carried out by the **processing unit**, which is some form of electrical circuit that is connected to the sensing device and provides a voltage at its output. If the voltage is small, then it may be necessary to amplify it. The voltage can then be used to control an output device.

The potential divider

A potential divider circuit (see page 187) enables a change in resistance of a sensing device to be converted into a voltage change. In Figure 14.8 the sensing device of resistance S is connected in series with a resistor of resistance R.

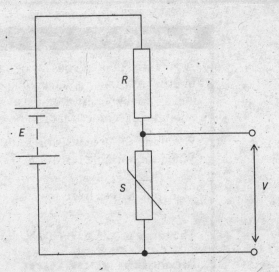

Figure 14.8 Potential divider circuit

The battery of e.m.f. E is assumed to have negligible internal resistance. The output voltage V is given by the expression

$$V = \frac{S}{S + R} \times E$$

The magnitude V of the potential difference (voltage) at any particular value S of the sensing device is dependent on the relative values of S and R. Note that, as the resistance S of the sensor increases, the output voltage V also increases. By monitoring the potential difference (voltage) across the resistor of fixed resistance R, as shown in Figure 14.9, the output voltage will be given by

$$V = \frac{R}{S + R} \times E$$

The output voltage V will then decrease as the sensor resistance S increases.

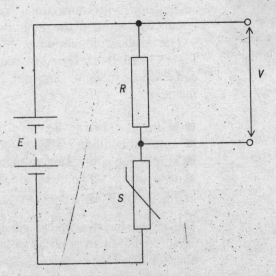

Figure 14.9 Alternative connections for a potential divider

Example

A potential divider consists of a battery of e.m.f. 6.00 V and negligible internal resistance connected in series with a resistor of resistance 120 Ω and a variable resistor of resistance $0 \rightarrow 200$ Ω. Determine the range of potential difference that can be obtained across the fixed resistor.

When the variable resistor is at 0 Ω, p.d. across fixed resistor = 6.0 V
When the variable resistor is at 200 Ω,

$$\text{p.d. across fixed resistor} = \frac{120}{120 + 200} \times 6.00 = 2.25 \text{ V}$$

The range is **2.25 V → 6.00 V**.

Now it's your turn

A potential divider consists of a battery of e.m.f. 7.5 V and negligible internal resistance connected in series with a resistor of resistance R and a variable resistor of resistance $0 \rightarrow 500$ Ω. Deduce how the potential divider may be arranged so as to provide a potential difference that may be varied between 0 and 3.0 V.

The operational amplifier (op-amp)

Once the output of the sensing device is in the form of a voltage, it may require further change (processing) before it is used to control an output device. The basis of many circuits used for processing the sensor voltage is the **operational amplifier** or **op-amp**.

An op-amp is an integrated circuit that contains about 20 transistors together with resistors and capacitors. It is referred to as an 'integrated circuit' because all the components are formed on a small slice of a semiconductor (e.g. silicon) with connections to enable the op-amp to be connected into a circuit. The whole of the integrated circuit is encapsulated.

The symbol for an op-amp and some of its connections are shown in Figure 14.10. It is referred to as an 'operational amplifier' because the circuit can easily be made to carry out different operations. These operations include:

- acting as a switch when a voltage reaches a certain level
- amplifying direct voltages
- amplifying alternating voltages
- comparing two voltages and giving an output that depends on the result of the comparison.

The op-amp has two inputs, called the inverting input (−) and the non-inverting input (+). For many applications, the positive and the negative power supplies are ±6 V or ±9 V.

a)

b)

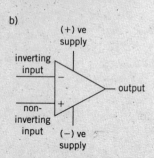

Figure 14.10 An op-amp and its symbol

Properties of an ideal op-amp

The ideal operational amplifier (op-amp) has the following properties:

- **infinite input impedance** (no current enters or leaves either of the inputs). Think about connecting an input to one of the resistors in a potential divider. The input would be in parallel with the resistor and therefore the resistance in the potential divider circuit would be changed. Connecting the op-amp to the potential divider would thus change the potential differences across the components in the potential divider. However, if the op-amp connection (the input) has infinite resistance (or, more strictly, for alternating voltages, infinite impedance) then connecting it to the potential divider will not affect the potential differences.
- **zero output impedance** (the whole of the output voltage is seen across the load connected to the output). If the output connection had some resistance, then the voltage produced in the op-amp (the output voltage) would be divided between the output and the resistor connected to the output. Zero output resistance (or zero output impedance) means that all of the output voltage will be seen across the resistor connected to the output.
- **infinite open-loop gain** (this means that when there is only a very small input voltage, the amplifier will **saturate** and the output will have the same value as the supply voltage). The voltage gain, or simply **gain** of an amplifier is a measure of how many times the output voltage is greater than the input voltage.

$$\text{voltage gain} = \frac{\text{output voltage}}{\text{input voltage}}$$

  When there are no components connected between the output and the input of the op-amp, the gain that is measured is said to be the **open-loop gain**. The output of the op-amp cannot be greater than the supply voltage (from energy conservation) and so, if a *very* small voltage is applied to the input and the gain is infinite, then the output will be at the supply voltage. The output cannot be any greater, even for a larger input signal. The amplifier is said to be saturated.
- **infinite bandwidth** (all frequencies are amplified equally). If an alternating voltage is applied to the input, then the output will have the same frequency but a larger amplitude. The range of frequencies that are amplified by the same amount (the input signals of different frequencies that all have the same gain) is known as the bandwidth.
- **infinite slew rate**. When the input signal is changed, then the output signal will also change. The slew rate is a measure of the time delay between the changes to the input and output. A high slew rate implies a short time delay. With an infinite slew rate there is no delay.

In reality, op-amps are not ideal. The input impedance is in the range 10^6–$10^{12}\ \Omega$ and the output impedance is about $10^2\ \Omega$. The open-loop gain is not constant but is the order of 10^5 for direct voltages. There is a finite bandwidth and the slew rate is about $10\ \text{V}\,\mu\text{s}^{-1}$.

The operational amplifier as a comparator

When an op-amp is used (incorporated) in an electrical circuit, it is usually connected to a dual power supply so that the output voltage can be either positive or negative. Such a power supply can be represented as two sets of batteries, as shown in Figure 14.11.

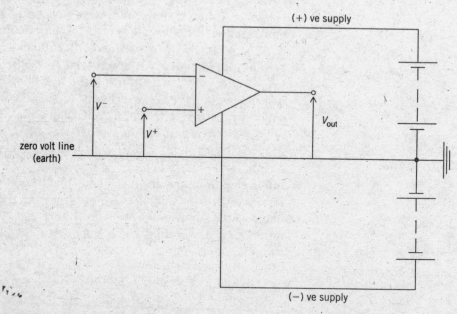

Figure 14.11 The op-amp used as a comparator

The link between the two batteries is referred to as the 'zero volt line' or 'earth' and it provides the reference from which the voltages on the inputs and the output are measured. The output V_{out} of this circuit is given by

$$V_{out} = A_0 (V^+ - V^-)$$

where V^+ is the voltage at the non-inverting input, V^- is the voltage at the inverting input, and A_0 is the open-loop gain of the amplifier.

Consider the case where the non-inverting input V^+ is 0.95 V and the inverting input V^- is 0.94 V, and $A_0 = 10^5$. The supply voltages are at ±6 V. By substituting into the equation, the output voltage V_{out} is given by $V_{out} = 10^5 \times (0.95 - 0.94) = +1000$ V. From energy considerations, this answer is impossible since the output voltage can never exceed the power supply voltage. The amplifier is saturated and the output voltage will be 6 V. So,

if $V^+ > V^-$ the output is $+V_{supply}$

if $V^- > V^+$ the output is $-V_{supply}$

The circuit of Figure 14.11 is called a **comparator** because it compares the voltages applied to the non-inverting and the inverting inputs and then gives an output that depends on whether $V^+ > V^-$ or $V^- > V^+$.

When a circuit incorporating an op-amp is used as a comparator, it is usual to connect each of the two inputs to a potential divider, as shown in Figure 14.12.

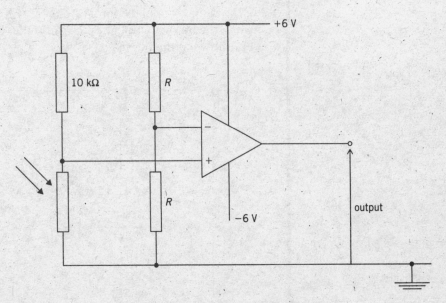

Figure 14.12 The op-amp used as a comparator to monitor illumination

The two resistors of equal resistance provide an input voltage of 3 V at the inverting input. When the light-dependent resistor (LDR) is in darkness, its resistance is greater than $10\,\text{k}\Omega$. The voltage at the non-inverting input will be greater than 3 V and the output will be at +6 V. In daylight, the resistance of the LDR will be less than $10\,\text{k}\Omega$ and the voltage at the non-inverting input will be less than 3 V. The output will be at −6 V.

The output therefore depends on the level of light illumination. This output can be made to operate an output device such as an LED (see Section 14.3). By changing the values of the resistors in the potential divider connected to the inverting input, the voltage at which the circuit switches from +6 V to −6 V can be altered. Thus, the level of illumination at which the circuit switches can be changed. In practice, one of these two resistors would be a variable resistor.

The circuit shown in Figure 14.12 switches from +6 V to −6 V when the level of illumination increases. By changing round the connections to the two inputs, the output could be made to switch from +6 V to −6 V when the level of illumination decreases.

Other devices could also be fitted into the comparator circuit. For example, a thermistor could be used so that the circuit provides a warning for either high or low temperatures.

Examples

1 State what is meant by a comparator circuit.

It is a circuit that usually incorporates an op-amp. The output voltage of the circuit is either positive or negative, depending on which of the two inputs to the circuit is at the higher potential.

2 Draw a diagram of a comparator that will give a positive output when the intensity of light incident on an LDR is high and a negative output when the intensity is low.

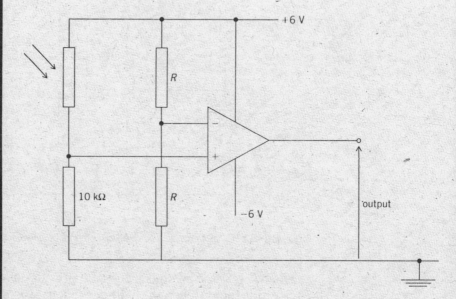

Figure 14.13

Now it's your turn

1 A thermistor has a resistance of 2500 Ω at 10 °C. Design a circuit that will switch from a positive output to a negative output when the thermistor temperature rises above 10 °C.

2 The resistance of the thermistor in question 1 is 1800 Ω at 20 °C. Show how your circuit in question 1 may be modified so that the comparator can be used *either* for temperatures rising above 10 °C *or* for temperatures rising above 20 °C.

Operational amplifiers and feedback

Feedback is a process whereby a fraction of the output of any device is fed back to the input, so as to assist in the control of the device. Much of the movement of humans is controlled by feedback. If a person wishes to pick up an object, for example, then the person stretches out a hand while, at the same time, looking at the hand and the object. The visual signal from the eye is fed back to the brain to provide control for the hand. This feedback is a continuous process of refining the position of the hand relative to the object.

For an amplifier circuit, the basic arrangement is as shown in Figure 14.14.

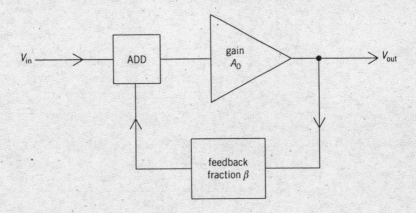

Figure 14.14 An amplifier circuit with feedback

A fraction β of the output signal V_{out} is fed back and added to the input signal V_{in}. The amplifier has open-loop gain A_0 and it amplifies whatever voltage there is at its input. So,

$$V_{out} = A_0 \times \text{(the amplifier input)}$$
$$= A_0 \times (V_{in} + \beta V_{out})$$

and this gives

$$V_{out}(1 - A_0\beta) = A_0 \times V_{in}$$

The overall **voltage gain** (or simply **gain**) V_{out}/V_{in} of the amplifier is given by

$$\frac{V_{out}}{V_{in}} = \frac{A_0}{(1 - A_0\beta)}$$

If the fraction β is negative, then $(1 - A_0\beta)$ is greater than unity and the overall gain of the amplifier circuit is less than the open-loop gain of the operational amplifier itself. This process is known as **negative feedback**.

Negative feedback seems, at first sight, to defeat the object of an amplifier. However, it does have a number of benefits:

- increased bandwidth (the range of frequencies for which the gain is constant)
- less distortion of the output
- greater stability.

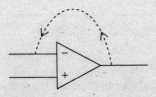

Figure 14.15 Negative feedback with an op-amp

Negative feedback can be achieved by feeding part of the output of the op-amp back to the *inverting* input, as illustrated in Figure 14.15.

The inverting amplifier

A circuit for an **inverting amplifier** incorporating an op-amp is shown in Figure 14.16. For simplicity, the power supplies are not shown.

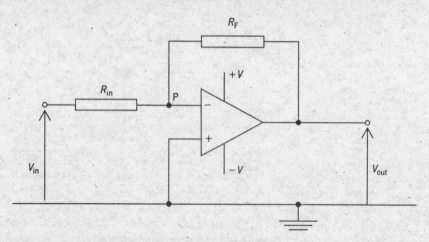

Figure 14.16 An inverting amplifier

Negative feedback is achieved by means of the resistor R_F. The resistors R_{in} and R_F act as a potential divider between the input and the output of the op-amp.

Since the open-loop gain of the op-amp is very large, then the input voltages at the non-inverting (+) and at the inverting (−) inputs must be almost the same. The non-inverting input is connected to the zero-volt line (the earth line) and therefore the input is zero. This means that the inverting input must be very nearly zero volts (or earth) and so point P in Figure 14.16 is known as a **virtual earth**.

The input impedance of the op-amp is very large which means there is no current in either the inverting or the non-inverting inputs. So all the current from, or to, the input signal to the circuit must pass through the feedback resistor to the output. This is illustrated in Figure 14.17.

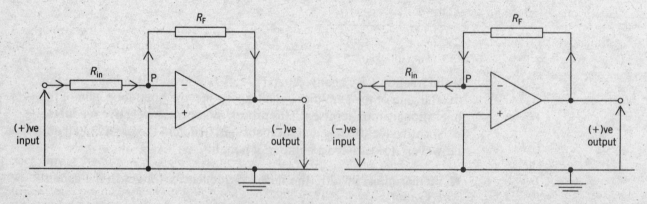

Figure 14.17 Feedback current

It can be seen that, because point P is a virtual earth, then a positive input signal gives rise to a negative output and vice versa. Consequently, the output is the inverse of the input and the amplifier is referred to as an **inverting amplifier**.

Referring to Figure 14.16, since the input impedance is infinite, current in R_{in} = current in R_F, So

$$\frac{\text{p.d. across } R_{in}}{R_{in}} = \frac{\text{p.d. across } R_F}{R_F}$$

The potential at P is zero and therefore

$$\frac{V_{in} - 0}{R_{in}} = \frac{0 - V_{out}}{R_F}$$

The overall voltage gain of the amplifier circuit is given by

$$\frac{V_{out}}{V_{in}} = -\frac{R_F}{R_{in}}$$

The negative sign indicates that there is a phase difference of π rad between the input and the output. That is, the input is negative when the output is positive, and the input is positive when the output is negative.

The non-inverting amplifier

Figure 14.18 is a circuit for a non-inverting amplifier.

Figure 14.18 A non-inverting amplifier

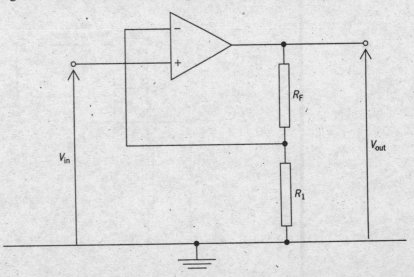

The input voltage V_{in} is applied directly to the non-inverting input. Negative feedback is achieved by means of the resistors R_F and R_1. If the amplifier has not saturated, the overall voltage gain of the amplifier is given by

$$\frac{V_{out}}{V_{in}} = 1 + \frac{R_F}{R_1}$$

The non-inverting amplifier produces an output that is in phase with the input. That is, if the input is positive, the output is positive, and if the input is negative, the output is negative.

Examples

1 What is meant by an inverting amplifier?

An electrical circuit where the output potential difference (voltage) is π rad out of phase with the input voltage.

2 The magnitude of the gain of an inverting amplifier is 25. The supply voltage to the op-amp is $\pm 9.0\,\text{V}$ and the non-inverting input is at earth potential. Calculate the output voltage of the amplifier circuit for an input voltage at the inverting input of:
(a) $+40\,\text{mV}$,
(b) $-80\,\text{mV}$,
(c) $-1.2\,\text{V}$.

$$V_{\text{out}} = -V_{\text{in}} \times \text{gain}$$

(a) $-1.0\,\text{V}$
(b) $+2.0\,\text{V}$
(c) $+9.0\,\text{V}$ (amplifier is saturated)

Now it's your turn

The circuit for an amplifier incorporating an op-amp is shown in Figure 14.19.

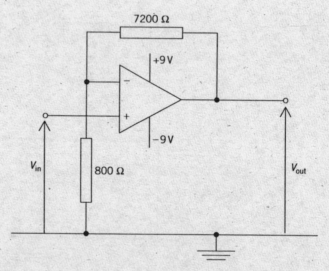

Figure 14.19

(a) State whether the amplifier is an inverting or a non-inverting amplifier.
(b) Calculate the gain of the amplifier circuit.
(c) Calculate the maximum input potential such that the amplifier does not saturate.

Section 14.2 Summary

- The processing unit is connected to the sensor and provides a voltage output that controls the output device.
- A potential divider provides a potential difference that changes as the resistance of the sensor changes.
- A comparator incorporating an operational amplifier (op-amp) enables a particular voltage level to be monitored, indicating whether the voltage is above or below a specified value.
- Low-value potential differences may be amplified using an amplifier incorporating an op-amp.
- Negative feedback reduces the gain of an amplifier but provides stability and increased bandwidth.
- The gain of an inverting amplifier is: $-R_F/R_{in}$
- The gain of a non-inverting amplifier is: $(1 + R_F/R_1)$

Section 14.2 Questions

1 A sensor may be connected to an output device via a processing unit. Explain the use of a potential divider as part of a processing unit.

2 You are provided with two identical cells and a light-dependent resistor that has a resistance of 950 Ω in normal lighting.
   (a) Design a circuit that will give an output that changes polarity from positive to negative as the light level changes from below normal to above normal.
   (b) Explain the action of your circuit.

14.3 Output devices

At the end of Section 14.3 you should be able to:
- show an understanding of the use of relays in electronic circuits
- show an understanding of the use of light-emitting diodes (LEDs) as devices to indicate the state of the output of electronic circuits
- show an understanding of the need for calibration where digital or analogue meters are used as output devices.

The processing units described in Section 14.2 produce an output voltage and, if this voltage is connected across some form of resistor, there will be a current from the output of the op-amp to the resistor.

This output current must not be larger than about 25 mA. With larger currents the op-amp would be destroyed. In fact, op-amps frequently have an output resistor as part of the integrated circuit so that, even if the output is 'shorted', the op-amp will not be damaged.

The sensing circuit may be required to switch on or off an appliance that requires a large current; for example, an electric motor. The switching on or off of a large current by means of a small current can be achieved using a relay.

The relay

A **relay** is an electromagnetic switch. When a small current passes through the coil of an electromagnet, the electromagnet operates a switch. This switch is used to switch on, or off, a much larger current. The symbol for a relay is shown in Figure 14.20.

Figure 14.21 shows one way in which a relay may be connected to the output of an op-amp.

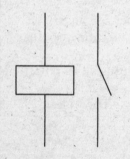

Figure 14.20 Symbol for a relay

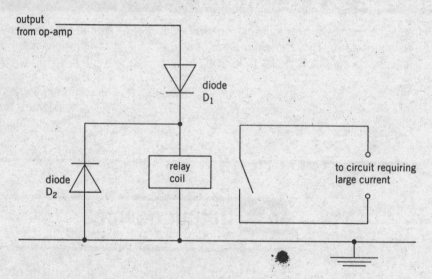

Figure 14.21 The connection of a relay to an op-amp

The output of the op-amp may be positive or negative. However, diode D_1 conducts only when the output is positive with respect to earth. Therefore, the relay coil is energised only when the output of the op-amp is positive.

When a current is switched off in an electromagnet, an e.m.f. is induced in the coil of the electromagnet. This e.m.f. could be large enough to damage the op-amp. A diode D_2 is connected across the coil to protect the op-amp from this e.m.f. Note that the diode is connected so that, when the output of the op-amp is positive, the diode D_2 is reverse-biased so that the current is in the coil, not the diode.

Relays are designed so that they can be used to switch on or switch off currents when the coil of the relay is energised.

The light-emitting diode (LED)

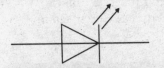

Figure 14.22 Symbol for a light-emitting diode (LED)

There are many uses of sensing devices where the purpose is to indicate whether something is too hot or too cold, switched on or off, etc. In these situations, all that is required for the output of the sensor is some form of visible indicator; that is, a lamp. Filament lamps not only dissipate a comparatively large amount of power but, also, they are not very robust and the filament is likely to break after being switched on and off a number of times. The **light-emitting diode (LED)** is a semiconductor device that is robust, reliable and dissipates much less power than a filament lamp. LEDs are available in different colours including red, green, yellow and amber. The symbol for an LED is shown in Figure 14.22.

The LED emits light only when it is forward-biased. A resistor is frequently connected in series with an LED so that, when the LED is forward-biased, the current in the diode is not so large that it would damage it. A typical maximum current for an LED is 20 mA. Also, the LED may be damaged if the reverse-bias voltage exceeds about 5 V.

Figure 14.23 shows two LEDs connected to the output of an op-amp to indicate whether the output is positive or negative.

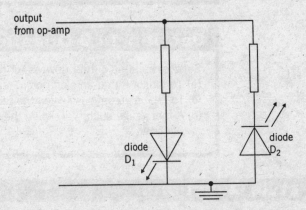

Figure 14.23 LEDs connected to indicate the state of an op-amp output

When the output is positive with respect to earth, diode D_1 will be forward-biased and will conduct, emitting light. Diode D_2 will be reverse-biased and will not emit light. When the polarity of the output changes, diode D_2 will be forward-biased and will conduct, emitting light. D_1 will be reverse-biased and will not emit light. The LEDs can be chosen so that they emit light of different colours.

Digital and analogue meters as output devices

LEDs are used to indicate the polarity of the output and may be used when the op-amp saturates (for example, using the op-amp as a comparator to detect whether a temperature is above or below a specific value). When the output of the op-amp does not saturate, the output voltage can be used to indicate the magnitude of whatever is being sensed (for example, the level of fuel in a tank). A digital or analogue voltmeter, connected between the op-amp output and earth, will indicate the output voltage.

The output voltage will be proportional to the input to the processing unit. It will also be dependent on the magnitude of whatever is being sensed. However, the voltmeter reads a potential difference, not the quantity that is being measured and therefore the voltmeter reading is unlikely to vary linearly with the change in this quantity. The voltmeter must be calibrated. The calibration of meters and the use of calibration curves is discussed in Section 2.1.

Section 14.3 Summary

- A relay is used to control large-value currents by means of the low current/voltage output of a processing unit.
- A light-emitting diode (LED) is used to indicate the state (high or low, on or off) of the output of a processing unit.
- Digital and analogue meters may be used to measure the output of a processing unit. Meters need to be calibrated in order to measure the quantity that is being monitored.

Section 14.3 Question

Suggest suitable output devices for the following applications.
(a) To indicate that an electrical circuit is switched on.
(b) To heat a greenhouse when the temperature falls to a particular setting.
(c) A fuel gauge such that, when the meter reads zero, there is reserve fuel in the tank.

Exam-style Questions

In all questions, assume that any operational amplifier (op-amp) is ideal and that the power supply to the op-amp is a dual $\pm 9.0\,V$ supply.

1 (a) State four properties of an ideal operational amplifier (op-amp).
(b) Figure 14.24 shows an amplifier incorporating an op-amp.

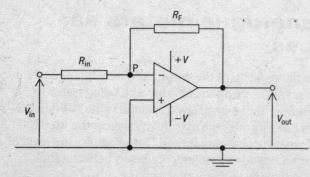

Figure 14.24

(i) The point P is referred to as a *virtual earth*. State what is meant by this.
(ii) Derive an expression, in terms of the resistances R_F and R_{in}, for the gain of the amplifier.
(c) The resistances R_F and R_{in} are $12\,k\Omega$ and $630\,\Omega$ respectively. Calculate:
  (i) the gain of the amplifier,
  (ii) the maximum input voltage such that the amplifier does not saturate.

2 A thermistor T is included in the circuit shown in Figure 14.25.
(a) Referring to the graph in Figure 14.5 (page 381), state the temperature at which the thermistor has a resistance of $3.00\,k\Omega$.
(b) Calculate the resistance of resistor R such that the output of the op-amp will change polarity at the temperature stated in (a).

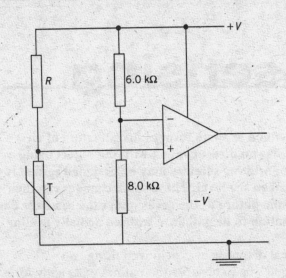

Figure 14.25

(c) Suggest and explain the direction of change of the polarity as the temperature is gradually increased from a low value.

(d) Two light-emitting diodes, one red and the other green, are to be used to indicate the temperature of the thermistor. The red LED is to emit light when the thermistor is below the temperature stated in (a) and the green LED is to emit light when the temperature is above that stated in (a). Sketch the circuit including the arrangement of the LEDs.

15. Remote sensing

Remote sensing may be described as some form of investigation of an object using equipment that has no direct contact with the object being investigated. For example, an orbiting satellite may be designed so that it can detect small changes in mean sea level. These small changes can then be interpreted to determine the nature of the rocks under the sea-bed. The investigation enables information to be gathered without actually drilling into the sea-bed.

Over a century ago, medical diagnosis was achieved using two techniques. The first technique was to observe the patient externally for signs of fever, changed pulse-rate, changed breathing, vomiting, skin condition, and so on. Diagnosis on this basis was part science and part art. The alternative was to carry out investigative surgery. Surgery always involves a risk and, in the past, many patients died either from the trauma of surgery or from resulting infection.

Diagnostic techniques have now been developed that enable externally placed equipment to obtain information from under the skin – this is a form of remote sensing. X-ray diagnosis was developed throughout the twentieth century. X-ray exposure is not without some risk, however small, and during the second half of the last century other techniques evolved. The use of ultrasound and magnetic resonance imaging are now commonplace.

Using these imaging techniques, it is possible to obtain detailed information about internal body structures without surgery, and consequently the techniques are referred to as being *non-invasive*. They are designed to cause as little risk as possible to the patient. It should be appreciated, however, that there is some risk with any procedure. The risks involved in imaging diagnoses are, in general, very small. These very small risks have to be weighed against the benefits of their use and the real finite risk that accompanies any disease.

15.1 The use of X-rays

At the end of Section 15.1 you should be able to:
- explain the principles of the production of X-rays by electron bombardment of a metal target
- describe the main features of a modern X-ray tube, including control of the intensity and hardness of the X-ray beam
- show an understanding of the use of X-rays in imaging internal body structures, including a simple analysis of the causes of sharpness and contrast in X-ray imaging

▶▶

- recall and solve problems by using the equation $I = I_0 e^{-\mu x}$ for the attenuation of X-rays in matter
- show an understanding of the purpose of computed tomography or CT scanning
- show an understanding of the principles of CT scanning
- show an understanding of how the image of an eight-voxel cube can be developed using CT scanning.

The production of X-rays

Figure 15.1 Modern X-ray machine

Whenever a charged particle is accelerated, electromagnetic radiation is emitted. This radiation is known as Bremmstrahlung radiation or 'braking (slowing down)' radiation. The frequency of the radiation depends on the magnitude of the acceleration. The larger the acceleration (or deceleration), the greater is the frequency of the emitted photon.

X-ray photons may be produced by the bombardment of metal targets with high-speed electrons. The electrons are first accelerated through a potential difference of several thousands of volts so that they have high energy and high speed. This acceleration is, however, not sufficient for X-ray radiation to be emitted. The high-speed electrons strike a metal target, which causes the electrons to change direction and to lose kinetic energy very rapidly. Large decelerations are involved that give rise to the emission of X-ray photons. It should be remembered that not all of the energy of the electrons is emitted as X-ray photons. The majority is transferred to thermal energy in the target metal.

A simplified design of an X-ray tube is shown in Figure 15.2. A metal filament is heated using a low-voltage supply and electrons are emitted from the filament – this is called the thermionic effect. The electrons are then accelerated towards the target, constructed of metal with a high melting point. The electrons are accelerated through a large potential difference of the order of 20–90 kV. When the electrons are stopped in the metal target, X-ray photons (Bremmstrahlung radiation) are emitted and the X-ray beam passes out of the tube through a window that is transparent to X-rays. The tube is evacuated and is made of a material that is opaque to X-rays. This reduces background radiation around the tube. The anode is earthed and the cathode is maintained at a high negative potential, so that the equipment is at earth potential. Since the majority of the energy of the electrons is transferred to thermal energy in the anode, the anode is either cooled or is made to spin rapidly so that the thermal energy is dispersed over a larger area of the target.

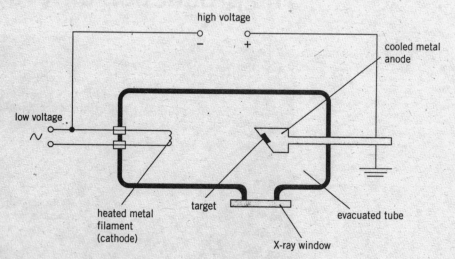

Figure 15.2 Design of an X-ray tube

Control of the X-ray beam

In order that the optimum X-ray image may be obtained, both the intensity and the hardness of the X-ray beam need to be controlled.

- The **intensity** is the wave power per unit area, and this affects the degree of blackening of the image.
- The **hardness** is the penetration of the X-ray beam, which determines the fraction of the intensity of the incident beam that can penetrate the part of the body being X-rayed. In general, the shorter the wavelength of the X-rays, the greater their penetration.

A typical X-ray spectrum showing the variation with wavelength of the intensity of an X-ray beam is shown in Figure 15.3.

The spectrum has two distinct components. First, there is a continuous distribution of wavelengths with a sharp cut-off at the shortest wavelength, λ_0. Second, sharp peaks may be observed. These sharp peaks correspond to the emission line spectrum of the target metal and are, therefore, a characteristic of the target.

The continuous distribution comes about because the electrons, when incident on the metal target, will not all have the same deceleration but will, instead, have a wide range of values. Since the wavelength of the emitted radiation is dependent on the deceleration, there will be a distribution of

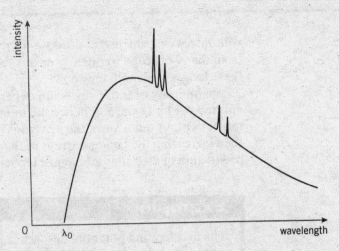

Figure 15.3 Typical X-ray spectrum

wavelengths. The cut-off wavelength corresponds to an electron that is stopped in one collision in the target so that all of its kinetic energy is given up as one X-ray photon.

The kinetic energy E_K of an electron is equal to the energy gained by the electron when it is accelerated from the cathode to the anode.

$$E_K = eV$$

where e is the charge on the electron and V is the accelerating potential difference.

The energy E of a photon of wavelength λ is given by

$$E = \frac{hc}{\lambda}$$

where h is the Planck constant (see page 241). Thus, at the cut-off wavelength,

$$eV = \frac{hc}{\lambda_0}$$

and so

$$\lambda_0 = \frac{hc}{eV}$$

The accelerating potential V thus determines the cut-off wavelength λ_0. The larger the potential difference, the shorter the wavelength. The hardness of the X-ray beam is therefore controlled by variation of the accelerating potential difference between the cathode and the anode.

The continuous distribution of wavelengths implies that there will be X-ray photons of long wavelengths that would not penetrate the person being investigated and so would not contribute towards the X-ray image. Such long-wavelength photons would add to the radiation dose received by the person without serving any useful purpose. For this reason the X-ray beam emerging from the X-ray tube frequently passes through aluminium filters that absorb these long-wavelength photons.

The intensity of the X-ray beam depends on the number of photons emitted per unit time and hence the number of electrons hitting the metal target per unit time. Since the electrons are produced by thermionic emission, then increasing the heater current in the cathode will increase the rate of production of electrons and hence increase the intensity of the X-ray beam.

Example

The accelerating potential difference between the cathode and the anode of an X-ray tube is 30 kV. Given that the Planck constant is 6.6×10^{-34} J s, the charge on the electron is 1.6×10^{-19} C and the speed of light in free space is 3.0×10^8 m s^{-1}, calculate the minimum wavelength of photons in the X-ray beam.

For the minimum wavelength,
energy gained by electron = energy of photon

$$eV = \frac{hc}{\lambda_0}$$

$$1.6 \times 10^{-19} \times 30 \times 10^3 = \frac{(6.6 \times 10^{-34} \times 3.0 \times 10^8)}{\lambda_0}$$

$$\lambda_0 = \mathbf{4.1 \times 10^{-11} \ m}$$

Now it's your turn

(a) State what is meant by the *hardness* of an X-ray beam.
(b) The Planck constant is 6.6×10^{-34} J s, the charge on the electron is 1.6×10^{-19} C and that the speed of light in free space is 3.0×10^8 m s^{-1}. Calculate the minimum wavelength of photons produced in an X-ray tube for an accelerating potential difference of 45 kV.

The X-ray image

An X-ray image is shown in Figure 15.4. This is not really an image in the sense of the real image produced by a lens. Rather, the image is like a shadow, as illustrated in Figure 15.5.

The X-ray beam is incident on the body part of the patient. The X-ray beam can penetrate soft tissues (skin, fat, muscle, etc.) with little loss of intensity and so photographic film, after development, will show a dark area corresponding to these soft tissues. Bone, however, causes a greater attenuation (reduces the intensity by a greater extent) than the soft tissues and therefore the photographic film will be lighter in colour in areas corresponding to the positions of bones. What is produced on the film is a two-dimensional shadow of the bone and the surrounding tissues.

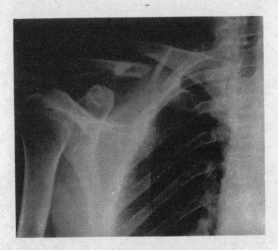

Figure 15.4 X-ray image showing a fractured collar bone

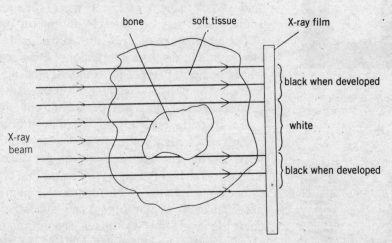

Figure 15.5 How an X-ray image is produced

The quality of the shadow image produced depends on its sharpness and contrast. **Sharpness** is related to the ease with which the edges of structures can be determined. A shadow image where the bones and other organs are clearly outlined is said to be a 'sharp image'. Although an image may be sharp, it may still not be clearly visible because there is little difference in the degree of blackening between, say, the bone and the surrounding tissue. An X-ray image having a wide range of degrees of blackening is said to have good **contrast**.

A sharp image requires a parallel X-ray beam. This can be achieved by reducing the area of the target anode in the X-ray tube, limiting the size of the aperture through which the X-ray beam passes, and reducing scattering of the emergent beam. Figure 15.6 illustrates the effect of changing the area of the target anode. The full shadow produces an area that is white on development of the photographic film. Where there is no shadow, the image will be black. There is a region of partial shadow, or greyness, where the image gradually changes from white to black. If the image is to be sharp, then this area of greyness must be reduced as much as possible. The area of the target anode should be kept to a minimum.

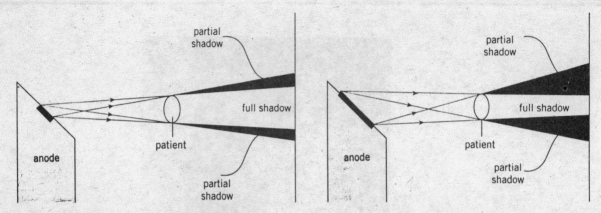

Figure 15.6 Effect of the area of the target anode on sharpness

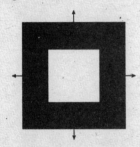

Figure 15.7 Limiting aperture size

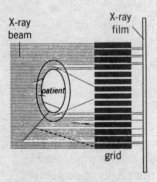

Figure 15.8

A reduction in the grey area at the edge of the image can also be achieved by limiting the size of the aperture through which the X-ray beam passes. This is achieved by using overlapping metal sheets, as shown in Figure 15.7. The aperture size may be varied by sliding the metal sheets over one another.

The emerging parallel X-ray beam should then give rise to a sharp image. However, as a result of interactions between photons and any substance through which the beam passes (this could even be air), some photons will be scattered. Scattered photons will result in a loss of sharpness. These scattered photons may be absorbed in a metal grid placed in front of the photographic film (Figure 15.8).

Good contrast is achieved when neighbouring body organs and tissues absorb the X-ray photons to very different extents. This is usually the case for bone and muscle. This is, however, not the case where, for example, the stomach is to be investigated. The patient is then asked to swallow a solution of barium sulphate – a 'barium meal' (Figure 15.9). The barium is a good absorber of X-ray photons. As a result, when the barium sulphate solution coats the inside of the stomach, the outline of the stomach will be shown up clearly on the image. Similarly, blood vessels can be made visible by injecting a radio-opaque dye into bloodstream.

Contrast also depends on other factors, such as exposure time. Contrast may be improved by backing the photographic film with a fluorescent material.

The attenuation of X-rays

When a beam of X-ray photons passes through a medium, absorption processes occur that reduce the intensity of the beam. The intensity of a parallel beam is reduced by the same fraction each time the beam passes through equal thicknesses of the medium. Consequently, the variation of the percentage of the intensity transmitted with thickness of absorber may be shown as in Figure 15.10.

It can be seen that the same thickness of medium is always required to reduce the transmitted beam intensity by 50%, no matter what starting point is chosen. This thickness of medium is called the **half-value thickness** (HVT) and is denoted by the symbol $x_{\frac{1}{2}}$.

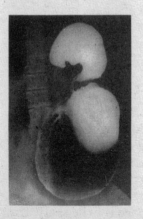

Figure 15.9 X-ray of stomach after a barium meal

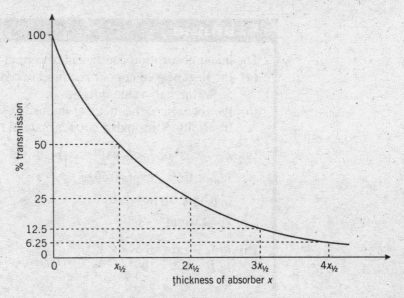

Figure 15.10 The percentage transmission of X-rays in a medium

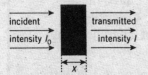

Figure 15.11 Absorption of X-rays in a medium

The decrease in transmitted intensity is an exponential decrease. Consider a parallel beam having an incident intensity I_0. The medium (the absorber) has thickness x and the transmitted intensity is I, as illustrated in Figure 15.11. The transmitted intensity is given by the expression

$$I = I_0\, e^{-\mu x} \quad \text{or} \quad I = I_0 \exp(-\mu x)$$

where μ is a constant that is dependent on the medium and on the energy of the X-ray photons, known as the **linear attenuation coefficient** or the **linear absorption coefficient** of the medium. The unit of μ is mm^{-1} or cm^{-1}.

For a thickness $x_{\frac{1}{2}}$ (the half-value thickness) of the medium, the intensity I will be equal to $\frac{1}{2}I_0$. Hence,

$$\tfrac{1}{2}I_0 = I_0 \exp(-\mu x_{\frac{1}{2}})$$

which gives

$$\ln 2 = \mu x_{\frac{1}{2}}$$

Approximate values of the linear attenuation (absorption) coefficient μ for some substances are given in Table 15.1.

Table 15.1 Some approximate values of linear attenuation (absorption) coefficient

substance	μ/cm^{-1}
copper	7
water	0.3
bone	3
fat	0.9

Example

The linear absorption coefficient of copper is $0.693\,\text{mm}^{-1}$. Calculate:

(a) the thickness of copper required to reduce the incident intensity by 50% (the half-value thickness),

(b) the fraction of the incident intensity of a parallel beam that is transmitted through a copper plate of thickness 1.2 cm.

(a) $I = I_0 e^{-\mu x}$ or $I = I_0 \exp(-\mu x_{\frac{1}{2}})$

$\quad I / I_0 = 0.50 = \exp(-0.693 x_{\frac{1}{2}})$

$\quad \ln 0.50 = -0.693 x_{\frac{1}{2}}$

$\quad x_{\frac{1}{2}} = \mathbf{1.0\,mm}$

(b) $I / I_0 = \exp(-0.693 \times 12)$

$\quad I / I_0 = \mathbf{2.4 \times 10^{-4}}$

Now it's your turn

1 For one particular energy of X-ray photons, water has a linear attenuation (absorption) coefficient of $0.29\,\text{cm}^{-1}$. Calculate the depth of water required to reduce the intensity of a parallel beam of these X-rays to 1.0×10^{-3} of its incident intensity.

2 The linear attenuation (absorption) coefficients of bone and of the soft tissues surrounding the bone are $2.9\,\text{cm}^{-1}$ and $0.95\,\text{cm}^{-1}$ respectively. A parallel beam of X-rays is incident, separately, on a bone of thickness 3.0 cm and on soft tissue of thickness 5.0 cm. Calculate the ratio:

$$\frac{\text{intensity transmitted through bone}}{\text{intensity transmitted through soft tissue}}$$

Computed tomography (CT scanning)

The image produced on X-ray film as outlined earlier is a 'shadow' or 'flat' image. There is little, if any, indication of depth. That is, the position of an organ within the body is not apparent. Also, soft tissues lying behind structures that are very dense cannot be detected. Tomography is a technique whereby a three-dimensional image or 'slice' through the body may be obtained. The image is produced by **computed tomography** using what is known as a CT scanner (Figure 15.12). An example of such an image is shown in Figure 15.13.

In this technique, a series of X-ray images is obtained. Each image is taken through the section or slice of the body from a different angle, as illustrated in Figure 15.14.

Data for each individual X-ray image and angle of viewing is fed into a high-power computer. The computer enables the image of each slice to be combined so that a complete three-dimensional image of the whole object is obtained which can then be viewed from any angle.

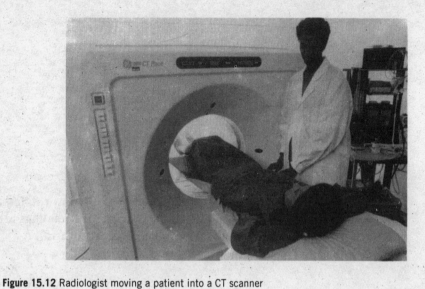

Figure 15.12 Radiologist moving a patient into a CT scanner

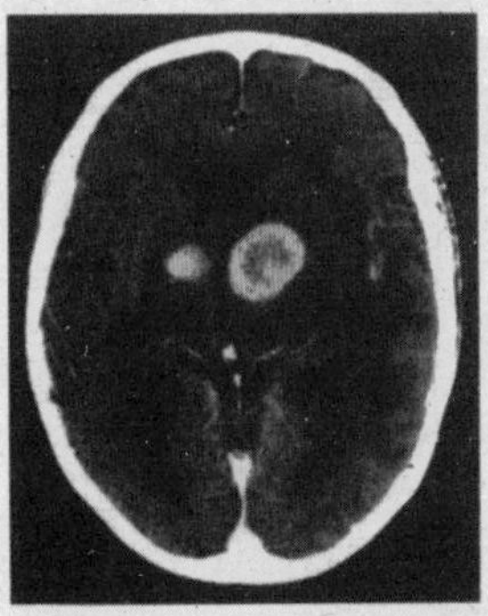

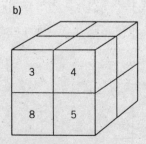

Figure 15.13 CT scan through the head of a patient with a cerebral lymphoma

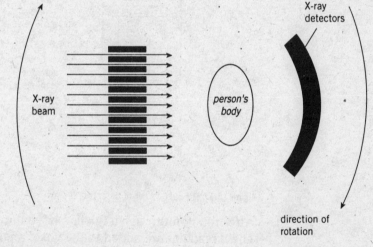

X-ray beam

person's body

X-ray detectors

direction of rotation

Figure 15.14 Arrangement for a CT scan

a)

b)

3	4
8	5

Figure 15.15 The first section showing the pixels

The basic principles of CT scanning may be illustrated using a simple cubic shape as shown in Figure 15.15. The aim of CT scanning is to produce an image of a section through the cube from measurements made about its axis. The section, or 'slice' through the cube is divided up into a series of small units, called **voxels**. Each voxel will absorb the X-ray beam to a different extent. The intensity transmitted through each voxel alone can be given a number, referred to as a **pixel**. The various pixels are built up from measurements of the X-ray intensity along different directions through the section or slice.

Suppose that the cube in Figure 15.15a is divided into eight voxels. The cube can be thought to consist of two slices or sections. For the first section, let the pixels be as shown in Figure 15.15b. The purpose of the CT scan is to reproduce these pixels in their correct positions.

When the X-ray beam is directed at the section from the left, as shown in Figure 15.16, the detectors will give readings of 7 and 13. The voxels will be partially completed as shown.

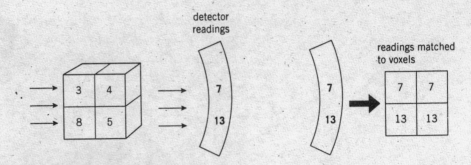

Figure 15.16 The first set of detector readings

The X-ray tube and the detectors are now rotated through 45°. The new detector readings are 4, 8 and 8. These readings are added to the readings already in the voxels, as shown in Figure 15.17.

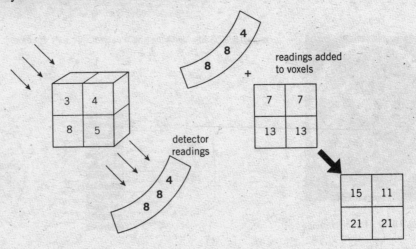

Figure 15.17 The second set of detector readings

After rotation through a further 45°, a third set of detector readings is taken. These readings are added to the voxel readings. The result is shown in Figure 15.18.

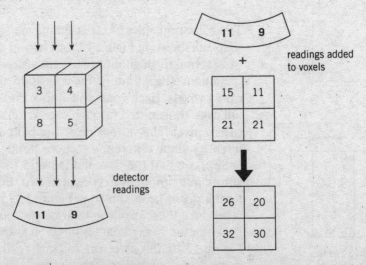

Figure 15.18 The third set of detector readings

After rotation through a further 45° a final set of readings is taken. Once again the readings are added to those already in the voxels, giving the result shown in Figure 15.19.

The resulting pattern of the pixels is shown in Figure 15.20a.

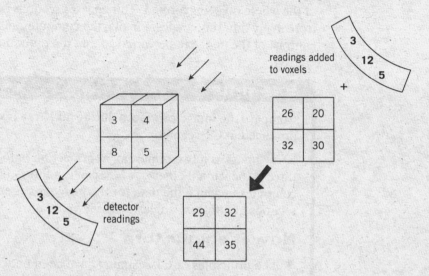

Figure 15.19 The fourth set of detector readings

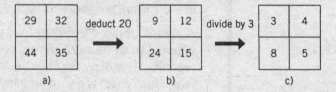

Figure 15.20 The final result

Having summed all the measurements, it remains to reduce these measurements to that of the original. This is achieved in two stages.

1 The background intensity must be removed. This background is equal to the sum of the detector readings for any one position. In this case, the sum is 20. This background is deducted from each pixel, as shown in Figure 15.20b.
2 Allowance must now be made for the fact that more than one view was made of the section. In this example, there were four sets of readings and consequently each pixel reading is divided by 3.

The final result is shown in Figure 15.20c – and note this is the same as Figure 15.15b.

Once the pattern of pixels for one section has been obtained, the CT scanner is moved relative to the object so that the next neighbouring section is analysed. This procedure is repeated until the whole of the object to be analysed has been scanned.

The analysis above is a very simple example. In practice, the image of each section is built up from a large number of units or voxels. The larger the number of voxels, the better the definition. This is similar, in principle, to the digital camera. The use of a large number of voxels implies that measurements must be made from a large number of different angles. The storage of the data for each angle and its construction into a final image on a screen requires a powerful computer. The reconstruction of the intensity in each voxel will involve more than one million separate computations. All the data for all the sections is stored in the computer memory so that a three-dimensional image of the whole object is formed. This enables sections of the image to be viewed from many different angles. Finally, the computer enables the brightness and contrast of the image to be varied so that the optimum image may be obtained.

Example

Compare the image produced during an X-ray investigation and that produced in CT scanning.

X-ray image is a two dimensional projection onto a flat screen of a three-dimensional object. A CT scan is a three-dimensional image. The computer in which the data for the image is stored enables different sections at different angles to be viewed.

Now it's your turn

1 The principles of CT scanning have been understood for some time. However, scanners could not be developed until large powerful computers were made available. By reference to the image produced in a CT scan, suggest why such a computer is necessary.

2 A simple object consists of four voxels. The object is scanned from four different directions, each at 45° to the next. The detector measurements for each individual voxel are summed and the result is shown in Figure 15.21.

40	49
34	31

Figure 15.21

The total of the readings of the detectors in any one position is 22. Determine the pattern of the pixels in the voxels.

Section 15.1 Summary

■ Remote sensing enables investigations to be made where there is no contact with the object under investigation.
■ X-rays are produced when high-speed electrons are stopped by a metal target.
■ An X-ray image is a 'shadow' of structures in which the X-ray beam is attenuated.
■ The attenuation in the intensity of an X-ray beam is given by:
$I = I_0 \, e^{-\mu x}$
■ Computed tomography (CT scanning) enables an image of a section through the body to be obtained by combining many X-ray images, each one taken from a different angle.

Section 15.1 Questions

1 (a) State what is meant by the *sharpness* of an
X-ray image.

(b) Explain how the sharpness of an X-ray
image may be improved.

2 (a) Outline how X-ray images are used to build
up the image produced in a CT scan.

(b) Explain why the radiation dose received
during a CT scan is greater than that for an
X-ray 'photograph'.

15.2 The use of ultrasound

At the end of Section 15.2 you should be able to:

■ explain the principles of the generation and detection of ultrasonic
waves using piezo-electric transducers

■ explain the main principles behind the use of ultrasound to obtain
diagnostic information about internal structures

■ show an understanding of the meaning of acoustic impedance and its
importance to the intensity reflection coefficient at a boundary.

The generation of ultrasound

Ultrasound waves may be generated using a **piezo-electric transducer**.
A transducer is the name given to any device that converts energy from one
form to another. In this case, electrical energy is converted into ultrasound
energy by means of a piezo-electric crystal such as quartz.

The structure of quartz is made up of a large number of tetrahedral silicate
units, as shown in Figure 15.22. These units build up to form a crystal of
quartz that can be represented, in two dimensions, as shown in Figure 15.23.

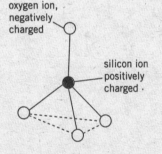

Figure 15.22 Tetrahedral
silicate unit

oxygen ion,
negatively
charged

silicon ion
positively
charged

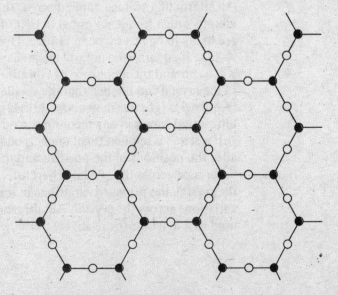

Figure 15.23 Two-dimensional representation of a quartz crystal

When the crystal is unstressed, the centres of charge of the positive and the negative ions in any one unit coincide, as shown in Figure 15.24a.

Electrodes may be formed on opposite sides of the crystal by depositing silver on its surfaces. When a potential difference is applied between the electrodes, an electric field is set up in the crystal. This field causes forces to act on the ions. The oxygen ions are negatively charged and the silicon ions have a positive charge. The ions are not held rigidly in position and, as a result, they will be displaced slightly when the electric field is applied across the crystal. The positive ions will be attracted towards the negative electrode and the negative ions will be attracted to the positive electrode. Dependent on the direction of the electric field, the crystal will become slightly thinner (Figure 15.24b) or slightly thicker (Figure 15.24c).

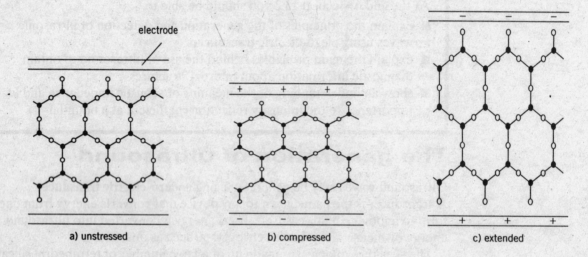

Figure 15.24 The effect of an electric field on a quartz crystal

An alternating voltage applied across the electrodes causes the crystal to vibrate with a frequency equal to that of the applied voltage. These oscillations are likely to have a small amplitude. However, if the frequency of the applied voltage is equal to the natural frequency of vibration of the crystal, resonance will occur and the amplitude of vibration will be a maximum. The dimensions of the crystal can be such that the oscillations are in the ultrasound range of frequencies (greater than about 20 kHz). These oscillations will give rise to ultrasound waves in any medium surrounding the crystal.

If a stress is applied to an uncharged quartz crystal, the forces involved will alter the positions of the positive and the negative ions, creating a potential difference across the crystal. Therefore, if an ultrasound wave is incident on the crystal, the pressure variations in the wave will give rise to voltage variations across the crystal. An ultrasound transducer may therefore also be used as a detector (or receiver).

A simplified diagram of a piezo-electric transducer/receiver is shown in Figure 15.25.

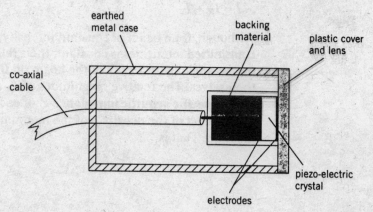

Figure 15.25 Piezo-electric transducer/receiver

A transducer such as this is able to produce and detect ultrasound in the megahertz frequency range, which is typical of the frequency range used in medical diagnosis.

The reflection and absorption of ultrasound

Ultrasound is typical of many types of wave in that, when it is incident on a boundary between two media, some of the wave power is reflected and some is transmitted. This is illustrated in Figure 15.26.

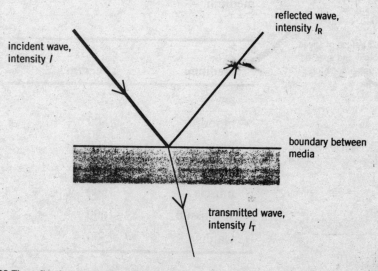

Figure 15.26 The reflection and transmission of a wave at a boundary

For a wave of incident intensity I, reflected intensity I_R and transmitted intensity I_T, by conservation of energy,

$$I = I_R + I_T$$

Although, for a beam of constant intensity, the sum of the reflected and transmitted intensities is constant, their relative magnitudes depend not only on the angle of incidence of the beam on the boundary but also on the media themselves. The relative magnitudes of I_R and I_T are quantified by reference to the **specific acoustic impedance** Z of each of the media. This is defined as the product of the density ρ of the medium and the speed c of the wave in the medium. That is,

$$Z = \rho c$$

For a wave incident normally on a boundary between two media having specific acoustic impedances of Z_1 and Z_2, the ratio of the reflected intensity I_R to the incident intensity I is given by

$$\frac{I_R}{I} = \frac{(Z_2 - Z_1)^2}{(Z_2 + Z_1)^2}$$

The ratio I_R/I is known as the **intensity reflection coefficient** for the boundary and is given the symbol α. As the above equation shows, α depends on the difference between the specific acoustic impedances of the two media. Some typical values of specific acoustic impedance are given in Table 15.2, together with the approximate speed of ultrasound in the medium.

Table 15.2 Values of speed of ultrasound and specific acoustic impedance for some media

medium	speed/m s^{-1}	specific acoustic impedance/ kg m^{-2}s^{-1}
air	330	430
blood	1600	1.6×10^6
bone	4100	$5.6 \times 10^6 - 7.8 \times 10^6$
fat	1500	1.4×10^6
muscle	1600	1.7×10^6
soft tissue	1600	1.6×10^6
water	1500	1.5×10^6

Example

Using data from Table 15.2, calculate the intensity reflection coefficient for a parallel beam of ultrasound incident normally on the boundary between:
(a) air and soft tissue,
(b) muscle and bone that has a specific acoustic impedance of $6.5 \times 10^6 \, \text{kg m}^{-2}\text{s}^{-1}$.

(a) $\alpha = (Z_2 - Z_1)^2 / (Z_2 + Z_1)^2$
$= (1.6 \times 10^6 - 430)^2 / (1.6 \times 10^6 + 430)^2$
$= \textbf{0.999}$
(b) $\alpha = (6.5 \times 10^6 - 1.7 \times 10^6)^2 / (6.5 \times 10^6 + 1.7 \times 10^6)^2$
$= \textbf{0.34}$

Now it's your turn

Using data from Table 15.2,
(a) suggest why, although the speed of ultrasound in blood and muscle is approximately the same, the specific acoustic impedance is different.
(b) calculate the intensity reflection coefficient for a parallel beam of ultrasound incident normally on the boundary between fat and muscle.

It can be seen that the intensity reflection coefficient for a boundary between air and soft tissue is approximately equal to unity. This means that when ultrasound is incident on the body, very little is transmitted into the body. In order that ultrasound may be transmitted into the body and also that the ultrasound may return to the transducer, it is important that there is no air between the transducer and the skin (soft tissue). This is achieved by means of a water-based jelly. This jelly has a specific acoustic impedance of approximately $1.5 \times 10^6 \, \text{kg m}^{-2}\text{s}^{-1}$.

Once the ultrasound wave is within the medium, the intensity of the wave will be reduced by absorption of energy as it passes through the medium. The medium is heated. In fact, the heating effect produced by ultrasound of appropriate frequencies is used in physiotherapy to assist recovery from sprains and similar injuries.

For a parallel beam, this absorption is approximately exponential (as with X-rays – see Section 15.1). For such a beam of ultrasound incident normally on a medium of thickness x, the transmitted intensity I is related to the incident intensity I_0 by the expression

$$I = I_0 \, e^{-kx} \quad \text{or} \quad I = I_0 \exp(-kx)$$

where k is a constant for the medium known as the linear absorption coefficient. The coefficient k depends not only on the medium itself but also on the frequency of the ultrasound. Some values of the linear absorption coefficient are shown in Table 15.3.

Table 15.3 Some values of linear absorption coefficient for ultrasound

medium	linear absorption coefficient / cm^{-1}
air	1.2
bone	0.13
muscle	0.23
water	0.0002

Example

A parallel beam of ultrasound is incident on the surface of a muscle and passes through a thickness of 3.5 cm of the muscle. It is then reflected at the surface of a bone and returns through the muscle to its surface. Using data from Tables 15.2 and 15.3, calculate the fraction of the incident intensity that arrives back at the surface of the muscle.

The beam passes through a total thickness of 7.0 cm of muscle. For the attenuation in the muscle,

$I = I_0 \exp(-0.23 \times 7.0) = 0.20 I_0$

Fraction reflected at muscle–bone interface = 0.34
(see part (b) of Example on page 417)

Fraction received back at surface $= 0.34 \times 0.20 = 0.068 = \dfrac{1}{15}$

Now it's your turn

A parallel beam of ultrasound passes though a thickness of 4.0 cm of muscle. It is then incident normally on a bone having a specific acoustic impedance of $6.4 \times 10^6 \, kg \, m^{-2} s^{-1}$. The bone is 1.5 cm thick. Using data from Table 15.3, calculate the fraction of the incident intensity that is *transmitted* through the muscle and bone.

Obtaining diagnostic information using ultrasound

The ultrasound transducer is placed on the skin, with the water-based jelly excluding any air between the transducer and the skin (Figure 15.27). Short pulses of ultrasound are transmitted into the body where they are partly reflected and partly transmitted at the boundaries between media in the body such as fat–muscle and muscle–bone. The reflected pulses return to the transducer where they are detected and converted into voltage pulses. These voltage pulses can be amplified and processed by electronic circuits such that the output of the circuits may be displayed on a screen as in, for example, a cathode-ray oscilloscope.

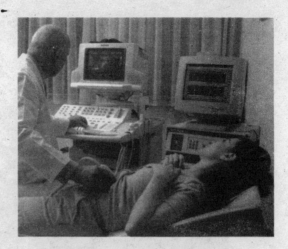

Figure 15.27 Ultrasound diagnosis

Pulses of ultrasound are necessary so that the reflected ultrasound pulses can be detected in the time intervals between the transmitted pulses. The time between the transmission of a pulse and its receipt back at the transducer gives information as to the distance of the boundary from the transducer. The intensity of the reflected pulse gives information as to the nature of the boundary. Two techniques are in common use for the display of an ultrasound scan.

In an **A-scan**, a short pulse of ultrasound is transmitted into the body through the coupling medium (the water-based jelly). At each boundary between media, some of the energy of the pulse is reflected and some is transmitted. The transducer detects the reflected pulses as it now acts as the receiver. The signal is amplified and displayed on a cathode-ray oscilloscope (c.r.o.). Reflected pulses (echoes) received at the transducer from deeper in the body tend to have lower intensity than those reflected from boundaries near the skin. This is caused not only by absorption of wave energy in the various media but also, on the return of the reflected pulse to the transducer, some of the energy of the pulse will again be reflected at intervening boundaries. To allow for this, echoes received later at the transducer are amplified more than those received earlier. A vertical line is observed on the screen of the c.r.o. corresponding to the detection of each reflected pulse. The time-base of the c.r.o. is calibrated so that, knowing the speed of the ultrasound wave in each medium, the distance between boundaries can be determined. An example of an A-scan is illustrated in Figure 15.28.

① transmitted pulse

② fat–muscle boundary

③ muscle–bone boundary

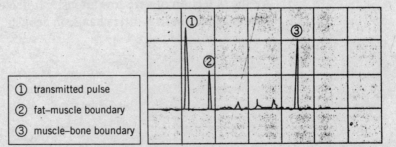

Figure 15.28 An A-scan

A **B-scan** consists of a series of A-scans, all taken from different angles so that, on the screen of the c.r.o., a two-dimensional image is formed. Such an image is shown in Figure 15.29.

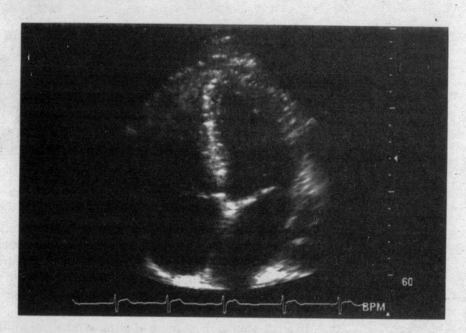

Figure 15.29 An image of a healthy heart produced from a B-scan

The ultrasound probe for a B-scan does not consist of a single crystal. Rather, it has an array of small crystals, each one at a slightly different angle to its neighbours. The separate signals received from each of the crystals in the probe are processed. Each reflected pulse is shown on the screen of the c.r.o as a bright spot in the direction of orientation of the particular crystal that gave rise to the signal. The pattern of spots builds up to form a two-dimensional image representing the positions of the boundaries within the body. The image may be either viewed immediately or photographed or stored in a computer memory.

The main advantage of ultrasound scanning compared to X-ray diagnosis is that the health risk to both the patient and to the operator is very much less. Also, ultrasound equipment is much more portable and is relatively simple to use.

Higher frequency ultrasound enables greater resolution to be obtained. That is, more detail can be seen. Furthermore, as modern techniques allow for the detection of very low intensity reflected pulses, boundaries between tissues where there is little change in acoustic impedance can be detected.

Section 15.2 Summary

- Ultrasound may be generated and detected by piezo-electric crystals.
- Ultrasound images are formed as a result of the detection and processing of ultrasound pulses that have been reflected from tissue boundaries.
- The acoustic impedance Z of a medium is: ρc
- The intensity reflection coefficient is: $(Z_2 - Z_1)^2 / (Z_2 + Z_1)^2$
- Two-dimensional scans may be obtained using a generator/detector consisting of many separate crystals all at different angles of orientation.

Section 15.2 Questions

1 Explain why, when obtaining an ultrasound scan,
  (a) the ultrasound is pulsed and is not continuous,
  (b) the reflected signal received from deeper in the body is amplified more than that received from near the skin.

2 The specific impedance of fat, muscle and bone are $1.4 \times 10^6 \, \text{kg m}^{-2} \text{s}^{-1}$, $1.6 \times 10^6 \, \text{kg m}^{-2} \text{s}^{-1}$ and $6.5 \times 10^6 \, \text{kg m}^{-2} \text{s}^{-1}$ respectively. The linear absorption coefficients in fat and in muscle are $0.24 \, \text{cm}^{-1}$ and $0.23 \, \text{cm}^{-1}$ respectively.

A parallel beam of ultrasound of intensity I is incident on the layer of fat. Discuss quantitatively, in terms of I, the reflection and the transmission of the beam of ultrasound as it passes through the layer of fat of thickness 4.0 mm, into the muscle of thickness 43.5 mm and finally into the bone.

15.3 The use of (nuclear) magnetic resonance imaging

At the end of Section 15.3 you should be able to:

- explain the main principles behind the use of magnetic resonance to obtain diagnostic information about internal structures
- show an understanding of the function of the non-uniform magnetic field, superimposed on the large constant field, in diagnosis using magnetic resonance.

Nuclear magnetic resonance

Many atomic nuclei have a property that is known as 'spin'. The 'spin' causes the nuclei to behave as if they were small magnets. Such nuclei have an odd number of protons and/or neutrons. Examples of these nuclei are hydrogen, carbon and phosphorus.

When a magnetic field is applied to these nuclei, they tend to line up along the field. However, this alignment is not perfect and the nuclei rotate about the direction of the magnetic field, due to their spin. The motion can be modelled as the motion of a top spinning about the direction of a gravitational field. This rotation is known as **precession**.

The spinning about the direction of the magnetic field has a frequency known as the frequency of precession or the **Larmor frequency**. The frequency depends on the nature of the nucleus and the strength of the magnetic field.

If a pulse of electromagnetic radiation of the same frequency as the Larmor frequency is incident on precessing nuclei, the nuclei will resonate, absorbing energy. This frequency is in the radio-frequency (RF) band with a wavelength shorter than about 10 cm. A short time after the incident pulse has ended, the nuclei will return to their equilibrium state, emitting RF radiation. The short time between the end of the RF pulse and the re-emitting of the radiation is known as the **relaxation time**. The whole process is referred to as **nuclear magnetic resonance**.

In practice, there are two relaxation processes and it is the time between these two that forms the basis of **magnetic resonance imaging (MRI)**. Since hydrogen is abundant in body tissues and fluids, hydrogen is the atom used in MRI.

Magnetic resonance imaging (MRI)

A schematic diagram of a magnetic resonance scanner is shown in Figure 15.30.

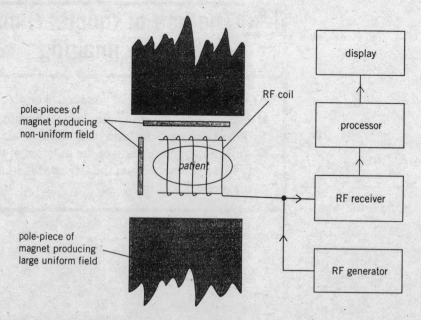

Figure 15.30 Schematic diagram of a magnetic resonance scanner

The patient who is to be investigated is positioned in the scanner (see Figure 15.30), between the poles of a large magnet that produces a very large uniform magnetic field (in excess of 1 tesla). The magnetic field causes all the hydrogen nuclei within the person to precess with the same Larmor frequency. In order that the hydrogen nuclei in only one small part of the body may be detected, a non-uniform magnetic field is also applied across the patient. This non-uniform field is accurately calibrated and results in a different magnitude of magnetic field strength at each point in the body of the patient. Since the Larmor frequency is dependent on the strength of the magnetic field, the Larmor frequency will also be different in each part of the patient. The particular value of magnetic field strength, together with the radio-frequency that is emitted, enables the hydrogen nuclei in the part under investigation to be located.

Radio-frequency (RF) pulses are produced in coils near the patient. These pulses pass into the patient. The emitted pulses produced as a result of de-excitation of the hydrogen nuclei are picked up by the coils. These signals are processed and displayed so as to construct an image of the number density of hydrogen atoms in the patient. As the non-uniform magnetic field is changed, atoms in different parts of the patient's body are detected and displayed. An image of a cross-section through the patient may be produced. One such image is shown in Figure 15.31.

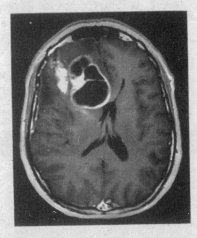

Figure 15.31 MRI scan showing a brain tumour

A whole series of images of sections through the patient can be produced and stored in a computer memory. As with CT scanning, this enables a three-dimensional image to be generated. Sections through the patient may be viewed from many different angles.

Example

The use of magnetic resonance imaging is now a common diagnostic tool in medicine. However, there may be disadvantages for some patients. Suggested why:
(a) some people may find the scanner to be claustrophobic,
(b) there are additional problems when scanning young children rather than adults.

▶▶

(a) In order to produce a large magnetic field through the patient, the patient must be placed inside a large cylindrical magnet. This can mean that the patient feels trapped.

(b) The total scan takes some time because separate scans of many slices through the patient must be taken. The patient must be very still whilst this is done and this is difficult for children.

Now it's your turn

During an MRI scan, the patient is placed in a large uniform magnetic field on to which is superimposed a highly calibrated non-uniform magnetic field. Explain the purpose of each of these two magnetic fields.

Section 15.3 Summary

- Magnetic resonance imaging (MRI) monitors the concentration of hydrogen nuclei in the body.
- Hydrogen nuclei have 'spin' which causes then to precess in an applied magnetic field.
- The frequency of precession (the Larmor frequency) depends on the magnetic field applied to the nuclei.
- A radio-frequency pulse at the Larmor frequency causes the nuclei to resonate
- On de-excitation, the nuclei give off radio-frequency waves that can be detected and analysed.

Section 15.3 Questions

1 Explain briefly what is meant by nuclear magnetic resonance.

2 (a) By reference to nuclear magnetic resonance imaging, explain what is meant by

(i) the Larmor frequency,
(ii) relaxation time.

(b) Suggest, with an explanation, a typical value for the Larmor frequency.

Exam-style Questions

1 CT scanning and MR scanning provide images of sections through the patient.
  (a) Describe, by reference to what is detected in each case, the differences between these two types of scan.
  (b) Suggest why both types of scan rely on the use of powerful computers.

2 (a) By reference to an X-ray image, explain what is meant by *sharpness* and *contrast*.
  (b) Describe how sharp X-ray images of soft organs, such as the stomach, may be obtained.
  (c) The linear attenuation (absorption) coefficients of X-rays in muscle and in bone are 0.89 cm^{-1}

and 3.0 cm^{-1} respectively. In a particular limb, a bone of thickness 1.5 cm is surrounded by muscle of thickness 3.0 cm. A parallel beam of X-rays is incident on the limb.
  Calculate the fraction of the incident intensity that is transmitted through the limb.

3 (a) Describe the principles behind the use of magnetic resonance to obtain diagnostic information about internal body structures.
  (b) Suggest three reasons why MR scanning would not be used to diagnose a simple bone fracture.

16. Communication

The ability to communicate is a gift that human children learn at a very early age. Human communication may be sophisticated speech or a simple smile. Animals can also communicate. For examples, whales and dolphins communicate using high- and low-pitched sounds that travel through water over many kilometres.

In the case of humans, the ability to communicate has been developed and extended so that messages and information may be sent over great distances in very short time intervals. This chapter looks briefly at some aspects of communication in the modern-day world.

16.1 Amplitude modulation and frequency modulation

At the end of Section 16.1 you should be able to:
- understand the term modulation and be able to distinguish between amplitude modulation (AM) and frequency modulation (FM)
- recall that a carrier wave, amplitude-modulated by a single audio frequency, is equivalent to the carrier wave frequency together with two sideband frequencies
- understand the term bandwidth
- demonstrate an awareness of the relative advantages of AM and FM transmissions.

Modulation

For any system of communication, there must be a transmitter and a receiver. A simple system could be one person speaking to another. One person is the transmitter, the other is the receiver, and the communication system is sound waves.

An alternative system, if the two people are in different rooms, is a microphone connected to a loudspeaker by means of twin wires. The microphone converts the sound wave into a varying electrical signal that is transmitted along the wires to the loudspeaker where it is converted back to sound waves.

Communication of speech could also be achieved using radio waves. The signal from the microphone would be amplified and applied to an aerial. The radio waves produced would then be picked up by a receiving aerial.

After amplification, the received signals would be passed to a loudspeaker. This simple system is shown in Figure 16.1.

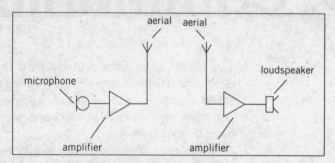

Figure 16.1 Simple radio communication system

This simple system has two serious disadvantages. First, only one system could operate in an area because the receiving aerial would not be selective – it would pick up all signals. Second, the aerial required for the transmission of the low frequencies of sound waves (about 20 Hz to 15 kHz) would be very long and the electrical power required for transmission over long distances would be very large.

These problems are solved by the process known as **modulation**. A high-frequency wave, known as the **carrier wave**, is transmitted. This high-frequency carrier wave has either its amplitude varied or its frequency varied so as to carry information. This is illustrated in Figure 16.2.

> In **amplitude modulation (AM)**, the carrier wave has constant frequency. The amplitude of the carrier wave is made to vary. These variations are in synchrony with the displacement of the information signal.

The rate at which the amplitude of the carrier wave varies is related directly to the frequency of the information signal.

Note: Amplitude modulation is not the same as superposition. Superposition involves the addition of displacements whereas amplitude modulation is achieved through the multiplication of displacements.

> In **frequency modulation (FM)**, the amplitude of the carrier wave remains constant. The frequency of the carrier wave is made to vary in synchrony with the displacement of the information signal.

The change in the frequency of the carrier wave is a measure of the displacement of the information signal. The rate at which the carrier-wave frequency varies is a measure of the frequency of the information signal.

The use of a carrier wave allows different radio stations in the same area to transmit at the same time. Each radio station has a different carrier wave frequency. The receiver is adjusted, or tuned, to the frequency of whichever transmitter is desired. The receiver accepts the signal transmitted on that particular carrier wave and rejects other carrier-wave frequencies.

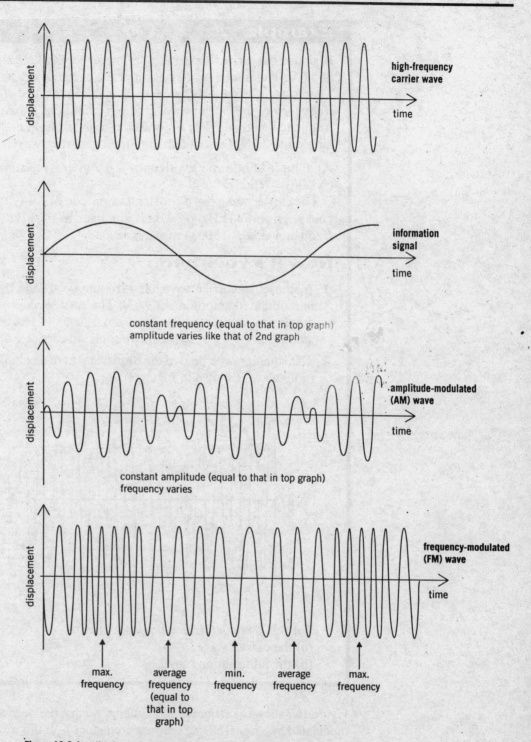

Figure 16.2 Amplitude modulation and frequency modulation of a carrier wave

Example

A sinusoidal carrier wave has a frequency of 800 kHz and an amplitude of 5.0 V. The frequency deviation of the carrier wave is 30 kHz V^{-1}. That is, for every 1.0 V change in displacement of the signal, the frequency of the carrier wave changes by 30 kHz. The carrier wave is frequency-modulated by a sinusoidal signal of frequency 10 kHz and amplitude 2.0 V. Describe the modulated carrier wave.

Amplitude of information signal = 2.0 V giving variation of $(2 \times 30) = 60$ kHz.

The carrier wave has a constant amplitude of 5.0 V. Its frequency changes from 740 kHz to 860 kHz and back to 740 kHz. This change of frequency occurs 10 000 times per second.

Now it's your turn

1 A sinusoidal carrier wave has a frequency of 750 kHz and an unmodulated amplitude of 4.0 V. The carrier wave is to be amplitude-modulated by a sinusoidal signal of frequency 3 kHz and amplitude 0.5 V. Describe the modulated carrier wave.

2 The variation with time of the displacement of an amplitude-modulated radio wave is shown in Figure 16.3.

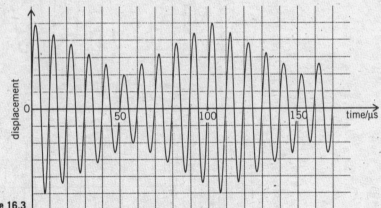

Figure 16.3

For this wave determine the frequency of:
(a) the carrier wave,
(b) the information signal.

Figure 16.3 shows a carrier wave, frequency f_C, that has been amplitude-modulated by a signal having only one frequency f_S. If this waveform is analysed, it is found to be made up of three frequencies $(f_C + f_S)$, f_C and $(f_C - f_S)$. The frequency spectrum of this modulated wave – that is, a graph showing the variation with frequency of the amplitudes of each component – is shown in Figure 16.4.

The central frequency f_C is the frequency of the carrier wave and it has the largest amplitude. The other two frequencies, $(f_C + f_S)$ and $(f_C - f_S)$, are known as **sidebands** or **sideband frequencies**. The frequency f_S is the frequency of the information signal.

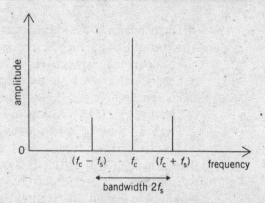

Figure 16.4 The spectrum of a carrier wave amplitude-modulated by a single frequency

> The **bandwidth** is the range of frequencies occupied by the amplitude-modulated waveform.

This bandwidth is equal to $2f_s$.

For the broadcast of music, the information signal will contain a wide range of frequencies, possibly from about 20 Hz to 15 kHz. A typical frequency spectrum of such an amplitude-modulated wave is shown in Figure 16.5.

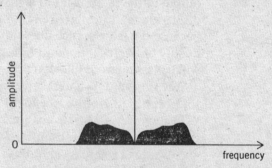

Figure 16.5 Typical spectrum for amplitude-modulated wave transmitting music

In practice, the very high frequencies in music may not be broadcast, so as to reduce the bandwidth of the transmitted signal. Some quality of reproduction will be lost but for normal broadcasting this loss would be minimal.

The frequency spectrum of a frequency-modulated (FM) waveform differs from that for an AM waveform. The FM waveform has additional side frequencies that are multiples of the information signal frequency, resulting in a greater bandwidth for the same range of broadcast frequencies.

Relative advantages of AM and FM transmissions

AM radio transmissions on the long-wave (LW), medium-wave (MW) and short-wave (SW) wavebands are broadcast over very large distances so that one transmitter can serve a large area. FM transmissions have a range of only about 30 km and this range is by line-of-sight. Consequently, many transmitters are required to broadcast over a large area. FM can be used to serve small local areas but it is simpler and cheaper to broadcast using AM.

The bandwidth of AM broadcasts on the LW and MW wavebands is 9 kHz. This means that the highest frequency that can be broadcast is 4.5 kHz. This is quite adequate for speech. However, the lack of quality for music is easily noticed. On the very-high-frequency (VHF) waveband, the bandwidth of an FM radio signal is about 200 kHz, giving a maximum frequency that can be broadcast of about 15 kHz. FM broadcasting does, therefore, offer higher quality.

The long-wave (LW) waveband has a range of frequencies from about 30 kHz to 300 kHz. If the bandwidth on each AM broadcast is 9 kHz, then, theoretically, $(300 - 30)/9 = 30$ transmitters could broadcast in the same area without causing interference between each other. In practice, the number would be less than this, so that 'space' in the frequency spectrum is left between transmitters. For FM broadcasting, $(300 - 30)/200 = 1$ transmitter could broadcast in the LW waveband. The number of transmitters that can share the same waveband is much larger for AM than for FM. For this reason, FM is broadcast only at frequencies in excess of 1 MHz.

Electrical equipment that produces sparks also produces electromagnetic waves. (For this reason, the ignition systems of petrol engines must be 'suppressed'.) These electromagnetic waves are detected as interference by the aerial in a radio. If the radio is tuned into an AM signal, the interference will add to the displacement of the AM signal and will appear as noise in the output of the radio. An FM signal is based on changes of frequency, not displacement. Interference is not picked up by the aerial of the FM radio because the interference does not alter the frequency of the signal. The quality of the reception of FM is generally better than that of AM, since there will be less noise or interference.

AM transmitters and receivers are much simpler, electronically, than those for FM. It is, therefore, cheaper to broadcast and receive AM.

In summary, FM transmissions are more expensive than AM and the area covered by one FM transmitter is much smaller. This difference in coverage is an advantage where local radio is concerned but a disadvantage as regards national radio. The bandwidth necessary for FM is greater. The quality of the received FM signal is much better because of the increased frequency spectrum, and it also suffers less noise.

Example

A particular transmitter is broadcasting an AM signal of frequency 200 kHz. The transmitter is broadcasting a programme of music with a maximum frequency of 4.5 kHz. Determine, for this AM signal:
(a) the wavelength,
(b) the bandwidth.

(a) speed of electromagnetic waves is $3 \times 10^8 \text{m s}^{-1}$
   wavelength = speed/frequency = $(3 \times 10^8)/(200 \times 10^3)$ = **1500 m**
(b) bandwidth = 2×4.5 = **9.0 kHz**

Now it's your turn

Discuss the relative advantages of AM and FM radio transmissions. Wherever appropriate, give numerical values to illustrate your answer.

Section 16.1 Summary

- Modulation is the process whereby either the amplitude (in AM) or the frequency (in FM) of a carrier wave is varied so as to carry information.
- Bandwidth is the range of frequencies that is used in any particular broadcast.
- Bandwidth is higher for FM broadcasts, giving better sound quality.

Section 16.1 Question

During the development of commercial radio, AM broadcasting preceded FM broadcasting. Explain why, in broadcasting:

(a) AM was developed before FM,

(b) there has been a change-over from AM to FM.

16.2 Digital transmission

At the end of Section 16.2 you should be able to:

- recall the advantages of the transmission of data in digital form
- understand that the digital transmission of speech or music involves analogue-to-digital conversion (ACD) on transmission and digital-to-analogue conversion (DAC) on reception
- show an understanding of the effect of sampling rate and the number of bits in each sample on the reproduction of an input signal.

Analogue and digital signals

An information signal that has the same variations with time as the information itself is known as an **analogue** signal. For example, the signal voltage produced by a microphone is analogous to the sound wave incident on the microphone. The voltage output from the microphone is an analogue signal.

Much of the information that we wish to transmit and communicate is analogue in nature (speech, music, television pictures, etc.). When any signal is transmitted over a long distance, it will pick up noise. Noise is not just unwanted sound, but any unwanted random signal that adds to the signal that is being transmitted. Also, the power of the signal becomes less; that is, the signal is attenuated. For long-distance transmission, the signal has to be amplified at regular intervals. The problem is that, on amplification of an analogue signal, the noise is also amplified. The signal becomes distorted or 'noisy'. This effect is shown in Figure 16.6.

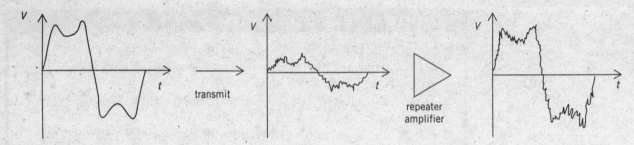

Figure 16.6 Amplification of a noisy analogue signal

A **digital** signal consists of a series of 'highs' and 'lows' with no values between the 'highs' and the 'lows'. The data in the signal is transmitted as a particular sequence of 'highs' and 'lows' or, effectively, a sequence of 1s and 0s. The signal will still suffer from noise and attenuation. However, on amplification, the noisy 1s and 0s can be re-shaped or regenerated to return the signal to its original form. Such amplifiers are known as regenerator amplifiers. These amplifiers 'filter out' any noise, and in so doing, restore the signal. This is illustrated in Figure 16.7.

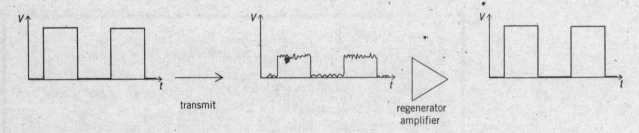

Figure 16.7 Amplification of a noisy digital signal

Unlike an analogue signal, a digital signal can be transmitted over a long distance with regular regenerations without the signal becoming degraded.

Modern digital electronic circuits are, in general, more reliable and cheaper to produce than analogue circuits. In the future, the vast majority of communications systems are likely to be digital.

An additional advantage of digital systems is that extra information, or data, can be added to the transmissions. These extra data are a code for the receiving system so that the transmitted signal may be checked and corrected before the signal is finally reproduced.

Binary numbers

A binary number is a number that has the base 2, whereas a decimal number has the base 10. Table 16.1 shows some decimal numbers and their equivalents in binary notation. A binary number consists of a number of digits, or **bits**. All the binary numbers shown are four-bit numbers. Binary numbers may also be called digital numbers.

Table 16.1

decimal number	binary number
0	0000
1	0001
2	0010
3	0011
4	0100
5	0101
6	0110
7	0111
8	1000
12	1100
13	1101
15	1111

Larger numbers would require digital numbers with more bits. When reading a digital number, the bit on the left-hand side of the number is called the most significant bit, or MSB. This bit has the highest value. The bit on the right-hand side has the least value and is known as the least significant bit, or LSB.

When the LSB is 1, this corresponds to decimal number 1. When the second bit shows 1, this corresponds to decimal number 2. When successive bits show 1 they correspond to decimal numbers 4, 8, 16, 32, 64, etc. Thus, the binary number 1101 corresponds to decimal number 8 + 4 + 0 + 1, or 13. Conversely, decimal number 11 which equals 8 + 0 + 2 + 1 corresponds to binary number 1011.

Example

Convert the following decimal numbers into five-bit digital numbers:
(a) 9,   (b) 23.

(a) **01001**   (b) **10111**

Now it's your turn

Convert the following five-bit numbers into decimal numbers:
(a) 11001,   (b) 01010.

The transmission of a signal

In an analogue signal such as speech or music, the generated voltage signal varies continuously. For digital transmission, this analogue signal must be converted into a digital signal. This is achieved using an **analogue-to-digital converter (ADC)**.

In an analogue-to-digital converter, the analogue voltage is sampled at regular intervals of time, at what is known as the **sampling frequency** or **sampling rate**. The value of the sample voltage measured at each sampling time is converted into a digital (binary) number that represents the voltage value.

For example, assuming that a four-bit number is being used, then the number representing a signal that is sampled as 5.0 V would be 0101. Note that when sampling, the number representing the sample would be the whole number below the actual value of the sampled voltage. If the signal were to be sampled as 11.4 V, then the 4-bit number would be 1011. A sampled signal of 11.8 V would also be 1011.

Figure 16.8a shows a sinusoidal analogue signal that is to be sampled at a sampling rate of 10 kHz. A four-bit system is used for the digital numbers generated.

The sample voltages are shown in Figure 16.8b. These sample voltages are converted into a digital signal, shown in Figure 16.8c, by the analogue-to-digital converter (ADC). After this digital signal had been transmitted, it is converted back into an analogue signal using a **digital-to-analogue converter (DAC)**.

In a digital-to-analogue converter, a digital signal is converted into an analogue signal.

Figure 16.8d shows the analogue signal that has been recovered. The recovered signal is very 'grainy', consisting of large 'steps'. The size of these steps, and hence the faithfulness of the reproduction of the initial analogue signal, can be improved by using more voltage levels and sampling at a higher frequency.

The number of bits in each digital number limits the number of voltage levels. In this example, there are four bits and $2^4 = 16$ levels. In practice eight or more bits would be used for sampling. An eight-bit number would give $2^8 = 256$ levels.

The choice of sampling frequency also determines the amount of information that can be transmitted. About 100 years ago, Nyquist showed that, in order to recover an analogue signal of frequency f, then the signal must be sampled at a frequency greater than $2f$. The greater the sampling frequency, the more faithful is the reproduction of the original signal. For example, for good quality reproduction of music, the higher audible frequencies must be present; that is, frequencies up to about 20 kHz. For compact discs (CDs) the sampling frequency is therefore 44.1 kHz. This quality of reproduction is not required for speech, and it would prove costly. In a telephone system, the sampling frequency is 8 kHz and the highest frequency to be transmitted is limited to 3.4 kHz.

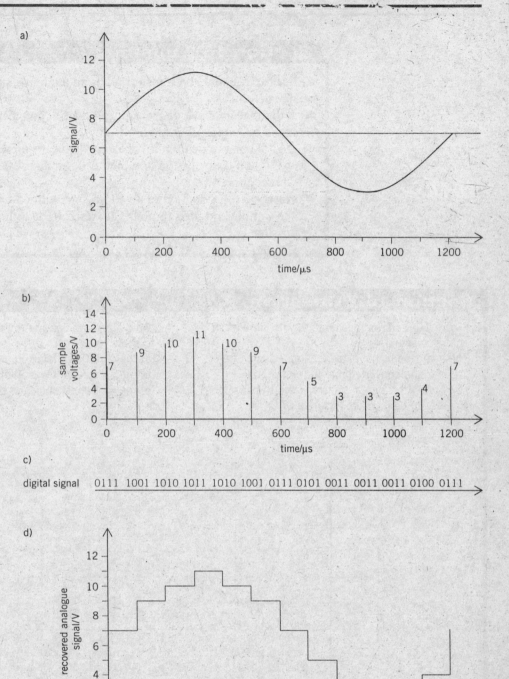

Figure 16.8 Analogue-to-digital and digital-to-analogue conversion

Section 16.2 Summary

- Information may be transmitted as an analogue signal or as a digital signal.
- A digital signal can be transmitted over a long distance without the signal becoming degraded.
- Digital communication involves analogue-to-digital conversion (ADC) on transmission and digital-to-analogue conversion (DAC) on reception.
- For good reproduction of an audio signal, the number of bits in a sample and the sampling rate need to be as high as possible.

Section 16.2 Question

Figure 16.9 shows the variation with time of an analogue signal.
The analogue signal is to be sampled at a rate of 1.0 kHz and converted into four-bit digital numbers. After digital transmission, the analogue signal is to be recovered using a DAC.

Draw diagrams (similar to those in Figure 16.8) to show
(a) the sample voltages in analogue form,
(b) the sample voltages in digital form,
(c) the recovered analogue signal.

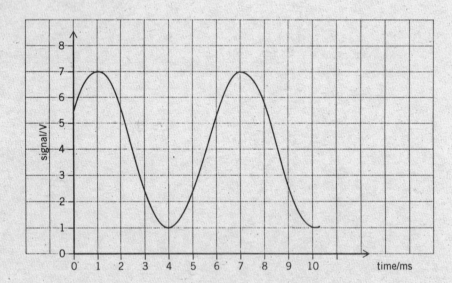

Figure 16.9

16.3 Channels of communication

At the end of Section 16.3 you should be able to:
- appreciate that information may be carried by a number of different channels, including wire-pairs, coaxial cables, radio and microwave links, and optic fibres
- discuss the relative advantages and disadvantages of channels of communication in terms of available bandwidth, noise, cross-linking, security, signal attenuation, repeaters and regeneration, cost and convenience
- recall the frequencies and wavelengths used in different channels of communication
- understand and use signal attenuation expressed in dB and dB per unit length
- recall and use the expression, number of dB = $10 \lg (P_1/P_2)$, for the ratio of two powers
- describe the use of satellites in communication
- recall the relative merits of both geostationary and polar-orbiting satellites for communicating information.

In the previous section, the conversion of signals between digital and analogue was considered. Between conversions, the signal has to be transferred from one place to another. This may be achieved in various ways using different channels of communication including wire-pairs, coaxial cables, radio and microwave links, and optic fibres.

Wire-pairs

Wire-pairs provide a very simple link between a transmitter of information and the receiver. In the early days of electrical communication using Morse code, the transmitter and receiver were connected directly to one another by means of two copper wires (or one copper wire and an 'earth return'). In modern communication systems, wire-pairs are used mainly for short-distance communication at relatively low frequencies, for example linking telephones to the nearest exchange. In the simplest of applications, they are used for linking a door bell in a house to the switch outside the door.

High-frequency electrical signals lose their energy over short distances in wire-pairs – the signals have a high attenuation. This high attenuation is partly due to the heating caused by the electrical resistance of the wires and partly due to the emission of radiation (radio waves), since the wires act as aerials. A signal going any distance in a wire-pair must be amplified at regular intervals along the length of the wires.

A further problem is that, since the wires acts as aerials, they pick up any electromagnetic waves and these unwanted signals, or noise, will cause deterioration of the signal. Wire-pairs close to one another will pick up each other's signals. This effect is called **cross-talk** or **cross-linking** and means that wire-pairs give rise to poor security since the signals can be 'tapped' easily.

The bandwidth of a wire-pair is only about 500 kHz and, as a result, wire-pairs are limited as to the amount of information that they can carry.

To summarise, wire-pairs:

- are used mainly for short-distance communication
- cause high attenuation of a signal
- easily pick up noise
- suffer from cross-talk and are of low security
- have limited bandwidth.

Coaxial cables

The construction of a coaxial cable is shown in Figure 16.10. It consists, basically, of two wire conductors. Any signal is transmitted along a central inner conductor that is covered by an insulator. The second conductor is in the form of thin wire braid that completely surrounds the insulator. This wire braid acts as the 'return' for the signal and is earthed. The braid is covered by a protective layer of insulation.

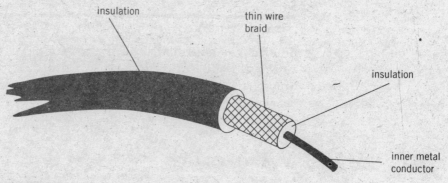

Figure 16.10 Coaxial cable

The earthed outer braiding shields the inner conductor from external interference. Coaxial cables are, therefore, far less noisy than wire-pairs and provide better security. Attenuation is also reduced. Consequently, repeater amplifiers can be further apart on coaxial cables.

The bandwidth of a coaxial cable is about 50 MHz. Much more information can be carried along a coaxial cable than along a wire-pair. However, coaxial cable is more costly. Coaxial cable is used to connect an aerial and a television.

To summarise, comparing a wire-pair and a coaxial cable, the coaxial cable:

- is more costly
- causes less attenuation
- is less noisy and is more secure
- has a larger bandwidth.

Radio waves

Alternating current in a wire acts as an aerial. Energy is radiated from the aerial in the form of electromagnetic waves. These waves travel outwards from the aerial with the speed of light. Electromagnetic waves in the frequency range 30 kHz to 3 GHz are generally referred to as radio waves.

The first radio waves to be used for communication were of very low frequency and very long wavelength. The radio waves were switched on and off so that communication was by Morse code. Later use of higher frequencies and the development of amplitude modulation enabled voice communication. Further developments, including FM broadcasts and the use of different carrier frequencies, enabled higher-quality communication and also more radio stations to operate in the same area.

The choice of aerial for broadcasting determines whether the radio waves are emitted in all directions (for broadcasting to an area) or, in one direction only (for point-to-point communication). Similarly, for the receiving of radio signals, the choice of aerial is determined by whether the signal from one direction, or all directions, is to be received. Aerials with dish reflectors enable the radio waves to be transmitted as a parallel beam.

The intensity of the radio waves will always be reduced (attenuated) as the distance from the transmitter increases. The degree of attenuation depends on the frequency of the waves. Some data on frequencies and ranges are given in Table 16.2.

Table 16.2 Radio waves

name of radio wave	frequency range	distance
space wave	greater than 30 MHz	line-of-sight between transmitter and receiver, plus satellite communication
sky wave	3 MHz → 30 MHz	worldwide, as a result of multiple reflections from the ionosphere and the ground or sea
surface wave	below 3 MHz	up to 1000 km

For simple radio communication, AM broadcasts on the medium-wave (MW) and long-wave (LW) wavebands are relatively cheap and technically less complex. They are transmitted as surface waves and provide coverage over large areas.

The wavelength of the radio waves determines the length of the aerial. For mobile phones, the aerial must be, for the sake of convenience, short and hence the wavelength must also be relatively short. High-frequency space waves are used. Remember that, as the frequency of the carrier wave increases, the bandwidth also increases.

The frequency bands used for radio communication are summarised in Table 16.3.

Table 16.3 Radio-frequency bands

communication type	frequency range	wavelength in air	frequency band
LW radio	30 kHz → 300 kHz	10 km → 1 km	low frequencies LF
MW radio	300 kHz → 3 MHz	1 km → 100 m	medium frequencies MF
SW radio	3 MHz → 30 MHz	100 m → 10 m	high frequencies HF
FM radio	30 MHz → 300 MHz	10 m → 1 m	very high frequencies VHF
TV broadcasting and mobile phones	300 MHz → 3 GHz	1 m → 10 cm	ultra-high frequencies UHF
microwave links	3 GHz → 30 GHz	10 cm → 1 cm	super-high frequencies SHF
satellite links	30 GHz → 300 GHz	1 cm → 1 mm	extra-high frequencies EHF

Microwaves

Microwaves are electromagnetic waves that can be considered alongside other radio waves, as shown in Table 16.3. They are in the frequency range 3 GHz to 30 GHz and are generally used for point-to-point communication since, for use on Earth, the range of the transmissions is limited to line-of-sight.

Reflecting parabolic dishes are used so that the transmission is in the form of a parallel beam and so that as much wave power as possible can be focused onto the receiving aerial, as shown in Figure 16.11.

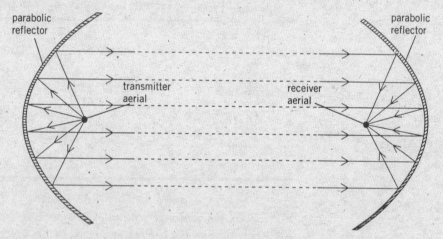

Figure 16.11 The use of parabolic reflectors with microwaves

It should be remembered that the reflecting parabolic dish is not the aerial. The aerial is found at the focus of the reflecting dish.

The bandwidth of a microwave link is of the order of 1 GHz. This large bandwidth means that the microwave beam has a large capacity for transmitting information.

Optic fibres

An optic fibre consists of a fine strand of very pure glass, surrounded by a protective covering. The glass fibre itself is thinner than a hair. Pulses of light or infra-red radiation travel along the fibre as a result of total internal reflection. These pulses carry digital information along the fibre.

The radiation pulses are provided by lasers and have very high frequencies, of the order of 10^8 MHz. In theory, a single pulse need only last for 10^{-14} s. However, lasers cannot be controlled at such high frequencies and the duration of a single pulse, or bit, is governed by the frequency at which the laser can be switched on and off. Technology is always improving and, at present, the frequency is about 800 MHz. Such high frequencies mean a large bandwidth, so many different phone calls can share the same optic fibre.

Fibre optic transmission is being installed in many different types of communication.

Figure 16.12 Optic fibres making up an optic cable

The advantages of optic fibres are summarised below:

- large bandwidth, giving rise to large transmission capacity
- much lower cost than metal wires
- diameter and weight of cable is much less than metal cable, hence easier handling and storage
- much less signal attenuation, so far fewer regenerator amplifiers are required, reducing the cost of installation
- do not pick up electromagnetic interference, so very high security and negligible cross-talk
- can be laid alongside existing routes such as electric railway lines and power lines.

Signal attenuation

When a signal passes along a wire or a fibre, it loses power. This loss of power is referred to as **attenuation**, and the amount of attenuation increases as the distance that the signal travels increases. In the case of an electrical signal in a metal wire, signal power is lost as heating of the wire. In optic fibres, light power is lost as a result of absorption in impurities in the glass and also scattering due to imperfections. A beam of electromagnetic waves travelling through air also loses power as a result of absorption and scattering.

In order that a signal may be detected, the power of the signal must be a minimum number of times greater than the noise power. This ratio, the **signal-to-noise ratio**, may be, for example, 50 to 100. Signals are amplified and the output of the amplifier is a certain number of times greater than the input. The **amplifier gain** could be 100 000, or 10^5. When a microwave signal is sent from Earth to a satellite, the signal power may be reduced by a factor of 10^{19}! It can be seen that in comparing power levels the ratio of two numbers is involved and that this ratio can be very large or very small. In order to condense the scale of such variations and to make the numbers more manageable, the power levels are compared on a logarithmic scale.

The result of this comparison gives the ratio in a unit known as the **bel**, which has the symbol B.

$$\text{number of bels} = \lg \left(\frac{P_2}{P_1} \right)$$

where P_2 and P_1 are the two powers that are being compared. Since the bel is a large unit, the ratio is usually expressed as the **decibel**, where 1 bel = 10 decibels or 10 dB.

$$\text{number of decibels (dB)} = 10 \lg \left(\frac{P_2}{P_1} \right)$$

It should be noted that if P_2 is greater than P_1, the dB number is positive and there has been amplification. Where P_2 is less than P_1, there is attenuation and the number of dB will be negative.

Example

A signal having a power of 2.4 μW is amplified in a two-stage amplifier. The first stage has a gain of 18 dB and the second stage provides a further amplification of 25 dB. Calculate:
(a) the total gain of the two-stage amplifier,
(b) the power of the output signal from the amplifier.

(a) For the first stage, the input is V_I and the output is V, giving a gain of V_I / V.

For the second stage, the input is V and the output is V_0, giving a gain of V_0 / V.

Total gain = $V_0 / V_I = V_0 / V \times V_I / V$
= gain of first stage × gain of second stage

Gain, when expressed in dB, is a logarithm. Therefore, when two gains are multiplied together, then the gains in dB must be added.

$\lg$ (total gain) = $\lg$ (gain of first stage) + $\lg$ (gain of second stage)
= 18 + 25 = **43 dB**

(b) Gain in dB = $10 \lg (P_2 / P_1)$
$43 = 10 \lg (P_2 / (2.4 \times 10^{-6}))$
$4.3 = \lg (P_2 / (2.4 \times 10^{-6}))$

Taking antilogs,

$10^{4.3} = P_2 / (2.4 \times 10^{-6})$
$P_2 = \textbf{0.048 W}$

Now it's your turn

A signal is amplified by two amplifiers A and B, connected in series. The input power to amplifier A is $17\,\mu W$. Amplifier A has a gain of 20 dB. The output power from amplifier B is 0.26 W.

Determine:
(a) the overall gain, in dB, of the two amplifiers,
(b) the gain of amplifier B.

Maths Note

The logarithm of a number, to the base 10, is the power to which the number ten must be raised in order to give that number.

For example,
$100 = 10^2$ and so the logarithm to the base ten of 100 (lg 100) is 2.00
and
$2.00 = 10^{0.301}$ and so the logarithm to the base ten of 2.00 (lg 2.00) is 0.301
also
$10^{1.699} = 50.0$ and so lg 50.0 = 1.699

Remember that when two numbers are multiplied together, if the numbers are expressed as numbers to the base ten, then the powers of the base ten are added. For division, the powers are subtracted.

For example, $75 \times 3.00 = 10^{1.875} \times 10^{0.477}$
Adding the powers, $75 \times 3.00 = 10^{2.352}$
$10^{2.352} = 225$
(We already know that $75 \times 3 = 225$!)

Also, $75 \div 3.00 = 10^{1.875} \div 10^{0.477}$
Subtracting the powers, $75 \div 3.00 = 10^{1.398}$
$10^{1.398} = 25$
(We already know that $75 \div 3 = 25$!)

The gain of an amplifier is usually expressed in dB.
Now, gain (in dB) $= 10\lg(P_{out}/P_{in})$ and so the gain, in dB, is a logarithm to the base ten.

When two amplifiers are connected in series, the combined gain is found *either* by multiplying together the two actual gains *or* by adding together the gains, when expressed in dB.

In a transmission line, the amount of attenuation is dependent on the length of the line. As a result, **attenuation per unit length** is normally quoted.

$$\text{attenuation per unit length} = \left(\frac{1}{L}\right) 10 \lg \left(\frac{P_2}{P_1}\right)$$

where L is the length of the transmission line. Attenuation per unit length is measured in dB km^{-1}.

An optic fibre may have an attenuation per unit length of between 1 and $3\,dB\,km^{-1}$, whereas for a coaxial cable, it varies with frequency, increasing from about $3\,dB\,km^{-1}$ at 10 MHz to about $40\,dB\,km^{-1}$ at 3 GHz.

Example

The signal input to an optic fibre is 7.0 mW. The average noise power in the fibre is 5.5×10^{-19} W and the signal-to-noise ratio must not fall below 24 dB. The fibre has an attenuation of $1.8\,dB\,km^{-1}$. Calculate:
(a) the minimum effective signal power in the cable,
(b) the maximum uninterrupted length of the optic fibre through which the signal can be transmitted.

(a) number of decibels (dB) = $10\lg (P_2/P_1)$
$$24 = 10\lg (P_{min}/5.5 \times 10^{-19})$$

where P_{min} is the minimum effective signal power.
$$10^{2.4} = P_{min}/5.5 \times 10^{-19}$$
minimum effective signal power $P_{min} = \mathbf{1.38 \times 10^{-16}\,W}$

(b) total attenuation of signal to reach minimum = $10\lg (P_{input}/P_{min})$
$$= 10\lg (7.0 \times 10^{-3})/(1.38 \times 10^{-16})$$
$$= 137\,dB$$

maximum uninterrupted length = total attenuation/attenuation per unit length
$$= 137/1.8 = \mathbf{76\,km}$$

Now it's your turn

A laser is used to provide the input to an optic fibre of length 65 m. The power input to the fibre is 6.20 mW and the output at the other end of the fibre is 6.07 mW.
(a) Calculate the power loss, in $dB\,km^{-1}$, in the optic fibre.
(b) The average noise power in the fibre is 2.4×10^{-18} W and the minimum acceptable signal-to-noise ratio is 22 dB. Calculate the maximum length of uninterrupted fibre through which the signal could be transmitted.

Communication satellites

Long-distance communication using radio waves is possible on the MW waveband (as surface waves) and the SW waveband (as sky waves). However, for modern communication systems, there are three major disadvantages.

1 Long-distance communication using sky waves is unreliable in that it depends on reflection from layers of ions in the upper atmosphere. These layers of ions vary in height and density, giving rise to variable quality of signal. Surface waves are also unreliable because there is poor reception in hilly areas.
2 The wavebands available on MW and SW are already crowded.
3 The bandwidths that are available are narrow and consequently unable to carry large amounts of information.

Satellite communication enables more wavebands to be made available and at much higher frequencies, thus giving rise to a much greater data-carrying capacity. The basic principle of satellite communication is shown in Figure 16.13.

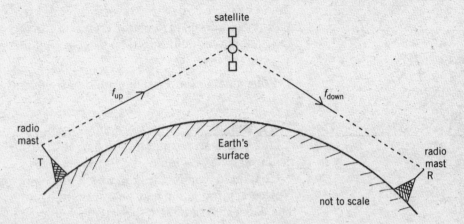

Figure 16.13 Satellite communication

A carrier wave of frequency f_{up} is sent from a transmitter T on Earth to a satellite. The satellite receives the greatly attenuated signal. The signal is amplified and the carrier frequency is changed to a lower value, f_{down}. The carrier wave is then directed back to a receiver R on Earth. The carrier wave frequencies f_{up} and f_{down} are different so that the very low power signal received from Earth is not swamped by (can be distinguished from) the high power signal that is transmitted back to Earth.

Values of the 'uplink' frequency f_{up} and the 'downlink' frequency f_{down} might be 6 GHz and 4 GHz respectively (the 6/4 GHz band). Other bands are 14/11 GHz and 30/20 GHz.

The communication satellite may be in **geostationary orbit**. In this case, the satellite orbits the Earth with a period of 24 hours at a height of 3.6×10^4 km above the Earth's surface. The satellite orbits in the same direction as the rotation of the Earth (from west to east) and the orbit is above the equator.

From the viewpoint of a person on Earth, the satellite remains above the same point on the Earth's surface. The transmitting aerial and the receiving aerial on Earth both have large parabolic reflectors. One advantage of geostationary satellites is that these aerials can be fixed in position since the satellite does not have to be tracked.

A geostationary satellite can have a permanent link with a transmitting ground station. Communication is maintained with any point on the Earth's surface that can receive the signal from the satellite. A number of satellites with overlapping areas, such that any area can receive a signal from one of the satellites, allows for long-distance communication. This removes the need for long-distance submarine cables. International television broadcasting is possible, allowing 'live' events in one country to be viewed in another.

Geostationary satellites are in equatorial orbits, which means that communication in polar regions may not be possible because a satellite will not be in line-of-sight. A further problem results from the height of the orbit.

Between transmission and receipt of the signal, the wave must travel at least twice the distance between the satellite and Earth; that is, 7.2×10^4 km. For the wave travelling at 3.0×10^8 m s^{-1}, the time taken is 0.24 s. This delay would be increased where several satellites were involved, and would not be acceptable during a telephone conversation. To avoid these problems, geostationary satellites may be used in conjunction with optic fibres.

Polar satellites are satellites that have low orbits and pass over the poles.

The orbital period of these satellites is about 100 minutes. Since the Earth rotates below such satellites, then in any period of 24 hours the satellite will pass over every region of the Earth's surface.

Continuous communication with a single polar satellite is not possible. However, information may be transmitted to the satellite while it is overhead. The data can be stored in the satellite and then transmitted back to Earth when the satellite is over the appropriate area. Continuous communication is possible using a number of polar satellites in orbits that are inclined to one another so that at least one satellite is always above the transmitter and receiver. In this case, the aerials must track the satellites in their orbits. The advantage of using such satellites is that their orbital height is only of the order of 10^5 m (a few hundred km) and thus delays in telephone conversations are not noticed.

Since polar satellites pass over the whole of the Earth in any 24-hour period, they are used for remote sensing. Such uses include military espionage, geological prospecting and weather forecasting. The Global Positioning System (GPS, or 'sat nav') uses the signals from a number of satellites that are not in geostationary orbits.

Section 16.3 Summary

- Information may be communicated using wire-pairs, coaxial cables, radio waves and microwaves, and optic fibres.
- The signal attenuation, the power gain on amplification, and the signal-to-noise ratio, are all given as a ratio of two powers.
- The ratio is expressed logarithmically and measured in decibels (dB): number of dB = $10 \lg (P_2/P_1)$
- Satellites in geostationary and in polar orbits are used for communication.

Section 16.3 Questions

1 The use of geostationary satellites and optic fibres for world-wide communication have, for many applications, replaced the use of sky waves.

   (a) Outline the use of sky waves for world-wide communication.

   (b) Suggest reasons why geostationary satellites and optic fibres have replaced sky waves.

   (c) Explain why a combination of geostationary satellites and optic fibres is preferable to geostationary satellites alone for communication over very long distances.

2 Both wire-pairs and coaxial cables are used in communication systems.
  (a) Suggest two uses of wire-pairs in communication systems.
  (b) Explain the structure of a coaxial cable.
  (c) Compare the use of wire-pairs and coaxial cables in communication systems.

3 (a) Outline the principles of communication using geostationary satellites.
  (b) Suggest one disadvantage of using several geostationary satellites for communication between opposite sides of the Earth.
  (c) Discuss the advantages and disadvantages of using polar, rather than geostationary, satellites for radio communication.

16.4 The mobile phone system

At the end of Section 16.4 you should be able to:

■ understand that in a mobile phone system, the public switched telephone network (PSTN) is linked to base stations via a cellular exchange

■ understand the need for an area to be divided into a number of cells, each cell served by a base station

■ understand the role of the base station and the cellular exchange during the making of a call from a mobile phone handset

■ recall a simplified block diagram of a mobile phone handset and understand the function of each block.

The public switched telephone network (PSTN)

In the early days of telephones, every caller was connected directly to all other callers. This was possible because there were few phones and all were local, for example, in one building. As the number of phones increased, the telephone 'exchange' was introduced. When a call was to be made, the caller would contact the exchange and the operator at the exchange would make the electrical connections necessary for the call to be made.

If the person to be called was not connected to the local exchange of the caller, the operator at the local exchange would contact the other person's local exchange via a 'trunk' exchange. Trunk exchanges were connected by trunk lines and so, from this, the expression 'to make a trunk call' developed.

In modern systems, the telephone operator has been replaced by electronic relays that carry out the switching operations. International exchanges, called **gateways**, have been introduced so that telephone communication may be worldwide. The **public switched telephone network (PSTN)** uses the principle of exchanges and is illustrated in Figure 16.14.

In the system shown, the caller is connected to the PSTN through a local exchange. Each caller has a 'fixed' line, either a wire-pair or an optic fibre, that links the 'subscriber' (the phone user) to the local exchange. The fixed line means that the user has limited mobility while making a call.

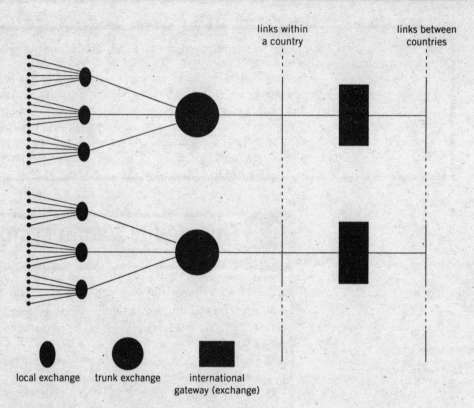

Figure 16.14 The public switched telephone network (PSTN)

During the latter part of the twentieth century, mobile phone systems were developed that did not require a permanent line to the local exchange. The system is shown in Figure 16.15.

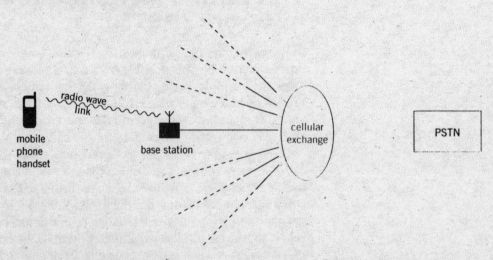

Figure 16.15 The mobile phone system

A mobile, or cellular (cell), phone is a small portable radio transmitter and receiver. When a call is made, the transmitter provides a radio link between the caller and a **base station**. There is a cable that links the base station to a **cellular exchange**. The cellular exchange provides access to the PSTN.

The mobile phone network

When a mobile phone is linked to a base station, this is achieved using a particular carrier-wave frequency. The range of carrier-wave frequencies is limited to a number that is far less than the number of mobile phones. Consequently, each mobile phone cannot have its own carrier frequency, and the same frequency must be shared with many other phones at the same time. This is made possible by having a network of base stations.

The carrier wave between the handset and the base station is in the UHF range of frequencies (see Table 16.3). Not only does this mean that the aerial on (or in) the handset is conveniently short but, importantly, the base station aerials have a limited range and operate on low power. The country to be covered by the mobile phone network is divided into **cells**, each cell having its own base station. The base station is usually near the centre of the cell so that the transmitter, sending out radio waves in all directions, covers the whole of the cell without overlapping significantly into neighbouring cells. The patchwork of cells covers the whole area and is illustrated in Figure 16.16. For convenience, each cell is shown as being hexagonal.

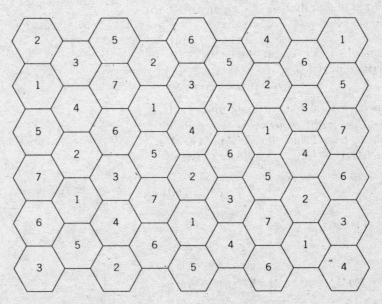

Figure 16.16 An arrangement of cells

In order that interference between phone calls does not occur near the boundary between two cells where the signals from the two base stations overlap, the band of carrier-wave frequencies allocated to neighbouring base stations is different. This is illustrated on Figure 16.16 by the number in each cell.

When a mobile phone handset is switched on, it transmits a short signal at regular intervals in order to identify itself. This signal is detected by one or more base stations. The base stations transfer the signal to the cellular exchange where a computer selects the base station with the strongest signal and also allocates a carrier-wave frequency for communication between the base station and the mobile phone.

During the time that the handset is switched on, or is actually in use, the signal strength is monitored at the cellular exchange. If the caller moves from one cell to another, then the relative strengths of the signal from base stations changes. The call from the handset is re-routed through the base station that provides the strongest signal.

The mobile phone handset

A simplified block diagram of a mobile phone handset is shown in Figure 16.17. Basically, it consists of a radio transmitter and receiver.

The caller speaks into the microphone. The microphone produces an analogue voltage of the sound wave that is amplified using the audio-frequency (a.f.) amplifier. The analogue voltage is converted into a digital signal using the analogue-to-digital converter (ADC). The parallel-to-series converter takes the whole of each digital number and emits it as a series of bits.

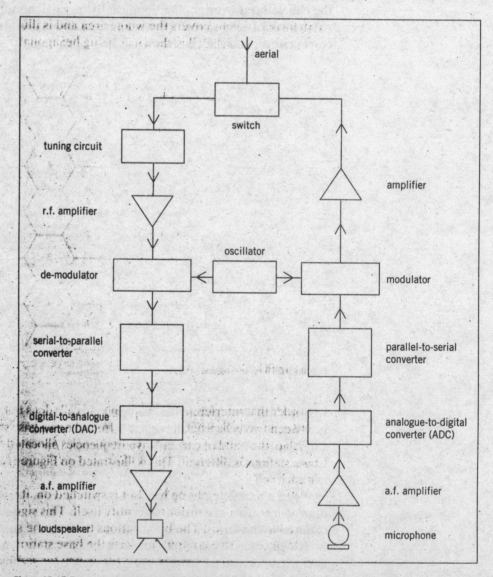

Figure 16.17 Block diagram of a mobile phone handset

The frequency of the oscillator is allocated by the computer at the cellular exchange. This carrier-wave frequency is modulated by the series of bits from the parallel-to-series converter. Finally, the modulated carrier wave is amplified and switched to the aerial where it is transmitted as a radio wave.

Any signal received at the aerial is switched to a tuning circuit. This circuit selects only the carrier-wave frequency that has been allocated to the handset by the computer at the cellular exchange. The selected signal is amplified by the radio-frequency (r.f.) amplifier. It is demodulated so that the information signal is in digital form. The series-to-parallel converter enables each sampled digital voltage to be separated and these digital numbers are then converted into an analogue waveform in the digital-to-analogue converter (DAC). Finally, the analogue signal is amplified before sound is produced in the loudspeaker.

Section 16.4 Summary

- In a mobile phone network, a region is divided into a number of cells, each with a base station.
- The mobile phone is connected to the public switched telephone network (PSTN) through a base station and a cellular exchange.
- A computer at the cellular exchange selects the carrier-wave frequency for each handset and also the base station that is to serve the handset.

Section 16.4 Questions

1 Outline the role of the computer at the cellular exchange in a mobile phone network.

2 (a) Draw a labelled block diagram of a mobile phone handset.
  (b) Outline the function of each block on the diagram.

Exam-style Questions

1 (a) Explain what is meant by:
   (i) an amplitude-modulated (AM) wave,
   (ii) a frequency-modulated (FM) wave.
  (b) Describe the relative advantages of broadcasting using AM and FM.
  (c) It is proposed to transmit the full range of audio frequencies using an AM broadcast. The carrier wave has a frequency of 400 kHz.

   (i) Suggest the range of audio frequencies to be transmitted and hence the bandwidth of the signal.
   (ii) Sketch a graph of the variation with frequency of the power of the signal at a time when all audio frequencies are being transmitted. Label the frequency axis with relevant values of frequency.

▶▶

2 Figure 16.18 shows the variation with time of the voltage produced by a microphone.

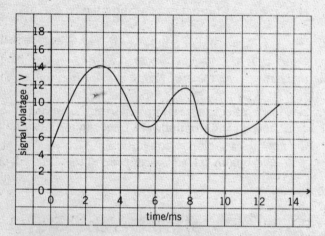

Figure 16.18

(a) Suggest how it is known that there is more than one frequency in the signal.

(b) The signal voltage is passed into a four-bit analogue-to-digital converter that samples the signal at a frequency of 500 Hz. The first sample is taken at time $t = 0$.
   (i) Calculate the time interval between samples.
   (ii) Copy the boxes below.

0101							

The first box shows the binary number for the sample corresponding to time $t = 0$. On your copy of the boxes,
■ underline the most significant bit (MSB) for time $t = 0$
■ complete the boxes for the next six samples.

(c) The binary numbers in (b) (ii) are passed through a digital-to-analogue converter (DAC). Draw a graph showing the variation with time of the recovered analogue signal.

3 The signal output of a microphone is 3.2 μW. This signal is amplified by an amplifier having a gain of 55 dB.

(a) Calculate the power of the signal from the amplifier.

(b) The amplified signal is passed along a wire-pair having an attenuation of 12 dB km^{-1} to a receiver. The constant noise power in the wire-pair is 0.032 μW. For the signal to be distinguished at the receiver, the signal-to-noise ratio must be at least 20 dB. Determine, for the maximum uninterrupted length of wire-pair:
   (i) the attenuation, in dB, of the signal,
   (ii) the length of the wire-pair.

4 The number of television programmes that are transmitted digitally, rather than in analogue form, is increasing.

(a) Discuss the advantages and disadvantages of transmitting digital, rather than analogue, signals.

(b) Television signals may be broadcast using terrestrial waves, geostationary satellites or optic fibres. Compare the relative merits of these three systems.

Answers

Chapter 1

Now it's your turn (page 4)

1 $10\,800\,cm^2 = 1.08 \times 10^4\,cm^2$
2 1.0×10^9

Now it's your turn (page 5)

1 $kg\,m^{-3}$
2 $kg\,m\,s^{-2}$

Now it's your turn (page 6)

1 **(a)** Yes   **(b)** Yes
2 $m^2 s^{-2}$

Now it's your turn (page 8)

(a) 100–150 g
(b) 50–100 kg
(c) 2–3 m
(d) 0.5–1.0 cm
(e) $0.5\,cm^3$
(f) $4 \times 10^{-3}\,m^3$
(g) $220\,m\,s^{-1}$
(h) 310 K

Section 1.1 (page 9)

1 **(a)** $6.8 \times 10^{-12}\,F$ **(b)** $3.2 \times 10^{-5}\,C$
   (c) $6.0 \times 10^{10}\,W$
2 800
3 4.6×10^4
4 **(a)** $kg\,m^2 s^{-2}$ **(b)** $m^2 s^{-2}\,K^{-1}$
5 $kg\,m^{-1} s^{-2}$
6 $kg\,m^2$

Now it's your turn (pages 10–11)

1 **(a)** scalar   **(b)** vector   **(c)** scalar
2 **(a)** 32 dollars, scalar   **(b)** 8 dollars credit, vector

Now it's your turn (page 13)

1 **(a)** $7\,km\,h^{-1}$ **(b)** $1\,km\,h^{-1}$
2 11 N at an angle to the 6.0 N force of 56°, in an
   anticlockwise direction

Now it's your turn (page 15)

1 **(a)** $250\,km\,h^{-1}$ **(b)** $180\,km\,h^{-1}$
2 **(a)** $1.0\,m\,s^{-1}$ **(b)** $9.1\,m\,s^{-1}$

Section 1.2 (page 16)

1 **(a)** vector   **(b)** scalar   **(c)** vector
2 Velocity has direction, speed does not. Velocity is
   speed in a certain direction.
3 Student is correct. Weight is a force which acts
   vertically downwards.
4 Direction of arrow gives direction of vector. Length
   of arrow drawn to scale represents magnitude of
   vector quantity.
5 **(a)** 690 N   **(b)** 210 N   **(c)** 510 N at an angle of
   28° to the 450 N force
6 upstream at 78° to the bank
7 120 N at an angle of 25° to the 50 N force in an
   anticlockwise direction

Exam-style Questions (pages 16–17)

3 **(b) (i)** 92 N   **(ii)** 77 N
   (c) (i) 59 N   **(ii)** 59 N
4 **(a) (i)** $18.3\,m\,s^{-1}$ **(ii)** 29° above horizontal
   (b) (i) $10\,m\,s^{-1}$ **(ii)** 33°

Chapter 2

Now it's your turn (page 23)

12.52 mm

Now it's your turn (page 28)

15 g; ±2 g

Now it's your turn (page 32)

10

Now it's your turn (page 34)

37.8 °C; about ±0.2 deg C

Now it's your turn (page 40)

(a) 28 mV
(b) 9.9 mV r.m.s.

Now it's your turn (page 44)

2 (a) ±0.06 A
  (b) (2.01 ± 0.09) A

Now it's your turn (page 45)

2 (a) ±0.2 mm
  (b) ±6%

Now it's your turn (page 47)

1 (a) end of ruler damaged; measure length from
    10 cm mark
  (b) parallax error; place pencil in contact with scale,
    over the graduations, and define the ends with
    set-squares
2 (a) micrometer screw gauge
  (b) zero error on drum
  (c) averaging reduces random errors; spiral readings
    allow for a non-circular cross-section, and moving
    along the length of the wire allows for any taper

Now it's your turn (page 48)

(1.2 ± 0.1) cm

Now it's your turn (page 50)

(870 ± 40) cm^3

Chapter 3

Now it's your turn (page 54)

1 5.3×10^{-11} m
2 3200 s or 53 min
3 6 m s^{-1}

Now it's your turn (page 58)

1 6.0 m s^{-2}
2 3.3 s

Sections 3.1–3.4 (page 59)

1 30 km
2 180 m s^{-1}
4 3.6 h; 610 km h^{-1}
5 −5.0 m s^{-2}

Now it's your turn (page 64)

1 2.5 m s^{-2}
2 7.5 s; 15 m s^{-1}
3 8.2 m s^{-1} upwards

Now it's your turn (page 66)

(a) 1.8 m s^{-2} (b) 770 m

Now it's your turn (page 70)

1 6.1 m s^{-1}
2 20 m

Sections 3.5–3.7 (page 71)

1 75 m
2 140 m s^{-1}
3 47 m s^{-1}
4 B; A
5 3.5°

Exam-style Questions (page 71)

2 8
3 (a) 61°; 2.8 m (b) 3.9 m s^{-1} (c) 1.4 s
4 (a) 9.4 m s^{-1} (b) +10%

Chapter 4

Now it's your turn (page 74)

1 (a) 180 J (b) 30 J
2 (a) 15.5 J (b) 7.6 J

Now it's your turn (page 76)

1 1800 J
2 75 J

Section 4.1 (page 77)

1 90 J; 0; 78 J; 0.15 m; 36°; 2.6×10^3 N
2 (a) 7.9 N (b) 0.19 J
3 8.66×10^4 J

Now it's your turn (page 80)

(a) 1.3×10^4 J gain (b) 12 J gain
(c) 2.4×10^9 J loss

Now it's your turn (page 82)

1 3.1×10^5 J
2 (a) 1000 J (b) 3000 J

Now it's your turn (page 83)

(a) kinetic energy at lowest point → potential energy at highest point → kinetic energy at lowest point, etc.

(b) potential energy of compressed gas → kinetic energy of spray droplets → heat when droplets have stopped moving

(c) kinetic energy when thrown → potential and kinetic energy at highest point of motion → heat and sound energy when clay hits ground

Now it's your turn (page 84)

1 The useful energy output of the heater is in the form of heat energy. All electrical energy is eventually changed to heat energy so the process is 100% efficient.

2 58%

Section 4.2 (page 85)

1 (a) chemical   (b) nuclear   (c) gravitational potential
(d) wind/kinetic   (e) sound, light, kinetic energy of gases, potential energy   (f) potential

2 (a) 0.51 m   (b) 180 J s^{-1}

3 (a) 11 J   (b) 14 m s^{-1}

4 (a) 16 MJ   (b) 1.1×10^2 MJ   (c) 2.1×10^2 MJ

5 15%

Now it's your turn (page 87)

1 400 N m^{-1}

2 (a) 3.0×10^4 N m^{-1} (b) 54 N

Now it's your turn (page 89)

1 5.4×10^{-2} J

2 4.2 cm

Now it's your turn (page 91)

1 (a) 1.29×10^7 Pa   (b) 1.17×10^{-4} (c) 0.16 mm

2 5.3×10^6 Pa

Section 4.3 (page 94)

1 (a) 500 N m^{-1} (b) 13.1 cm

2 7.9 J

3 0.36 mm

Now it's your turn (page 96)

1 (a) 1400 J gain   (b) 41 kJ loss

2 4200 J kg^{-1} K^{-1}

Now it's your turn (page 97)

1 (a) (i) 2.5×10^5 J (ii) 4.6×10^4 J   (b) 0.84

2 1100 J K^{-1}

Now it's your turn (page 99)

(a) (i) 16.5 kJ (ii) 113 kJ

(b) 6.8

Now it's your turn (page 100)

1 480 °C

2 41 g

Section 4.4 (page 101)

1 −20 °C

2 (a) 40 J K^{-1}
   (b) 38 °C

3 (a) 22.6 g

4 (a) 2020 J kg^{-1} K^{-1}
   (b) 91 s

Now it's your turn (page 103)

1 4.32×10^5 J

2 120 kN

3 1.5 kW

Now it's your turn (page 104)

1 6.7 cents

2 0.96 cents

3 5.3×10^{10}

Section 4.5 (page 105)

1 2100 cents

2 (a) (i) 16.5 kW (ii) 3.2 kN (iii) 96 kW
   (b) (i) 1.32 MJ (ii) 3.84 MJ

Now it's your turn (page 107)

5.6 N m

Now it's your turn (page 108)

4.5 N

Now it's your turn (page 109)

13.5 N

Section 4.6 (page 111)

1 (a) 4.2 N m   (b) 9.1 N

2 (a) 2.9 N m   (b) 8.0 N

3 (a) 80 N   (b) 120 N

4 67 N

Exam-style Questions (pages 111–112)

1 (c) (i) 1950 J   **(ii)** 780 kW
2 (c) 152 deg C
3 (c) (i) 147 s   **(ii)** $1.02 \, \text{g s}^{-1}$ **(d)** 280 cents
4 (a) (i) 60 W   **(b) (i)** $900 \, \text{J kg}^{-1} \, \text{K}^{-1}$

Chapter 5

Now it's your turn (page 117)

1 $1.7 \, \text{m s}^{-2}$
2 5.3 N

Now it's your turn (page 120)

(a) 61 N   **(b)** 94 N

Sections 5.1–5.4 (page 121)

1 50 kg
2 (a) $7.7 \, \text{m s}^{-1}$ **(b)** $45.2 \, \text{m s}^{-2}$ **(c)** 2040 N
4 $T/2$

Now it's your turn (page 125)

1 $6.8 \times 10^{-24} \, \text{kg m s}^{-1}$
2 $1.2 \, \text{m s}^{-1}$

Now it's your turn (page 127)

600 N

Now it's your turn (page 130)

1 (a) $u_A/2$ **(b)** 0.5

Sections 5.5–5.8 (page 131)

1 $3.3 \times 10^4 \, \text{kg m s}^{-1}$
2 $3.6 \times 10^7 \, \text{N}$
3 $2.7 \times 10^{-4} \, \text{N}$
4 $1.03 \times 10^5 \, \text{m s}^{-1}$
5 Heavy particle's speed is practically unchanged; light particle moves with speed $2u$, in same direction as the incident heavy particle.
6 the lighter body

Exam-style Questions (page 132)

1 (a) (i) $2.16 \, \text{kg m s}^{-1}$ **(ii)** $1.07 \, \text{m s}^{-1}$ **(iii)** 1.16 J
   (b) 0.059 m
2 $3080 \, \text{m s}^{-1}$
4 (a) $3u$ to left   **(b) (i)** $3u$ to right   **(ii)** $3t_1$
   (c) $3u/2$
5 (a) $3m$ **(b)** 0.25

Chapter 6

Now it's your turn (page 137)

$103.6 \, \Omega$

Now it's your turn (page 138)

$1.001 \times 10^3 \, \text{Pa}$

Sections 6.1–6.2 (page 140)

1 $-43.9 \, ^\circ\text{C}$
2 133

Now it's your turn (page 146)

1 $3.3 \times 10^{-9} \, \text{m}$
2 $1.29 \, \text{kg m}^{-3}$
3 1.1 mol; 6.7×10^{23}; $2.1 \times 10^{25} \, \text{m}^{-3}$
4 $2.5 \times 10^{-5} \, \text{m}^3$
5 $8.7 \times 10^{-6} \, \text{m}^3$

Section 6.3 (page 147)

1 212 kPa
2 1.28
3 $5.8 \times 10^3 \, \text{N}$

Now it's your turn (page 152)

1 $1.0 \times 10^{-20} \, \text{J}$
2 $3.1 \times 10^3 \, \text{m s}^{-1}$

Section 6.4 (page 153)

1 $6100 \, \text{m s}^{-1}$

Now it's your turn (page 156)

200 J of work is done *by* the system; constant temperature means constant internal energy

Sections 6.5–6.6 (page 157)

1 0; 250 J
2 20 J increase

Now it's your turn (page 159)

$3 \times 10^{-10} \, \text{m}$

Now it's your turn (page 162)

$1.4 \times 10^4 \, \text{Pa}$

Section 6.8 (Page 163)

$1.5 \times 10^5 \, \text{Pa}$

Exam-style Questions (page 163)

1 **(a)** 15.8 Ω **(b)** 149.0 °C
2 **(a)** **(i)** $2.69 \times 10^{25} \text{m}^{-3}$ **(ii)** 3.0×10^{-9} m
(b) 4×10^{-4}
4 **(a)** 939 °C (1212 K) **(b)** 239 kg
5 2.4×10^5 Pa
7 **(a)** 16 **(b)** 1
(c) 16 (in each part, the ratio is hydrogen : oxygen)
8 **(a)** 0 **(b)** 500 J increase
(c) temperature rises

Chapter 7

Now it's your turn (page 166)

1 2.0 A
2 4.83×10^5 s
3 **(a)** 60 C **(b)** 3.75×10^{20}

Now it's your turn (page 168)

1 1.6×10^{-13} J
2 **(a)** 0.25 C **(b)** 2.2 J

Now it's your turn (page 169)

100 Ω

Now it's your turn (page 170)

1 current in both is 2.4 A
2 **(a)** 9.2 A **(b)** 26 Ω

Section 7.1 (page 171)

1 **(a)** 760 C **(b)** 1000 W **(c)** 57 Ω
2 **(a)** **(i)** 0.20 A **(ii)** 0.60 W **(b)** 5400 J
3 3.5×10^6 J
4 2.0 kW

Now it's your turn (page 175)

1 18 m
2 0.97 mm

Section 7.2 (page 176)

1 6.7 m
2 **(b)** I/A 0.20 0.40 0.60 0.80 1.00 1.20 1.40
R/Ω 0.95 1.20 2.45 3.65 4.56 5.47 6.21
3 **(a)** 0.62 Ω **(b)** 4.3×10^{-7} Ω m

Now it's your turn (page 178)

1 0.25 A
2 2.0 A with the current entering the junction

Now it's your turn (page 183)

1 0.56 Ω
2 3.0 A when short circuited; 1.1 W when load resistance equals internal resistance

Section 7.3 (page 184)

1 **(a)** 0.05 Ω **(b)** 0.3 Ω
2 **(a)** 0.25 A **(b)** 1.6 Ω **(c)** 12 J

Now it's your turn (page 187)

1 4.3 Ω
2 25 Ω

Now it's your turn (page 190)

1 **(a)** 12 V; **(b)** 0.57 V; **(c)** 5 kΩ
2 **(a)** 0.8 V; **(b)** 7.7 V

Section 7.4 (page 191)

1 **(a)** 169 Ω **(b)** 13 Ω
2 **(a)** 5 Ω **(b)** 3.0 A
3 **(a)** 25 Ω

Now it's your turn (page 194)

(a) 1.4 kW **(b)** 2.9 kW, 0 kW

Section 7.5 (page 197)

1 $V_{rms} = 20$ V, $I_{rms} = 2.5$ A
2 **(a)** 340 V **(b)** 4.8 mA

Exam-style Questions (pages 197–198)

1 **(a)** 4.5 V
(b) **(i)** 50 Ω **(ii)** 0.090 A **(iii)** 0.90 V
3 ten resistors, each of resistance 12 kΩ and power rating 0.5 W, connected in parallel
4 **(a)** 4 Ω **(b)** 8 Ω **(c)** 3 Ω **(d)** 1.0 A
5 **(a)** **(i)** 1.6×10^{-2} Ω **(ii)** 1.1×10^{-3} Ω
(iii) 28 W
(b) **(i)** 4.4 s **(ii)** 4.4×10^{21}
(c) **(i)** 11.7 V **(ii)** 307 W
6 **(a)** 1.02 V, 1.22 W
(b) **(ii)** 7.53 m **(iii)** 1.41 W

8 (a) 1.4 kW
  (b) 2.9 kW
  (c) 0
The graph of power against time is a $\sin^2$ curve with maximum instantaneous power (2.9 kW) at 0.25 of a cycle and 0.75 of a cycle, and minimum instantaneous power (0 kW) at 0, 0.50 of a cycle and 1.00 cycle.

Chapter 8

Now it's your turn (page 204)

1 $0.40 \, \text{m s}^{-1}$
2 0.68 m
3 $4.6 \times 10^{14} \, \text{Hz}$
4 1.4

Section 8.1 (page 208)

1 (a) 200 Hz   **(b)** $5.0 \times 10^{-3} \, \text{s}$
2 (a) $7.5 \times 10^{14} \, \text{Hz}$ to $4.3 \times 10^{14} \, \text{Hz}$ **(b)** 1.2 m
3 1/4
4 (a) $5.7 \times 10^{-8} \, \text{J}$ **(b)** $4.9 \times 10^{-5} \, \text{W m}^{-2}$

Now it's your turn (page 218)

1 625 nm
2 63 m

Section 8.3 (pages 219–220)

2 (c) 3.5 mm

Now it's your turn (page 228)

1 880 Hz; 1320 Hz
2 0.38 m
3 340 Hz
4 128 Hz

Section 8.4 (page 229)

2 (b) 8.5 cm
3 (a) $108 \, \text{m s}^{-1}$ **(b)** 45 Hz

Now it's your turn (page 232)

1 56°
2 22 cm

Now it's your turn (page 234)

1 652 nm
2 28.2°, 70.7°
3 2

Section 8.5 (page 235)

1 (a) $2.5 \times 10^{-6} \, \text{m}$ **(b)** 10.2°   **(c)** 5, 3
3 0.73° (1.26×10^{-2} rad)

Exam-style Questions (page 236)

1 $2.0 \times 10^9 \, \text{W m}^{-2}$
2 (b) π rad (180°)   **(c)** 5:1   **(d)** $224 \, \text{m s}^{-1}$
4 26.6 cm
6 (b) 212 mm   **(c)** 449 mm

Chapter 9

Now it's your turn (page 243)

2 $5.8 \times 10^{14} \, \text{Hz}$

Section 9.1 (page 244)

1 3.1 eV to 1.8 eV
2 (a) 340 nm   **(b)** $2.5 \times 10^{-19} \, \text{J}$
3 $5.6 \times 10^{-19} \, \text{J}$ (3.5 eV)

Now it's your turn (page 245)

1 $1.3 \times 10^{-11} \, \text{m}$
2 $1.2 \times 10^{-10} \, \text{m}$

Section 9.2 (page 245)

1 $3.9 \times 10^{-11} \, \text{m}$
2 $2.6 \times 10^3 \, \text{m s}^{-1}$

Now it's your turn (page 251)

1 $3.9 \times 10^{-8} \, \text{m}$
2 $3.19 \times 10^{15} \, \text{Hz}$

Section 9.3 (page 252)

1 430 mm
2 91 nm

Exam-style Questions (pages 252–253)

2 8.0×10^{14}
3 (a) 500 nm
   (b) (i) $3.9 \times 10^{-19} \, \text{J}$ **(ii)** $9.3 \times 10^5 \, \text{m s}^{-1}$
4 (a) $4.4 \times 10^{-35} \, \text{m}$ **(b)** $2.6 \times 10^{-35} \, \text{m}$
   (c) $2.4 \times 10^{-11} \, \text{m}$ **(d)** $1.3 \times 10^{-13} \, \text{m}$
6 A: 480 nm; B: 1.8 μm; C: 660 nm

Chapter 10

Now it's your turn (page 255)

1 $9.1 \times 10^{22} \, \text{m s}^{-2}$
2 (a) $7.8 \, \text{km s}^{-1}$ **(b)** $5.3 \times 10^3 \, \text{s}$

Now it's your turn (page 258)

(a) $0.22 \, \text{m s}^{-1}$ **(b)** $0.63 \, \text{rad s}^{-1}$ **(c)** $0.14 \, \text{m s}^{-2}$

Now it's your turn (page 261)

(a) 590 m   (b) 4200 N

Section 10.1 (page 262)

2 (a) 3.3 rad s^{-1}
  (b) 0.53 revolutions per second

Now it's your turn (page 267)

(a) $x = 0.10 \cos \pi t$ (b) 0.20 s

Now it's your turn (page 270)

(a) 0.25 m   (b) 2.5 s

Sections 10.2–10.4 (page 271)

1 (a) 220 Hz   (b) 1.4×10^3 rad s^{-1} (c) 42 m s^{-1}
  (d) 5.9×10^4 m s^{-2} (e) 7.1 m s^{-1}
2 2.8 m s^{-1}
3 9.826 m s^{-2}
4 (a) 5.8 N m^{-1} (b) 15 mm
  (c) 0.58 s   (d) −8.2 mm

Now it's your turn (page 273)

(a) 0.87 J   (b) 1.97 J   (c) 2.84 J

Section 10.5 (page 274)

1 3.2 : 1
2 (a) 0.87 x_0 (b) 0.71 x_0

Section 10.6 (page 279)

(a) 2.0 Hz
(b) Suggestion is incorrect; driving frequency will act against natural frequency every half-cycle.

Exam-style Questions (page 280)

1 (b) 2.72×10^{-3} m s^{-2}
2 (a) (i) 25 rad s^{-1} (ii) 4.0 revolutions per second
  (b) (i) 0.84 N
3 (b) (i) $2A\rho g\Delta x$ (ii) $2g\Delta x/L$
  (iv) $(1/2\pi) \sqrt{(2g/L)}$
4 (a) 1.3 s   (b) 16 N
5 (a) 1.28×10^{14} Hz

Chapter 11

Now it's your turn (page 284)

2.3×10^{-8} N

Now it's your turn (page 288)

(a) 5.0×10^4 V m^{-1} (b) 2.4×10^{-12} C

Now it's your turn (page 289)

1.0×10^6 N C^{-1}

Now it's your turn (page 290)

0.45 J

Section 11.1 (page 292)

3 (a) 2.1×10^{21} N C^{-1} (b) 650 N
  (c) 1.7×10^7 V   (d) 5.3×10^{12} J

Now it's your turn (page 293)

1.0 mF

Now it's your turn (page 294)

0.203 J

Now it's your turn (page 296)

1.4×10^{-9} F

Now it's your turn (page 299)

(a) 9.4×10^{-3} C
(b) (ii) 940 μF   (iii) 10 V   (iv) 4.7×10^{-3} C

Now it's your turn (page 302)

(a) 6.0×10^{-5} C   (b) 2.7×10^{-5} C   (c) 5.4 V

Section 11.2 (page 303)

1 (a) 1.1×10^{-10} F
  (b) (i) 1.3×10^{-9} C   (ii) 2.4×10^4 V m^{-1}
2 (a) 2.0 μF   (b) 1.2 μF
3 (a) (i) 4.1×10^{-4} J   (ii) 9.0×10^{-5} C
  (b) (i) 6.0×10^{-5} A   (ii) 1.5 s   (iii) 0.77 s

Now it's your turn (page 306)

25 N kg^{-1}

Now it's your turn (page 307)

-9.4×10^9 J; 1.1×10^4 m s^{-1}

Now it's your turn (page 310)

1.43 hours

Section 11.3 (page 312)

1 5.5×10^3 kg m^{-3}
2 7.78×10^8 km
3 −0.50%; −0.25%

Exam-style Questions (pages 312–313)

1 (a) 1.0×10^4 V m^{-1} (c) 5.0×10^{-5} N
2 2.2×10^{39}; 2.2×10^{39}

3 (a) 2.8×10^{-2} N   **(c)** 1.9×10^{-7} C
4 (a) 4.4×10^{-7} F
5 (a) 5.7×10^{-11} N
6 (c) 1.5 m to left of A
7 (b) (ii) 8.86×10^{4} km

Chapter 12

Now it's your turn (pages 317–318)

1 Diagram should be as in Figure 12.4 but with direction of field lines reversed.

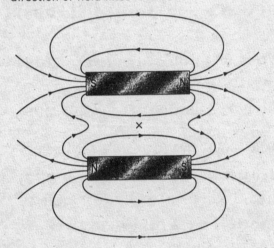

2

Now it's your turn (page 319)

1

2

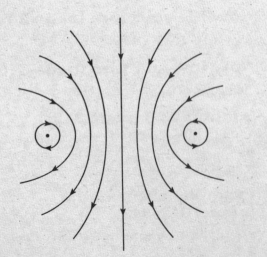

Now it's your turn (page 324)

1 3.4×10^{-7} N
2 (a) 2.8×10^{-2} N m^{-1} **(b)** 2.0×10^{-2} N m^{-1}

Now it's your turn (page 328)

The sphere moves downwards with an initial acceleration of 1.6 m s^{-2}.

Now it's your turn (page 330)

(a) 8.8×10^{6} m s^{-1} **(b)** 9.3 cm

Now it's your turn (page 332)

4.2×10^{6} m s^{-1}

Now it's your turn (page 333)

1 (a) 5.0×10^{-14} N   **(b)** 8.8 cm
2 Deflections in opposite directions because the particles are oppositely charged; electron has smaller radius of path because q/m is larger.

Section 12.2 (pages 334–335)

1 (a) (i) east–west   **(ii)** 2.5×10^{4} A
   (b) No, large current would melt the wire.
2 (a) (i) 1.5×10^{-4} T   **(ii)** 2.3×10^{-3} N m^{-1}
   (b) Forces not large enough to move copper wire.
3 (a) 0.15 T
   (b) uniform field, no field outside poles, wire normal to the field
4 (a) left to right using $r = mv/Bq$
   (b) Positive, with argument based on Fleming's left-hand rule.

Now it's your turn (pages 340–341)

1 (a) (i) 1600 m^2 **(ii)** 0.064 Wb   **(iii)** 0.064 V
   (b) wing pointing west
2 86 mV
3 (i) 2.7 m^2 **(ii)** 0.41 Wb   **(iii)** 0.41 V

Now it's your turn (pages 343–344)

1 (a) 45   **(b)** 56 mA
2 (a) 30   **(b)** 7500   **(c)** 450 mA

Section 12.3 (pages 345–346)

1 (a) (i) 1.4 T   **(ii)** 4.0 mWb   **(iii)** 1.6 Wb
   (b) 150 V across switch
   (c) Large voltage induced which could be dangerous.

2 (a) 0   (b) 50 mV   (c) 29 mV
3 (a) 0.034   (b) 4100
   (c) Thicker wire reduces heating effect caused by the larger current.

Exam-style Questions (page 346)

1 (a) north-south
   (b) (i) east-west   (ii) 26 μT   (iii) 55°
   (c) 2.0×10^{-4} N m^{-1}
2 (a) (i) 4.5 mT   (ii) 1.1×10^{-4} Wb
   (b) 7.1 mV
3 $E = 0$ when current constant. Spikes in opposite directions at times t_{on} and t_{off}.

Chapter 13

Now it's your turn (page 349)

19, 40, 21

Sections 13.2–13.4 (page 358)

3 $^{a-4}_{b-2}$Y; $^{a-4}_{b-1}$Z, $^{0}_{0}$γ

Now it's your turn (page 361)

(a) 1/4   (b) 1/1024   (c) 3/4   (d) 15/16

Now it's your turn (page 363)

(a) 5.6×10^{12} Bq
(b) 2.5×10^{6} Bq

Now it's your turn (page 364)

(a) 3.3×10^{12} (b) 1.3×10^{12}

Now it's your turn (page 365)

1 (a) 1.6×10^{-4} s   (b) 1.7×10^{11} s
2 (a) 1.2×10^{19} s^{-1} (b) 4.6×10^{-2} h^{-1}

Sections 13.5–13.6 (pages 365–366)

1 (b) 6 minutes
2 2.5×10^{9} Bq
3 (a) X: 3.5×10^{5} Bq; Y: 2.0×10^{5} Bq
4 1.2×10^{-10} kg
5 (a) 2.2 h   (b) 1.3×10^{21}

Now it's your turn (page 367)

0.109906 u

Now it's your turn (page 368)

101 MeV

Now it's your turn (page 370)

7.3 MeV per nucleon

Sections 13.7–13.11 (page 374)

1 (a) 0.00857 u, 7.98 MeV, 2.66 MeV/nucleon
   (b) 0.687 u, 640 MeV, 6.60 MeV/nucleon
   (c) 1.79 u, 1660 MeV, 7.50 MeV/nucleon
2 (a) 236.132 u; 235.918 u
   (b) 0.214 u   (c) 208.5 MeV
   (d) 8.13×10^{10} J   (e) 74 g
3 (a) 4.96 MeV   (b) 14 MeV

Exam-style Questions (pages 377–378)

1 15.8 years
2 (a) $P = 131$; $Q = 54$
   (b) (i) 9.2×10^{20} (ii) 0.032 g
3 (a) 0.046 h^{-1}
   (b) 6000 cm^{3}
4 (a) 3.27 MeV
5 (c) 3.0×10^{5} m s^{-1}
6 (a) about 50 mm; 3300 mm^{-1}

Chapter 14

Now it's your turn (page 386)

Connections for the potential difference are made across the variable resistor. Fixed resistor has resistance 750 Ω.

Now it's your turn (page 390)

1

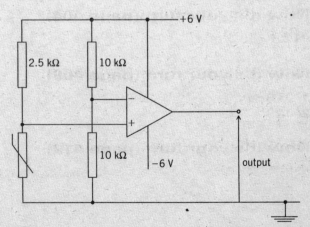

2

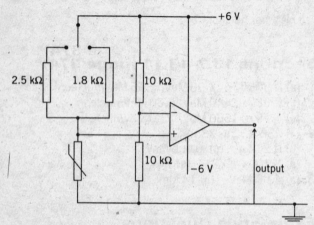

Now it's your turn (page 394)

(a) non-inverting
(b) 10
(c) 0.9 V

Section 14.3 (page 398)

(a) a light-emitting diode
(b) a relay switching on/off a heater
(c) An analogue meter with a systematic error such that it reads zero when there is a small input.

Exam-style Questions (pages 398–399)

1 (c) (i) −19
   (ii) 0.47 V
2 (a) (6 ±1) °C
   (b) 2.25 kΩ

Chapter 15

Now it's your turn (page 404)

(b) 2.8×10^{-11} m

Now it's your turn (page 408)

1 23.8 cm
2 0.19

Now it's your turn (page 412)

2

6	9
4	3

Now it's your turn (page 417)

(b) 9.4×10^{-3}

Now it's your turn (page 418)

0.11

Section 15.2 (page 421)

2 0.908 I incident on fat–muscle boundary
  0.904 I transmitted through fat–muscle boundary
  0.332 I received at muscle–bone boundary
  0.121 I transmitted into bone

Exam-style Questions (page 424)

2 (c) 5.3×10^{-5}

Chapter 16

Now it's your turn (page 428)

1 The signal has constant frequency of 750 kHz and the amplitude varies from 3.5 V to 4.0 V. The frequency of this variation in amplitude is 3 kHz.
2 (a) 1.0×10^5 Hz
  (b) 1.0×10^4 Hz

Now it's your turn (page 433)

(a) 25
(b) 10

Now it's your turn (page 443)

(a) 42 dB
(b) 22 dB

Now it's your turn (page 444)

(a) 1.4 dB km^{-1}
(b) 94 km

Exam-style Questions (page 451–452)

1 (c) (i) range 20 Hz–15 kHz, bandwidth 30 kHz
2 (b) (i) 2.0 ms
   (ii)

0101	1101	1011	0111	1011	0110	1000

3 (a) 1.0 W
  (b) (i) 55 dB
     (ii) 4.6 km

Index

Acknowledgements

The Publishers would like to thank the following for permission to reproduce copyright material:

Picture credits

p.2 Fig. 1.2*r* Rosenfeld Images Ltd/Science Photo Library; Fig. 1.2*l* Royal Observatory Edinburgh/Science Photo Library; Fig. 1.3*l* Andrew Lambert Photography; Fig. 1.3*c* M&C Denis-Hoot/Still Pictures; Fig. 1.3*r* Museum of Flight/Corbis; **p.3** Fig. 1.4 Alfred Pasieka/Science Photo Library; **p.7** Fig. 1.6*b* Hodder Education; Fig. 1.6*t* Yves Lefevre/Still Pictures; **p.10** Fig. 1.8 Action Plus; **p.15** Fig. 1.18 Action Plus; **p.19** Fig. 2.1 Chad Ehlers/Alamy; **p.21** Fig. 2.4 Leslie Garland Picture Library/Alamy; **p.22** Fig. 2.7 Andrew Lambert Photography/Science Photo Library; **p.26** Fig. 2.15 Martyn F. Chillmaid/ Science Photo Library; **p.27** Fig. 2.16 Lew Robertson/Corbis; Fig. 2.17 Michael Dalton/Fundamental Photos/Science Photo Library; Fig. 2.18 Alex Bartel/Science Photo Library; **p.30** Fig. 2.21 Martyn Chillmaid; Fig. 2.22 Synthetic Alan King/ Alamy; **p.31** Fig. 2.23 Martyn F. Chillmaid/Science Photo Library; **p.35** Fig. 2.30 DK Limited/Corbis; **p.36** Fig. 2.32 Andrew Lambert Photography/Science Photo Library; Fig. 2.33 sciencephotos/Alamy; **p.37** Fig. 2.35 Noel Toone, photographersdirect.com; **p.38** Fig. 2.36 Andrew Lambert Photography/Science Photo Library; **p.40** Fig. 2.41 Andrew Lambert Photography; **p.61** Fig. 3.7 Martyn Chillmaid; **p.62** Fig. 3.8 Archivo Iconografico, SA/Corbis; Fig. 3.9 Tony Craddock/Science Photo Library; Fig. 3.10 John McAnulty/Corbis; **p.67** Fig. 3.17 Glyn Kirk/Action Plus; Fig. 3.19 Barry Mayes/Life File; **p.72** Fig. 4.1 Action Plus; **p.73** Fig. 4.3 Paul Ridsdale/Alamy; **p.75** Fig. 4.6*all* Alex Bartel/Science Photo Library; **p.76** Fig. 4.8 Chrysler; **p.78** Fig. 4.10 Andrew Lambert Photography/Science Photo Library; **p.79** Fig. 4.11 Lester Lefkowitz/Corbis; **p.81** Fig. 4.13 Dr Jeremy Burgess/Science Photo Library; **p.84** Fig. 4.14 J-L Charmet/Science Photo Library; **p.104** Fig. 4.27 Clynt Garnham/Alamy; **p.107** Fig. 4.33 Stefan Sollfors/Alamy; **p.115** Fig. 5.2 Bill Sanderson/ Science Photo Library; **p.116** Fig. 5.4 Corbis; **p.126** Fig. 5.13 Edward Kinsman/Science Photo Library; **p.127** Fig. 5.14 Ping Europe Ltd; **p.139** Fig. 6.7 Andrew Lambert Photography; **p.141** Fig. 6.11 Science Photo Library; **p.167** Fig. 7.3 Wendy Brown; **p.172** Fig. 7.7 Science Photo Library; **p.188** Fig. 7.33*both* Andrew Lambert Photography; **p.207** Fig. 8.13 Bruce Coleman Inc./Alamy; **p.211** Fig. 8.18*both* Polaroid; Fig. 8.19 Peter Aprahamian/Sharpless Stress Engineers/Science Photo Library; **p.218** Fig. 8.30 Andrew Lambert Photography; **p.219** Fig. 8.31 Andrew Lambert Photography; **p.220** Fig. 8.33 Bob Winsett/Corbis; **p.221** Fig. 8.37 Andrew Lambert Photography; **p.231** Fig. 8.50 Andrew Lambert Photography; **p.241** Fig. 9.4 Lucien Aigner/Corbis; **p.244** Fig. 9.5 WA Steer/Daresbury Lab/UCL; Fig. 9.6 Andrew Lambert Photography/Science Photo Library; **p.247** Fig. 9.8 Imperial College/Science Photo Library; **p.249** Fig. 9.12 Imperial College/Science Photo Library; **p.250** Fig. 9.13 Imperial College/Science Photo Library; Fig. 9.14*all* Mike Guidry, University of Tennessee; **p.260** Fig. 10.9 Tek Image/Science Photo Library; Fig. 10.10 Novosti/Science Photo Library; Fig. 10.11 The Travel Library/Rex Features; **p.275** Fig. 10.27 Andrew Lambert Photography; **p.277** Fig. 10.32 Andrew Lambert Photography; **p.278** Fig. 10.33 Kit Kittle/Corbis; Fig. 10.34 AP/PA Photos; **p.310** Fig.11.25 David Ducros/ Science Photo Library; **p.315** Fig. 12.1 Hodder Education; **p.317** Fig. 12.7 Science Photo Library; **p.319** Fig. 12.16 Pete Saloutos/Corbis; **p.327** Fig. 12.25 Lawrence Berkeley Laboratory/Science Photo Library; **p.329** Fig. 12.29 Steve Allen/ Science Photo Library; **p.343** Fig. 12.42 Basil Booth/GeoScience Features Picture Library; **p.351** Fig. 13.2 N Feather/ Science Photo Library; **p.356** Fig. 13.5 Martyn F Chillmaid/Science Photo Library; **p.373** Fig. 13.16 European Space Agency/Science Photo Library; **p.374** Fig.13.17 Prof Erwin Mueller/Science Photo Library; **p.376** Fig. 13.21 Dr David Wexler/Science Photo Library; **p.380** Fig. 14.2a Martyn F. Chillmaid/Science Photo Library; **p.381** Fig. 14.4a Martyn F. Chillmaid/Science Photo Library; **p.386** Fig. 14.10a picture supplied by RS Components (rswww.com); **p.401** Fig.15.1 Gustoimages/Science Photo Library; **p.405** Fig. 15.4 Sovereign, Ism/Science Photo Library; **p.406** Fig. 15.9 Sovereign, Ism/Science Photo Library; **p.409** Fig. 15.12 Mauro Fermariello/Science Photo Library; Fig. 15.13 Zephyr/Science Photo Library; **p.419** Fig. 15.27 Aj Photo/Hop Americain/Science Photo Library; **p.420** Fig. 15.29 Sovereign, Ism/Science Photo Library; **p.423** Fig. 15.31 Zephyr/Science Photo Library; **p.441** Fig. 16.12 TEK Image/Science Photo Library

Text credits

Material in Chapters 14–16 adapted from *A Level Science Applications Support Booklet: Physics*, 2005. Reproduced by permission of the University of Cambridge Local Examinations Syndicate.

Every effort has been made to trace all copyright holders for text and artwork used in this publication. However, if any have been inadvertently overlooked the Publishers will be pleased to make the necessary arrangements at the first opportunity.